Larry Bull, Tim Kovacs (Eds.)

Foundations of Learning Classifier Systems

T0135241

Studies in Fuzziness and Soft Computing, Volume 183

Editor-in-chief
Prof. Janusz Kacprzyk
Systems Research Institute
Polish Academy of Sciences
ul. Newelska 6
01-447 Warsaw
Poland
E-mail: kacprzyk@ibspan.waw.pl

Further volumes of this series
can be found on our homepage:
springeronline.com

Vol. 167. Y. Jin (Ed.)
Knowledge Incorporation in Evolutionary Computation, 2005
ISBN 3-540-22902-7

Vol. 168. Yap P. Tan, Kim H. Yap,
Lipo Wang (Eds.)
Intelligent Multimedia Processing with Soft Computing, 2005
ISBN 3-540-22902-7

Vol. 169. C.R. Bector, Suresh Chandra
Fuzzy Mathematical Programming and Fuzzy Matrix Games, 2005
ISBN 3-540-23729-1

Vol. 170. Martin Pelikan
Hierarchical Bayesian Optimization Algorithm, 2005
ISBN 3-540-23774-7

Vol. 171. James J. Buckley
Simulating Fuzzy Systems, 2005
ISBN 3-540-24116-7

Vol. 172. Patricia Melin, Oscar Castillo
Hybrid Intelligent Systems for Pattern Recognition Using Soft Computing, 2005
ISBN 3-540-24121-3

Vol. 173. Bogdan Gabrys, Kauko Leiviskä,
Jens Strackeljan (Eds.)
Do Smart Adaptive Systems Exist?, 2005
ISBN 3-540-24077-2

Vol. 174. Mircea Negoita, Daniel Neagu,
Vasile Palade
Computational Intelligence: Engineering of Hybrid Systems, 2005
ISBN 3-540-23219-2

Vol. 175. Anna Maria Gil-Lafuente
Fuzzy Logic in Financial Analysis, 2005
ISBN 3-540-23213-3

Vol. 176. Udo Seiffert, Lakhmi C. Jain,
Patric Schweizer (Eds.)
Bioinformatics Using Computational Intelligence Paradigms, 2005
ISBN 3-540-22901-9

Vol. 177. Lipo Wang (Ed.)
Support Vector Machines: Theory and Applications, 2005
ISBN 3-540-24388-7

Vol. 178. Claude Ghaoui, Mitu Jain,
Vivek Bannore, Lakhmi C. Jain (Eds.)
Knowledge-Based Virtual Education, 2005
ISBN 3-540-25045-X

Vol. 179. Mircea Negoita,
Bernd Reusch (Eds.)
Real World Applications of Computational Intelligence, 2005
ISBN 3-540-25006-9

Vol. 180. Wesley Chu,
Tsau Young Lin (Eds.)
Foundations and Advances in Data Mining,
2005
ISBN 3-540-25057-3

Vol. 181. Nadia Nedjah,
Luiza de Macedo Mourelle
Fuzzy Systems Engineering, 2005
ISBN 3-540-25322-X

Vol. 182. John N. Mordeson,
Kiran R. Bhutani, Azriel Rosenfeld
Fuzzy Group Theory, 2005
ISBN 3-540-25072-7

Vol. 183. Larry Bull, Tim Kovacs (Eds.)
Foundations of Learning Classifier Systems,
2005
ISBN 3-540-25073-5

Larry Bull
Tim Kovacs
(Eds.)

Foundations of Learning Classifier Systems

 Springer

Larry Bull
School of Computer Science
University of the West of England
Bristol BS16 1QY
U.K.
E-mail: larry.bull@uwe.ac.uk

Tim Kovacs
Department of Computer Science
University of Bristol
Bristol BS8 1UB
U.K.
E-mail: kovacs@cs.bris.ac.uk

ISBN 978-3-642-06413-5

e-ISBN 978-3-540-32396-9

ISSN print edition: 1434-9922
ISSN electronic edition: 1860-0808

Springer is a part of Springer Science+Business Media
springeronline.com
© Springer-Verlag Berlin Heidelberg 2005
Softcover reprint of the hardcover 1st edition 2005

Cover design: E. Kirchner, Springer Heidelberg

Contents

Introduction
L. Bull and T. Kovacs 1

Section I – Rule Discovery

Population Dynamics of Genetic Algorithms
J.E. Rowe 21

Approximating Value Functions in Classifier Systems
L.B. Booker 45

Two Simple Learning Classifier Systems
L. Bull 63

Computational Complexity of the XCS Classifier System
M.V. Butz et al. 91

**An Analysis of Continuous-Valued Representations
for Learning Classifier Systems**
C. Stone and L. Bull 127

Section II – Credit Assignment

Reinforcement Learning: A Brief overview
J. Wyatt 179

**A Mathematical Framework for Studying Learning
in Classifier Systems**
J.H. Holland 203

Rule Fitness and Pathology in Learning Classifier Systems
T. Kovacs 219

**Learning Classifier Systems: A Reinforcement Learning
Perspective**
P.L. Lanzi 267

Learning Classifier System with Convergence and Generalization
A. Wada et al. 285

Section III – Problem Characterization

On the Classification of Maze Problems
A.J. Bagnall and Z.V. Zatuchna 307

What Makes a Problem Hard for XCS?
T. Kovacs and M. Kerber 317

Foundations of Learning Classifier Systems:
An Introduction

Larry Bull[1] & Tim Kovacs[2]

[1]School of Computer Science
University of the West of England
Bristol BS16 1QY, U.K.
Larry.Bull@uwe.ac.uk

[2]Department of Computer Science
University of Bristol
Bristol BS8 1UB, U.K.
Kovacs@cs.bris.ac.uk

[Learning] Classifier systems are a kind of rule-based system with general mechanisms for processing rules in parallel, for adaptive generation of new rules, and for testing the effectiveness of existing rules. These mechanisms make possible performance and learning without the "brittleness" characteristic of most expert systems in AI.

Holland et al., *Induction*, 1986

1. Introduction

Learning Classifier Systems (LCS) [Holland, 1976] are a machine learning technique which combines evolutionary computing, reinforcement learning, supervised learning or unsupervised learning, and heuristics to produce adaptive systems. They are rule-based systems, where the rules are usually in the traditional production system form of "IF state THEN action". An evolutionary algorithm and heuristics are used to search the space of possible rules, whilst a credit assignment algorithm is used to assign utility to existing rules, thereby guiding the search for better rules. The LCS formalism was introduced by John Holland [1976] and based around his more well-known invention – the Genetic Algorithm (GA)[Holland, 1975]. A few years later, in collaboration with Judith Reitman, he presented the first implementation of an LCS [Holland & Reitman, 1978]. Holland then revised the framework to define what would become the standard system [Holland, 1980; 1986a]. However, Holland's full system was somewhat complex and practical experience found it difficult to realize the envisaged behaviour/performance [e.g., Wilson & Goldberg, 1989] and interest waned. Some years later, Wilson presented the "zeroth-level" classifier system, ZCS [Wilson, 1994] which "keeps much of Holland's original framework but simplifies it to increase understandability and performance" [ibid.]. Wilson then introduced a form of LCS which altered the way in which rule fitness is calculated – XCS [Wilson,

L. Bull and T. Kovacs: *Foundations of Learning Classifier Systems: An Introduction*, Stud-
Fuzz **183**, 1–17 (2005)
www.springerlink.com

1995]. The following decade has seen resurgence in the use of LCS as XCS in particular has been found able to solve a number of well-known problems optimally. Perhaps more importantly, XCS has also begun to be applied to a number of hard real-world problems such as data mining, simulation modeling, robotics, and adaptive control (see [Bull, 2004] for an overview) and where excellent performance has often been achieved. Further, given their rule-based nature, users are often able to learn about their problem domain through inspection of the produced solutions, this being particularly useful in areas such as data mining or safety-critical control for example. However their combination of two machine learning techniques and potentially many heuristics means that formal understanding of LCS is non-trivial. That is, current formal understanding of, for example, Genetic Algorithms and Reinforcement Learning is significant but understanding of how the two interact within Learning Classifier Systems is severely lacking. The purpose of this volume is to bring together current work aimed at understanding LCS in the hope that it will serve as a catalyst to a concerted effort to produce such understanding.

The rest of this contribution is arranged as follows: Firstly, the main forms of LCS are described in some detail. A number of historical studies are then reviewed before an overview of the rest of the volume is presented. See [Barry, 2000] for more on early LCS.

2. Holland's LCS

Holland's Learning Classifier System [Holland, 1986] receives a binary encoded input from its environment, placed on an internal working memory space - the blackboard-like message list (Figure 1). The system determines an appropriate response based on this input and performs the indicated action, usually altering the state of the environment. Desired behaviour is rewarded by providing a scalar reinforcement. Internally the system cycles through a sequence of performance, reinforcement and discovery on each discrete time-step.

The rule-base consists of a population of N condition-action rules or "classifiers". The rule condition and action are strings of characters from the ternary alphabet $\{0,1,\#\}$. The # acts as a wildcard allowing generalisation such that the rule condition 1#1 matches both the input 111 and the input 101. The symbol also allows feature pass-through in the action such that, in responding to the input 101, the rule IF 1#1 THEN 0#0 would produce the action 000. Both components are initialised randomly. Also associated with each classifier is a fitness scalar to indicate the "usefulness" of a rule in receiving external reward. This differs from Holland's original implementation [Holland & Reitman, 1978], where rule fitness was essentially based on the accuracy of its ability to predict external reward (after [Samuel, 1959]).

On receipt of an input message, the rule-base is scanned and any rule whose condition matches the external message, or any others on the message list, at each position becomes a member of the current "match set" [M]. A rule is selected from those rules comprising [M], through a bidding mechanism, to become the system's external action. The message list is cleared and the action string is posted to it ready for the next cycle. A number of other rules can then be selected through bidding to fill any remaining spaces on the internal message list. This selection is performed by a simple stochastic roulette wheel scheme. Rules' bids consist of two components, their fitness and their specificity, that is the proportion of non-# bits they contain. Further, a constant (here termed β) of "considerably" less than one is factored in, i.e., for a rule C in [M] at time t:

$$Bid(C,t) = \beta \cdot specificity\,(C) \cdot fitness(C,t)$$

Reinforcement consists of redistributing bids made between subsequently chosen rules. The bid of each winner at each time-step is placed in a "bucket". A record is kept of the winners on the previous time step and they each receive an equal share of the contents of the current bucket; fitness is shared amongst activated rules. If a reward is received from the environment then this is paid to the winning rule which produced the last output. Holland draws an economic analogy for his "bucket-brigade" algorithm (BBA), suggesting each rule is much like the middleman in a commercial chain; fitness is seen as capital. The reader is referred to [Sutton & Barto, 1998] for an introduction to reinforcement learning.

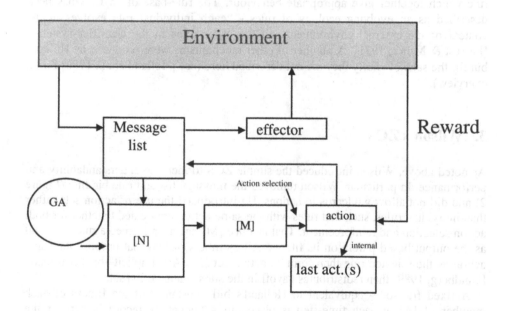

Fig. 1: Schematic of Holland's Learning Classifier System.

The LCS employs a steady-state Genetic Algorithm operating over the whole rule-set at each instance. After some number of time-steps the GA uses roulette wheel selection to determine two parent rules based on their fitness relative to the total fitness of the population:

$$Probabilty\_Selection(\,C,t\,) = fitness(\,C,t\,) \,/\, \Sigma \; fitnesses(\,t\,)$$

The effect of this scheme is to bias reproduction towards those rules which appear to lead to higher reward from the environment. Copies are made of the chosen rules which are then subjected to two genetic operators: mutation and crossover. Mutation is applied probabilistically at a per-locus rate (e.g., 1/100) along the length of the rule and upon satisfaction the value at that locus is altered – typically, a locus becomes one of the other two possible values with equal probability. For example, if the above mentioned rule 1#1:0#0 experiences a mutation event on its last locus it could become 1#1:0#1 or 1#1:0##. Crossover begins by randomly choosing a position within the rules and then swaps them from that point to their end. For example, the two rules 000:000 and 111:111 which experience crossover at position two would become 001:111 and 110:000 respectively. The purpose of the genetic operators is to introduce new rules into the population based on known good rules with the aim of discovering better rules. The new rules then replace two existing rules, often chosen using roulette wheel selection based on the reciprocal of fitness. The reader is referred to [Eiben & Smith, 2004] for a recent introduction to evolutionary computing.

It is important to note that the role of the GA in LCS is to create a cooperative set of rules which together solve the task. That is, unlike a traditional optimisation scenario, the search is not for a single fittest rule but a number of different types of rule which together give appropriate behaviour. The rule-base of an LCS has been described as an evolving ecology of rules - "each individual rule evolves in the context of the external environment and the other rules in the classifier system." [Forrest & Miller, 1991]. A number of other mechanisms were proposed by Holland but for the sake of clarity they are not described here (see [Holland et al., 1986] for an overview).

3. Wilson's ZCS

As noted above, Wilson introduced the simple ZCS to increase understandability and performance. In particular, Wilson removed the message list and rule bidding (Figure 2) and did not allow wildcards in actions. He introduced the use of action sets rather than individual rules, such that rules with the same action are treated together for both action selection and reinforcement. That is, once [M] has been formed a rule is picked as the output based purely on its fitness. All members of [M] that propose the same action as the selected rule then form an action set [A]. An "implicit" bucket brigade [Goldberg, 1989] then redistributes payoff in the subsequent action set.

A fixed fraction - equivalent to Holland's bid constant - of the fitness of each member of [A] at each time-step is placed in a bucket. A record is kept of the previous action set [A], and if this is not empty then the members of this action set each receive an equal share of the contents of the current bucket, once this has been reduced by a pre-determined discount factor γ (a mechanism used in temporal difference learning to encourage solution brevity [e.g., Sutton & Barto, 1998]). If a reward is received from the environment then a fixed fraction of this value is distributed evenly amongst the members of [A] divided by their number. Finally, a tax is imposed on the members of [M] that do not belong to [A] on each time-step in order to encourage exploitation of the fitter classifiers. That is, all matching rules not in [A] have their fitnesses reduced by factor τ thereby reducing their chance of being selected on future cycles. Wilson considered this technique provisional and suggested there were better approaches to controlling exploration. The effective update of action sets is thus:

$$\text{fitness ([A])} \leftarrow \text{fitness ([A])} + \beta \, [\, \text{Reward} + \gamma \, \text{fitness([A]}_{+1}) - \text{fitness([A])} \,]$$

where $0 \leq \beta \leq 1$ is a learning rate constant. Wilson noted that this is a change to Holland's formalism since specificity is not considered explicitly through bidding and pay-back is discounted by 1-γ on each step. ZCS employs two discovery mechanisms, a steady state GA and a covering operator. On each time-step there is a probability p of GA invocation. When called, the GA uses roulette wheel selection to determine two parent rules based on fitness. Two offspring are produced via mutation and crossover. The parents donate half their fitness to their offspring who replace existing members of the population. The deleted rules are chosen using roulette wheel selection based on the reciprocal of fitness. The cover heuristic is used to produce a new rule with an appropriate condition to the current state and a random action when a match-set appears to contain low quality rules, or when no rules match an input.

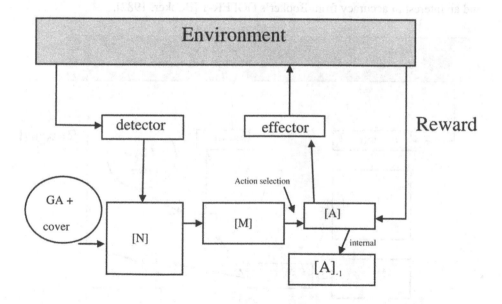

Fig. 2: Schematic of ZCS.

When ZCS was first presented, results from its use indicated it was capable of good, but not optimal, performance [Wilson, 1994][Cliff & Ross, 1995]. More recently, it has been shown that ZCS is capable of optimal performance, at least in a number of well-known test problems, but appears to be particularly sensitive to some of its parameters [Bull & Hurst, 2002]. It should be noted that ZCS has two closely related forerunners, namely BOOLE [Wilson, 1987] and NEWBOOLE [Bonelli et al., 1990].

4. Wilson's XCS

The most significant difference between XCS (Figure 3) and most other LCS (e.g., ZCS) is that rule fitness for the GA is not based on payoff received (P) by rules but on the accuracy of predictions (p) of payoff. Hence, XCS has been termed an accuracy-based LCS, in contrast to earlier systems which were for the most part strength-based (also called payoff-based systems). The intention in XCS is to form a complete and accurate mapping of the problem space (rather than simply focusing on the higher payoff niches in the environment) through efficient generalizations. In RL terms, XCS learns a value function over the complete state/action space. In this way, XCS makes the connection between LCS and reinforcement learning clear and represents a way of using traditional RL on complex problems where the number of possible state-action combinations is very large (other approaches have been suggested, such a neural networks – see [Sutton & Barto, 1998] for an overview).

XCS shares many features with ZCS, and inherited its niche GA, deletion scheme and an interest in accuracy from Booker's GOFER-1 [Booker, 1982].

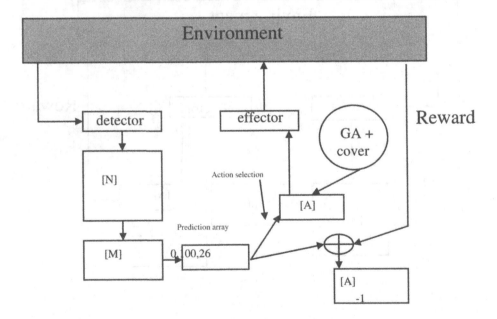

Fig. 3: Schematic of XCS.

On each time step a match set is created. A system prediction is then formed for each action in [M] according to a fitness-weighted average of the predictions of rules in each [A]. The system action is then selected either deterministically or randomly (usually 0.5 probability per trial). If [M] is empty covering is used.

Fitness reinforcement in XCS consists of updating three parameters, ε, p and F for each appropriate rule; the fitness is updated according to the relative accuracy of the rule within the set in five steps:

i) Each rule's error is updated: $\varepsilon_j = \varepsilon_j + \beta(\ |P - p_j| - \varepsilon_j)$ where as in ZCS $0 \le \beta \le 1$ is a learning rate constant.

ii) Rule predictions are then updated: $p_j = p_j + \beta(P - p_j)$

iii) Each rule's accuracy κ_j is determined:
$\kappa_j = \alpha(\varepsilon_0/\varepsilon)^v$ or $\kappa = 1$ where $\varepsilon < \varepsilon_0$
where v, α and ε_0 are constants controlling the shape of the accuracy function.

iv) A relative accuracy κ_j' is determined for each rule by dividing its accuracy by the total of the accuracies in the action set.

v) The relative accuracy is then used to adjust the classifier's fitness F_j using the moyenne adaptive modifee (MAM) procedure: If the fitness has been adjusted $1/\beta$ times, $F_j = F_j + \beta(\kappa_j' - F_j)$. Otherwise F_j is set to the average of the values of κ' seen so far.

In short, in XCS fitness is an inverse function of the error in reward prediction, with errors below ε_0 not reducing fitness. The maximum $P(a_i)$ of the system's prediction array is discounted by a factor γ and used to update rules from the previous time step. Thus XCS exploits a form of Q-learning [Watkins, 1989] in its reinforcement procedure, whereas Holland's 1986 system and ZCS both use a form of TD(0) (as noted in [Sutton & Barto, 1998]).

The GA acts in action sets [A], i.e., niches. Two rules are selected based on fitness from within the chosen [A]. Rule replacement is global and based on the estimated size of each action set a rule participates in with the aim of balancing resources across niches. The GA is triggered within a given action set based on the average time since the members of the niche last participated in a GA (after [Booker, 1989]).

XCS is more complex than ZCS but results from its use in a number of areas have been impressive. Wilson originally demonstrated results on the Boolean multiplexer function and a maze problem [Wilson, 1995]. Early on Kovacs emphasised its ability to learn complete, accurate, and minimal representations of Boolean functions [Kovacs, 1997]. XCS has since shown good performance on data mining tasks [e.g., Bernado, Llora & Garrell, 2002] and has been widely adopted in the LCS community; the majority of contributions to a recent volume on applications of LCS [Bull, 2004] used XCS. An algorithmic description of XCS can be found in [Butz & Wilson, 2001], while further details of XCS and an example execution cycle can be found in [Kovacs, 2004]. A brief overview of selected theoretical works now follows. We concentrate on pre-ZCS and XCS systems in order to complement the remaining chapters of this text, and on formal studies rather than experimental ones.

5. Previous Research on the Foundations of LCS

Since Learning Classifier Systems combine two machine learning algorithms, previous studies of their behaviour from a theoretical standpoint have tended to focus on one aspect over the other. The following historical review is divided to reflect this. Further related material is available in [Wilson & Goldberg, 1989; Lanzi & Riolo, 2001].

5.1 Rule Discovery: Evolutionary Algorithms

The term Evolutionary Algorithm denotes a family of stochastic problem solvers based on a population of solutions being manipulated by the neo-Darwinian processes of selection, recombination and mutation. The Genetic Algorithm, as briefly described above, is the most commonly used approach but recent work has included parameter self-adaptation [e.g., Bull et al., 2000] normally associated with Evolution Strategies [Rechenberg, 1973] and the later forms of Evolutionary Programming [e.g., Fogel, 1992], and the use of LISP S-expressions to represent rules [e.g., Lanzi, 1999b] as found in Genetic Programming [Koza, 1992]. Until the early 1990's, Holland's Schema Theorem [Holland, 1975] was the most widely used theoretical tool for understanding GAs and thus it was also used as a basis for some of the earliest work on Learning Classifier Systems.

Smith and Valenzuela-Rendon [1989] presented a simple proportion vector form of the canonical GA through which they considered the propagation of the set of eighteen rules with two-bit conditions and one-bit actions where there was no pass-through in the latter, i.e., the rules 00:0 to ##:1. Roulette wheel selection and single-point crossover were included in this infinite population generational model - a model based on the traditional scenario of replacing the whole population per reproduction cycle. The LCS was assigned a stimulus-response task, that is, a task under which each response from the LCS is rewarded immediately by the environment; three Boolean functions of varying difficulty were used. Initial results showed how a standard GA is unable to converge to a solution containing a full set of rules required to solve the given tasks. That is, the GA operated as it does in the standard function optimization scenario and simply sought solutions/rules which typically led to the highest fitness/reward only. They then examined the effects of fitness sharing in their model. Fitness sharing was highlighted by Booker [1982] as a way to prevent the GA population from clustering around such solutions. Simply, individuals are said to share the reward received with those who are similar to them in some way. In GA function optimization similarity is traditionally based on Hamming distance, i.e., on how many loci are of the same value, with all those within a predefined neighborhood being included [e.g., Goldberg & Richardson, 1987]. Using this scheme, Smith and Valenzuela-Rendon [1989] found complete rule sets were maintained in two of the three cases and the failure in the third was identified as being due to the disruptive actions of crossover. That is, rules which were individually useful always produced rules which were not useful through their recombination. Booker [1982] also suggested that mating restrictions could be used such that sufficiently dissimilar rules do not recombine. Using a simple mating restriction scheme Smith and Valenzuela-Rendon [1989] found the previously unsolved problem benefited but that another no longer maintained a full solution. They concluded by suggesting that the combination of both schemes may be beneficial.

The Schema Theorem has been somewhat criticized for reasons such as the difficulty in using it to explain the dynamical or limit behaviour of GAs. Goldberg and Segrest [1987] presented a Markov chain for a simple finite population generational GA and the use of such models has remained widespread [e.g., Vose, 1999] as they enable more predictive analysis. Holland [1986b] was the first to consider using Markov chains to model LCS, the BBA in particular (reprinted in this volume). Horn et al. [1994] presented a version of Goldberg and Segrest's model to examine fitness sharing in LCS and the effects of varying the amount of interaction between two rule classes. Their model enabled them to vary the fitness ratio between the two rule types and the degree of overlap in their generalizations of the input space. By calculating the expected time to absorption of the Markov chain, they were able to show that rule maintenance times are very large even for relatively small population/rule base sizes but that this niching pressures reduces as the degree of overlap increases. Horn et al. also calculated the steady state distributions during the maintenance of both rule classes through a well-known manipulation of the absorbing Markov chain to create an ergodic chain. The degree of overlap in the generalization space was again shown to be important, causing a decrease in the probability of achieving the coverage/constitution expected from the given fitness ratio.

As noted above, these two studies used models of generational GAs as their basis. However, as described in Section 4, LCS use a steady-state GA whereby only a small percentage of the rule base is replaced per GA invocation which means that the selection pressure can be very different for example [e.g., Chakraborty et al., 1997]. Bull [e.g., 2002] presented a steady-state GA version of Goldberg and Segrest's [1987] Markov model to examine aspects of accuracy-based fitness as presented by Wilson in XCS. In comparison to a traditional strength-based fitness scheme (that is, of Holland-style systems) without fitness sharing, it was shown that XCS-type accuracy-based fitness maintained selective pressure against an incorrect rule regardless of the degree of its incorrectness, whereas the strength-based fitness scheme selected for the incorrect rule in certain cases. That is, without fitness sharing, it was shown that under strength-based fitness, a rule whose average payoff is higher than that of a correct rule can lead to the extinction of the correct rule. This phenomenon has been termed "overgeneralization" [e.g., Wilson, 1995]. Using a simple set of difference equations, Bull and Hurst [2002] showed how fitness sharing has the potential to avoid overgeneralization in both single and multi-step scenarios. Bull [2002] also included mutation into his model and showed how the accuracy-based fitness scheme appears more sensitive to the mutation rate than the strength-based scheme, a result which was previously suggested in his work on self-adaptation [Bull et al., 2000]. A simple two-step problem was also examined with the Markov chain which indicated that, under certain relationships between the rewards given for each route to the goal state, selection pressure can disappear depending upon the constituency of the rule-base. That is, using roulette-wheel selection, the effective selection pressure can vary significantly over time due to the coevolutionary nature of LCS.

5.2 Credit Assignment

The first implemented classifier system, CS-1, [Holland & Reitman, 1978] used an epochal credit assignment scheme partly inspired by Samuel's work on checkers [Samuel, 1959]. This scheme found little subsequent use (see e.g., [Smith et al., 2004] for a recent example) as it was supplanted by the Bucket-Brigade Algorithm (BBA) introduced in section 2. However, many difficulties with the BBA were soon found and alternatives suggested, e.g., [Wilson & Goldberg, 1989; Riolo, 1989; Liepins et al., 1991; Huang, 1989]. The most common form is the implicit bucket brigade described above for ZCS and XCS, wherein matching rules do not bid for control of the system, and, instead, credit is apportioned between all rules proposing a given action. After [Holland, 1986b], Westerdale [e.g., 1991; 1999] has developed a general Markov chain model for a learning entity approximating the payoff (value) of states within a given transition matrix/environment via the BBA. The aforementioned closer connection between the BBAs of ZCS and XCS and the temporal difference algorithms of the reinforcement learning literature have put credit assignment in recent LCS on firmer ground than their predecessors. For example, as noted in Section 4, XCS evolves complete maps of the entire state/action space to an estimate of value, unlike earlier systems which aim only to form a best action map, mapping each state to an action and estimate of value. The difference is significant as the more complete map potentially allows better exploration control and proper propagation of credit through the state space in the manner of reinforcement learners [Kovacs, 2004]. Indeed, convergence proofs for reinforcement learning methods require infinite updates to the estimated value of *all* state/action pairs.

Some early work also considered the use of tools emerging from the field of complexity/non-linear systems to examine LCS. Forrest and Miller [1991] cast the internal processes of Holland's LCS, in particular with a message list, as a Random Boolean network [Kauffman, 1984]. Here each node of the network is a rule and connections are formed between nodes/rules if the antecedent of one satisfies the condition of the other. By varying the specificity of rules, they show a phase transition-like dynamic exists for the emergence of self-sustaining/long inductive chains; too much or too little generalization and the LCS is unable to sustain "appropriate" internal activity. Compiani et al. [e.g., 1991] considered the fact that rule discovery and credit assignment operate over different timescales. As such, they present models of the dynamics of rule updating, for a message list of a given size, as rule discovery occurs. They find "random regimes" exist which temporarily disrupt system performance, to a significant degree, if a careful balance is not maintained against exploring newly introduced rules and exploiting existing ones.

Yates and Fairley [1994] used aspects of Evolutionary Game Theory [Maynard Smith, 1986] to show that LCS under the BBA are "evolutionary stable." That is, the rule-base of the LCS will be optimally configured for the learning task. After identifying commonalities between the features required for an evolutionary stable learning rule, i.e., one capable of finding an evolutionary stable state (ESS), and the BBA, they show a simplified LCS without a GA solving a well-known two-player game to its ESS. However, akin to the findings of Compiani et al. [1991], they note the GA is likely to disrupt the ESS, even if only temporarily.

5.3 Other Early Considerations

As noted in Section 2, LCS typically use a ternary alphabet {0,1,#} to represent rule conditions. Rule conditions are minterms, and sets of rule conditions are in Disjunctive Normal Form. This simple syntax, very similar to the binary strings used with genetic algorithms, was chosen by Holland as it was thought to be most suitable for genetic search. In particular, it was argued that the lower the cardinality of the alphabet, the higher the number of schemata and the higher the degree of implicit parallelism [Booker et al., 1989; Goldberg, 1989].

A consequence of the limited expressive power of individual rules is that sets of rules are required to represent solutions for non-trivial tasks, which introduces issues concerning the interaction of rules (i.e., competition and cooperation). Under some fitness schemes the system becomes co-evolutionary (as the fitness of one rule depends on what others exist), which complicates credit assignment and hence adaptation.

Although sets of rules using the ternary alphabet are capable of representing complex information (indeed, Holland's LCS is computationally complete [Forrest, 1985]), concerns have been raised regarding the utility of this language [e.g., Belew & Forrest, 1988; Carbonell, 1989; Grefenstette, 1989; Schuurmans & Schaeffer, 1989]. In response, Booker [1991] demonstrated a number of more expressive languages using the ternary alphabet, claiming it was the syntax of the language, rather than the cardinality of the alphabet, which was often at fault.

Whilst many continued to advocate the use of low-cardinality alphabets, the application of evolutionary methods to more complex data structures such as trees and graphs, with Genetic Programming being perhaps the best-known approach, has become widespread. Wilson suggested the use of LISP S-expressions in a classifier system [Wilson, 1994], and Lanzi subsequently studied the use of messy encodings [Lanzi, 1999a] and then S-expressions [Lanzi, 1999b] (see also [Ahluwalia & Bull, 1999]). Other representations include fuzzy logic [e.g., Valenzuela-Rendon, 1991] and neural network rules [e.g., Bull & O'Hara, 2002]. Our view is that LCS are rule-based systems, and that the vast array of possible rule languages gives them considerable expressive power, comparable to other learning methods. As always, the representation (and inductive methods) used must suit the task at hand.

Default Hierarchies (DHs) are sets of rules in which exception rules override the action of default rules (see [Holland et al, 1986] for discussions). A typical example consists of an overgeneral default rule and a set of more specific exception rules. It seems plausible that a default rule, which provides better than random performance, might be found first and that the exception rules would then be found and gradually refine the performance of the rule population as a whole. In addition to allowing such gradual refinement of representation, DHs have been seen as a means of increasing the number of solutions to a problem without increasing the size of the search space. A final advantage of DHs is that they allow more compact representations of the solution [e.g., Valenzuela-Rendon, 1989a, 1989b]. Despite these potential advantages, it has proved difficult to form and retain DHs in practice due to the complex co-evolutionary dynamics they introduce, and interest in them waned in the early 1990s. The most advanced work on the subject remains that by Smith and Goldberg [1991].

6. Foundations of Learning Classifier Systems: An Overview

The rest of this book, in keeping with the distinct areas of formal enquiry which have emerged from the field, is divided into three main sections: rule discovery, credit assignment, and problem characterization.

6.1 Rule Discovery

Jon Rowe – Population Dynamics of Genetic Algorithms. As noted above, formal understanding of evolutionary optimization techniques has progressed significantly in recent years. This contribution introduces one of the more commonly used models by which this has been done, that of Michael Vose [1999], and then introduces some extensions which are relevant to LCS thereby indicating a potentially fruitful way forward.

Lashon Booker – Approximating value functions in classifier systems. The quality of a solution to a reinforcement learning problem depends on the quality of the value function approximation (assuming one is used). This chapter notes the similarity between tile coding and a classifier system with a fixed rule population and compares the quality of the value function approximation made by the two approaches. Although the standard approach, minus the genetic algorithm, performs poorly compared to tile coding, a new hyperplane coding is introduced and the best of several variations on it is found comparable to tile coding. This represents a promising new direction for function approximation with LCS.

Larry Bull – Two Simple Learning Classifier Systems. LCS are complex and as such the production of meaningful executable models is non-trivial. This chapter presents canonical forms of each basic type, i.e., strength and accuracy-based systems, with which to examine the underlying features of each through such models.

Martin Butz et al. – Computational Complexity of the XCS Classifier System. Based on experimental results with Boolean multiplexer problems, Wilson [1998] hypothesized that the difficulty of a problem for XCS (in terms of population size and amount of experience needed) grows as a low order polynomial of the problem complexity. This chapter presents an overview of the authors' previous work to examine these, and other, aspects of XCS formally, and establishes that k-DNF functions are PAC-learnable by XCS.

Christopher Stone and Larry Bull – An Analysis of Continuous-Valued Representations for Learning Classifier Systems. For a number of applications, particularly data mining [e.g., Wilson, 2000] and adaptive control [e.g., Hurst et al., 2002], an interval encoding has been used. This contribution considers the biases inherent within such an encoding, for both panmictic and niche-based GAs.

6.2 Credit Assignment

Jeremy Wyatt – Reinforcement Learning: A Brief Overview. LCS are now clearly identified as reinforcement learners. This contribution presents the basic mathematical framework used in the formal understanding of such techniques and discusses the various forms and extensions built from it.

John Holland – A Mathematical Framework for Studying Learning Classifier Systems. Shortly after presenting the most well-known instantiation of his LCS framework, Holland published this vision of a path to a more formal understanding of LCS ([Holland, 1986b] Reprinted with kind permission from Elsevier).

Pier-Luca Lanzi – Learning Classifier Systems: A Reinforcement Learning Perspective. This contribution demonstrates the direct connection between XCS and traditional reinforcement learning. Further, it suggests that a GA is exactly the right sort of approach to build generalizations over the input-output space of such techniques (see also [Hartmann, 1994] for a similar conclusion but from the perspective of learning difficulty in LCS).

Tim Kovacs – Rule Fitness and Pathology in Learning Classifier Systems. This chapter considers the conditions in which undesireable types of rules may prosper. Specifically, the concepts of strong overgeneral and fit overgeneral rules are introduced and linked to the structure of the value function. The prospects for such rules are investigated in both strength and accuracy-based systems, and it is suggested that accuracy-based systems have an advantage in dealing with them. This work demonstrates the existence of the above rule types using very simple tasks, to which any reinforcement learner could be applied. In doing so it demonstrates one way in which complex tasks and learners can be analysed.

Atsushi Wada et al. – Learning Classifier Systems with Convergence and Generalization. LCS for reinforcement learning incorporate function approximation through the use of rules which generalize over (aggregate) states. This chapter takes steps toward integrating LCS and standard formulations of linear function approximation in reinforcement learning. The chapter also considers convergence results. Convergence proofs exist for a number of tabular reinforcement learning methods, but no such proofs for LCS appear in the literature. As a first step, this chapter introduces a variant of ZCS to which an existing convergence proof extends. Although this version of ZCS generalizes over states, it is limited to a fixed rule population.

6.3 Problem Characterization

Anthony Bagnall and Zhanna Zatuchna – On the Classification of Maze Problems. Surprisingly, despite the many papers and many maze problems which have been presented, no overarching categorization of such tasks has been presented to date. This paper highlights features of such problems and how they can be used to group previously presented mazes and design new ones.

Tim Kovacs and Manfred Kerber – What Makes a Problem Hard for XCS? This contribution identifies four dimensions of problem complexity for XCS in the domain of Boolean functions. It suggests functions which bound the complexity of the space of functions of a given string length, and discusses how to measure the complexity of a function for XCS. Finally, it proposes a scalable Boolean test suite and argues for its use. Interested readers are referred to related work in Bernado and Ho [to appear].

7. Summary

Almost thirty years after Holland presented the Learning Classifier System paradigm, the ability of LCS to solve complex real-world problems is becoming clear. In particular, the XCS system Wilson presented ten years ago has sparked renewed interest in LCS. This article has given a brief introduction to LCS and previous formal studies of their behaviour. The rest of the book brings together work by a number of individuals who are contributing to the current formal understanding of how they achieve good performance. Future work must build on these insights to produce a coherent picture of how LCS work.

Acknowledgements

Thanks to everyone involved in this edited collection: Professor Kacprzyk for agreeing to publish the book in his series and the authors without whose efforts there would be no book. Dr. Bull would also like to thank his fellow members of the Learning Classifier Systems Group at UWE for so many interesting discussions.

References

Ahluwalia, M. & Bull, L. (1999) A Genetic Programming-based Classifier System. In W. Banzhaf, J. Daida, A.E. Eiben, M.H. Garzon, V. Honavar, M. Jakiela & R.E. Smith (eds) *GECCO-99: Proceedings of the Genetic and Evolutionary Computation Conference*, pp. 11-18. Morgan Kaufmann.

Barry, A. (2000) XCS Performance and Population Structure within Multiple-Step Environments. Ph.D. Thesis, Queens University Belfast.

Belew, R.K., & Forrest, S. (1988) Learning and Programming in Classifier Systems. *Machine Learning*, 3:193-223.

Bernado, E., Llora, X., Garrell, J.M. (2002) XCS and GALE: A Comparative Study of Two Learning Classifier Systems on Data Mining. In Lanzi, Stolzmann & Wilson (eds) Advances in Learning Classifier Systems, pp. 115-132, LNAI 2321, Springer.

Bernado, E. & Ho, T. (to appear) Domain of Competence of XCS Classifier System in Complexity Measurement Space. *IEEE Transactions on Evolutionary Computation*.

Bonelli, P., Parodi, A., Sen, S. & Wilson, S.W. (1990) NEWBOOLE: A Fast GBML System. In *International Conference on Machine Learning*, pp. 153-159. Morgan Kaufmann.

Booker, L. (1989) Triggered Rule Discovery in Classifier Systems. In Schaffer (ed.) *Proceedings of the International Conference on Genetic Algorithms*, pp. 265-274. Morgan Kaufmann.

Booker, L. (1982) Intelligent Behavior as an Adaptation to the Task Environment. Ph.D. Thesis, the University of Michigan.

Booker, L. (1991) Representing Attribute-based Concepts in a Classifier System. In Rawlins (ed.) *Proceedings of the First Workshop on the Foundations of Genetic Algorithms*, pp. 115-127. Morgan Kaufmann.

Booker, L., Goldberg, D.E. & Holland, J.H. (1989) Classifier Systems and Genetic Algorithms. *Artificial Intelligence*, 40: 235-282.

Bull, L. (2002) On Accuracy-based Fitness. *Soft Computing* 6(3-4): 154-161.

Bull, L. (2004)(ed.) *Applications of Learning Classifier Systems*. Springer.

Bull, L. & Hurst, J. (2002) ZCS Redux. *Evolutionary Computation* 10(2): 185-205.

Bull, L. Hurst, J. & Tomlinson, A. (2000) Self-Adaptive Mutation in Classifier System Controllers. In J-A. Meyer, A. Berthoz, D. Floreano, H. Roitblatt & S.W. Wilson (eds) *From Animals to Animats 6 - The Sixth International Conference on the Simulation of Adaptive Behaviour*, pp. 460-470. MIT Press.

Bull, L. & O'Hara, T. (2002) Accuracy-based Neuro and Neuro-Fuzzy Classifier Systems. In W.B. Langdon, E. Cantu-Paz, K. Mathias, R. Roy, D. Davis, R. Poli, K. Balakrishnan, V. Honavar, G. Rudolph, J. Wegener, L. Bull, M.A. Potter, A.C. Schultz, J. F. Miller, E. Burke & N.Jonoska (eds) *GECCO-2002: Proceedings of the Genetic and Evolutionary Computation Conference*, pp. 905-911. Morgan Kaufmann.

Butz, M. & Wilson, S.W. (2001) An Algorithmic Description of XCS. In *Advances in Learning Classifier Systems: Proceedings of the Third International Conference – IWLCS2000*. Springer, pp. 253-272.

Carbonell, J.G. (1989) Introduction: Paradigms for Machine Learning. *Artificial Intelligence* 40:1-9.

Chakraborty, U., Deb, K. & Chakraborty, M. (1997) Analysis of Selection Algorithms: A Markov Chain Approach. *Evolutionary Computation* 4(2): 133-167.

Compiani., M., Montanari, D. & Serra, R. (1991) Learning and Bucket Brigade Dynamics in Classifier Systems. *Physica D* 42: 202-212.

Cliff, D. & Ross, S. (1995) Adding Temporary Memory to ZCS. *Adaptive Behavior* 3(2): 101-150.

Eiben, A. & Smith, J. (2003) *Introduction to Evolutionary Computing*. Springer.

Fogel, D.B. (1992) *Evolving Artificial Intelligence*. PhD Thesis, University of California.

Forrest, S. (1985) A Study of Parallelism in the Classifier System and its Application to Classification in KL-ONE Semantic Networks. Ph.D. Thesis, University of Michigan, Ann Arbor.

Forrest, S. & Miller, J. (1991) Emergent Behavior in Classifier Systems. *Physica D* 42: 213-217.

Goldberg, D.E. (1989) *Genetic Algorithms in Search, Optimization, and Machine Learning*. Addison-Wesley.

Goldberg, D.E. & Richardson, J. (1987) Genetic Algorithms with Sharing for Multimodal Function Optimization. In J.J. Grefenstette (ed) *Proceedings of the 2nd International Conference on Genetic Algorithms*, pp. 41-49. Lawrence Erlbaum.

Goldberg, D.E. & Segrest P. (1987) Finite Markov Chain Analysis of Genetic Algorithms. In J.J. Grefenstette (ed) *Proceedings of the 2nd International Conference on Genetic Algorithms*, pp. 1-7. Lawrence Erlbaum.

Grefenstette, J. (1989) A System for Learning Control Strategies with Genetic Algorithms. In Schaffer (ed.) *Proceedings of the 3rd International Conference on Genetic Algorithms*, pp. 183-190. Morgan Kaufmann.

Hartmann, U. (1994) On the Complexity of Learning in Classifier Systems. In Y. Davidor, H-P. Schwefel & R. Manner (eds) *Parallel Problem Solving from Nature – PPSN III*. Springer, pp. 280-289.

Holland, J.H. (1975) *Adaptation in Natural and Artificial Systems*. University of Michigan Press.

Holland, J.H. (1976) Adaptation. In Rosen & Snell (eds) *Progress in Theoretical Biology*, 4. Plenum.

Holland, J.H. (1980) Adaptive Algorithms for Discovering and using General Patterns in Growing Knowledge Bases. *International Journal of Policy Analysis and Information Systems* 4(3): 245-268.

Holland, J.H. (1986a). Escaping brittleness: the possibilities of general-purpose learning algorithms applied to parallel rule-based systems. In Michalski, Carbonell, & Mitchell (eds) *Machine learning, an artificial intelligence approach*. Morgan Kaufmann.

Holland, J.H. (1986b) A Mathematical Framework for Studying Learning in Classifier Systems. *Physica D* 2(1-3): 307-317.

Holland, J.H. & Reitman, J.H. (1978) Cognitive Systems Based in Adaptive Algorithms. In Waterman & Hayes-Roth (eds) *Pattern-directed Inference Systems*. Academic Press.

Holland, J.H., Holyoak, K.J., Nisbett, R.E. & Thagard, P.R. (1986) *Induction: Processes of Inference, Learning and Discovery*. MIT Press.

Horn, J., Goldberg, D.E. & Deb, K. (1994) Implicit Niching in a Learning Classifier System: Nature's Way. *Evolutionary Computation* 2(1) 37-66.

Huang, D. (1989) The Context-Array Bucket-Brigade Algorithm: An Enhanced Approach to Credit-Apportionment in Classifier Systems. In Schaffer (ed.) *Proceedings of the 3rd International Conference on Genetic Algorithms*, pp. 311-316. Morgan Kaufmann.

Hurst, J., Bull, L. & Melhuish, C. (2002) TCS Learning Classifier System Controller on a Real Robot. In J.J. Merelo, P. Adamidis, H-G. Beyer, J-L. Fernandez-Villacanas & H-P. Schwefel (eds) *Parallel Problem Solving from Nature - PPSN VII*, pp. 588-600. Springer Verlag.

Kauffman, S. (1984) Emergent Properties of Randomly Complex Automata. *Physica D* 10: 145-156.

Kovacs, T. (1997) XCS Classifier System Reliably Evolves Accurate, Complete and Minimal Representations for Boolean Functions. In Roy, Chawdhry, & Pant (eds) *Soft Computing in Engineering Design and Manufacturing*, pp. 59-68. Springer-Verlag.

Kovacs, T. (2004) *Strength or Accuracy: Credit Assignment in Learning Classifier Systems*. Springer.

Koza, J. (1992) *Genetic Programming*. MIT Press.

Lanzi, P.L. (1999a) Extending the Representation of Classifier Conditions Part I: From Binary to Messy Coding. In Banzhaf et al. (eds.) *Proceedings of the Genetic and Evolutionary Computation Conference*, pp. 337-344. Morgan Kaufmann.

Lanzi, P.L. (1999b) Extending the Representation of Classifier Conditions Part II: From Messy Coding to S-Expressions. In Banzhaf et al. (eds.) *Proceedings of the Genetic and Evolutionary Computation Conference*, pp. 345-352. Morgan Kaufmann.

Lanzi, P.L. & Riolo, R.L. (2000) A Roadmap to the Last Decade of Learning Classifier System Research. In Lanzi, Stolzmann & Wilson (eds.) *Learning Classifier Systems. From Foundations to Applications*, pp. 33-62, LNAI 1813. Springer.

Liepins, G.E., Hillard, M.R., Palmer, R. & Rangarajan, G. (1991) Credit Assignment and Discovery in Classifier Systems. *International Journal of Intelligent Systems* 6:55-69.

Rechenberg, I. (1973) *Evolutionstrategie: Optimierung Techniser Systeme nach Prinzipien des Biologischen Evolution*. Fromman-Hozlboog Verlag.

Riolo, R.L. (1989) The Emergence of Coupled Sequences of Classifiers. In Schaffer (ed.), *Proceedings of the 3rd International Conference on Genetic Algorithms*, pp. 256-264. Morgan Kaufmann.

Samuel, A.L. (1959) Some Studies in Machine Learning using the Game of Checkers. *IBM Journal of Research and Development* 3: 211-229.

Schuurmans, D. & Schaeffer, J. (1989) Representational Difficulties with Classifier Systems. In Schaffer (ed.), *Proceedings of the 3rd International Conference on Genetic Algorithms*, pp. 328-333. Morgan Kaufmann.

Smith, R.E. & Valenzuela-Rendon, M. (1989) A Study of Rule Set Development in a Learning Classifier System. In Schaffer (ed.) *Proceedings of the International Conference on Genetic Algorithms*, pp. 265-274. Morgan Kaufmann.

Smith, R.E. & Goldberg, D.E. (1991) Variable Default Hierarchy Separation in a Classifier System. In Rawlins (ed.) *Proceedings of the First Workshop on the Foundations of Genetic Algorithms*, pp. 148-170. Morgan Kaufmann.

Smith, R.E., El-Fallah, A., Ravichandran, B., Mehra, R.K. & Dike, B.A. (2004) The Fighter Aircraft LCS: A Real-World, Machine Innovation Application. In L. Bull (ed) *Applications of Learning Classifier Systems*, pp. 113-142. Springer.

Sutton, R. & Barto, A. (1998) *Reinforcement Learning*. MIT Press.

Valenzuela-Rendon, M. (1989a) Two Analysis Tools to Describe the Operation of Classifier Systems. PhD Thesis, University of Alabama.

Valenzuela-Rendon, M. (1989b) Boolean Analysis of Classifier Sets. In Schaffer (ed.), *Proceedings of the 3rd International Conference on Genetic Algorithms*, pp. 346-353. Morgan Kaufmann.

Vose, M. (1999) *The Simple Genetic Algorithm* MIT Press.

Watkins, C.J. (1989) Learning from Delayed Rewards. Ph.D. Thesis, Cambridge University.

Westerdale, T. (1991) Quasimorphisms or Queasymorphisms? Modeling Finite Automaton Environments. In Rawlins (ed.) *Proceedings of the First Workshop on the Foundations of Genetic Algorithms*, pp. 128-147. Morgan Kaufmann.

Westerdale, T. (1999) An Approach to Credit Assignment in Classifier Systems. *Complexity* 4:49-52.

Wilson, S.W. (1987) Classifier Systems and the Animat Problem. *Machine Learning* 2: 219-228.

Wilson, S.W. (1994) ZCS: A Zeroth-level Classifier System. *Evolutionary Computation* 2(1):1-18.

Wilson, S.W. (1995) Classifier Fitness Based on Accuracy. *Evolutionary Computation* 3(2): 149-76.

Wilson, S.W. (1998) Generalization in the XCS Classifier System. In Koza et al. (eds.) *Genetic Programming 1998: Proceedings of the Third Annual Conference*, pp. 322-334. Morgan Kaufmann.

Wilson, S.W. (2000) Get real! XCS with Continuous-valued Inputs. In Lanzi, P. L., Stolzmann, W., and Wilson, S. W., (eds.) *Learning Classifier Systems. From Foundations to Applications*. Springer-Verlag.

Wilson, S.W. & Goldberg, D.E. (1989) A critical review of classifier systems. In *Proceedings of the 3rd International Conference on Genetic Algorithms*, pp. 244-255, Morgan Kauffman.

Yates, D. & Fairley, A. (1994) Evolutionary Stability in Simple Classifier Systems. In T. Fogarty (ed) *Evolutionary Computing*, pp. 28-37. Springer.

Section I

Rule Discovery

Population Dynamics of Genetic Algorithms

Jonathan E. Rowe

School of Computer Science, University of Birmingham, Birmingham B15 2TT, Great Britain
J.E.Rowe@cs.bham.ac.uk

1 Introduction

The theory of evolutionary algorithms has developed significantly in the last few years.
A variety of techniques and perspectives have been brought to bear on the analysis and
understanding of these algorithms. However, it is fair to say that we are still some way
away from a coherent theory that explains and predicts behaviour, and can give guidance to applied practitioners. Theory has so far developed in a fragmented, piecemeal
fashion, with different researchers applying their own perspectives, and using tools with
which they are familiar. This is beginning to change, as the research community develops and individual insights become shared. Consequently, the work presented in this
chapter is a somewhat biased selection of results. However, I hope that other researchers
will appreciate this material, even if they would themselves have concentrated on a different approach. Readers who are interested in a survey of current theory are referred to
the books *Genetic algorithms: principles and perspectives* [15] and *Theoretical aspects
of evolutionary computation* [9].

We begin, then, by considering the basic framework for studying genetic algorithms,
laid out by Michael Vose in his book *The Simple Genetic Algorithm* [25]. We will introduce only the basic concepts: other theoretical approaches map easily onto this framework at this level. Genetic algorithms are Markov processes, and we describe, in quite
general terms, how they may be described as such, as well as their relationship to the so-
called infinite population model. We will then concentrate on a particular example: the
"simple" genetic algorithm, comprising proportional selection, mutation and crossover
(by masks) acting on fixed-length strings (over some alphabet). Some results relating to
the fixed-points of these operators (in various combinations) will be described, as well
as the variant known as *genepool* crossover, which has recently been investigated.

The second half of the chapter then looks at some extensions to the basic model. Firstly,
we will look at the possibility of having variable-sized structures in the search space.
This kind of thing arises, for example, when considering rules (in which the action
part might be of arbitrary length), grammars (to represent developmental encodings,
for example), and programs (as in Genetic Programming). Much of the theory here has
been developed by Riccardo Poli and colleagues [10]. We will relate a few results which
fit nicely into the Vose framework.

Secondly, we will consider what happens when the fitness function (or the *environment*) changes with time [27, 3]. A simple case is when the fitness function is periodic,

J.E. Rowe: *Population Dynamics of Genetic Algorithms*, StudFuzz **183**, 21–43 (2005)
www.springerlink.com

in which case the fixed-point analysis for the stationary case generalises in a straightforward way. Another situation in which the fitness changes with time, is when it is, in fact, a function of the population itself. This leads us to our third extension: co-evolutionary systems. Such algorithms have been used, for example, in cooperative problem solving with parallel populations, as well as with learning classifier systems, in which the fitness of a rule depends on the context in which it is used. Such systems have been studied extensively in theoretical biology using evolutionary game theory [6]. We present a simple discrete-time version, which maps directly into our mathematical framework.

Clearly, there is a lot of theoretical work that will not be covered in this chapter. I would like to mention some, and give pointers to the interested reader. Staying within the framework developed by Michael Vose, there is a considerable amount of more advanced material. Some of this is covered in Vose's book. More recent work by Vose, and collaborators, generalises this approach to arbitrary finite search spaces, with particular emphasis on algebraic properties of the genetic operators [20, 21]. Related work has been done by Christopher Stephens, and co-workers [24].

In parallel to this approach, are a number of models of specific systems which make use of techniques from statistical physics [23, 1]. These techniques are useful in helping to understand the effects of finite population sizes on the underlying dynamics. A further parallel development is the application of techniques from algorithmic analysis to evolutionary algorithms applied to different optimisation problem classes [8]. These consider evolutionary algorithms as being "black-box" function optimisers, and ask what the expected running time is to finding the optimal solution. Finally, in the case when the search space comprises real Euclidean space (that is, the individuals are vectors of real numbers), a considerable amount of impressive theoretical work has been done [2]. Readers who are keen to understand the current state of the art in evolutionary computation theory are strongly advised to study all these areas.

2   Genetic algorithms as Markov processes

We will be considering a generational genetic algorithm, in which, at each generation, the entire population is updated.[1] If we are working, as is usual, with a fixed population size, then the *state* of the algorithm at a given time step is simply the current population. The behaviour of a genetic algorithm can therefore be traced through time as a (random) sequence of populations. Of course, the population that we see at any one generation depends rather heavily on the population of the previous generation, and also on the particular genetic operators and fitness function that were applied to that population. In fact, given the knowledge of these things, the probability of obtaining any particular population in the next generation is completely determined. A genetic algorithm is therefore an instance of a *Markov process*, since the state at any time step depends only on the previous time step. If the search space is finite (or countable), then it is a Markov chain.

---

[1] Some of the theory has been extended to *steady-state* algorithms — see [28].

Markov chains can be characterised by their corresponding *transition matrix*. This matrix contains the probabilities that the chain will move from one state j to another i as follows:

$$Q_{i,j} = \text{Prob}\,[i|j]$$

(that is, the probability of state i given state j).

Markov chains come in various types (see [7] for an introduction to the theory). The first important type (for our purposes) is when the chain is *irreducible*. This means that given any two states (that is, populations), there is always a non-zero probability of going from one to the other in a finite number of generations. Typically, this happens in a genetic algorithm if there is mutation present: there is always a chance (however small) of all the strings of one population mutating to those of another. One of the key properties of an irreducible Markov chain is that is visits every possible state infinitely often. Some states may be much more likely to occur than others, however, and we would like to be able to characterise these states. One characterisation comes from the transition matrix: the vector v that satisfies the equation

$$Qv = v$$

contains the probabilities of seeing each state over an infinitely long run. That is, state k will occur with frequency v_k if the algorithm is run for long enough. v is referred to as the *stationary distribution*.

If a Markov chain is not irreducible, then it might have *absorbing states*. These are states which, should the process ever arrive in one of them, it will remain stuck there forever. This is typically the case for genetic algorithms without mutation. A subset of states may also be absorbing in the sense that, having arrived at one state in the subset, the system remains forever within that subset (even though it might move around within it). An example here would be a genetic algorithm with mutation, but also with elitism (so that the best individual of a population is preserved to the next generation untouched). It is fairly straightforward to see that the fitness of the best of the population cannot decrease, due to the elitism, and in fact that this fitness will converge to the optimum (see [22] for a formal proof). This means that the algorithm is not irreducible, since it can never move to a population with a worse best fitness. Once it finds the optimum, it can never lose it. However, the rest of the population is free to be mutated into any other individual. Therefore, the set of all populations containing the optimum comprises an absorbing subset.

When the system has absorbing states, we know immediately, of course, where the system will end up (if run for long enough). The question of interest is: how long will this take. Theoretically, this question can also be answered by examining the transition matrix. However, as with finding the stationary distribution, this is intractible in practice, as the number of populations associated with a genetic algorithm is very large. Approximate methods can sometimes be employed here (as in the work on analysing the time complexity of evolutionary algorithms [8]).

In the following section, we will show how the transition probabilities of the Markov process are related to the underlying operators of selection, mutation and crossover.

First, we will describe the general setting: the *random heuristic search* framework of Michael Vose.

To start with, we need a way to describe populations (which are the states of the Markov chain) mathematically. We will represent a population with a corresponding *population vector*

$$p = (p_0, p_1, \ldots, p_{n-1})$$

where p_k represents the proportion of item k in the population, and n is the size of the search space. Notice that we associate the search space with the set $\Omega = \{0, 1, \ldots, n - 1\}$ by applying some arbitrary ordering. If the population size is N, then we can find the number of copies of item k simply by multiplying: Np_k. Population vectors are a subset of the following set, the *simplex*:

$$\Lambda = \left\{ x \in \mathbb{R}^n \mid \sum_k x_k = 1 \text{ and } x_k \geq 0 \text{ for all } k \right\}$$

We can therefore think of the genetic algorithm as mapping out a random sequence of points in the simplex. This sequence will arise in the following way. Any element of the search space has a certain probability of being in the next generation, given the current population. This probability depends on the correct combination of selection, mutation and crossover happening, in just the right way, so as to produce the element under consideration. Let us write $\mathcal{G}(p)_k$ to be the probability that item k is produced by the genetic operators, starting with population vector p. If we can write down these probabilities for all elements $k \in \Omega$ then we define a map:

$$\mathcal{G} : \Lambda \to \Lambda$$

Notice that Λ is here serving a dual purpose of representing probability distributions over the search space Ω.

The random process of the genetic algorithm is now exactly equivalent to the following:

1. Start with an initial population vector p.
2. Calculate the probability distribution $\mathcal{G}(p)$.
3. Sample this distribution N times (with replacement) to form the next population q
4. Set $p := q$ and go to 2.

The function $\mathcal{G}$ is referred to as the *heuristic* of the search process, and has to be appropriately defined to take into account the effects of the genetic operators (see the following section).

Even at this level of generality (that is, without going into the details of $\mathcal{G}$) it is possible to say something about the behaviour of the system. In the first place, we know that, given our current population, the next will be chosen according to some probability distribution over all possible populations (of size N). These probabilities form the contents of the Markov transition matrix, which we can write down as follows (theorem

3.4 of [25]).

$$\text{Prob}\,[q|p] = N! \prod_{k \in \Omega} \frac{(\mathcal{G}(p)_k)^{Nq_k}}{(Nq_k)!}$$

We can ask what the *expected* next population is (that is, the mean of this distribution). Theorem 3.3 of [25] tells us that if the current population vector is p then the expected next population is in fact simply $\mathcal{G}(p)$. In other words, the map $\mathcal{G}$ not only tells us the probability distribution from which the next generation is sampled, it also tells us the average result, over all possible populations.

Further, we can also ask what the *variance* of the distribution is. Theorem 3.5 of [25] tell us that it is:

$$\frac{1 - \mathcal{G}(p)^T \mathcal{G}(p)}{N}$$

(where x^T indicates the transpose of the column vector x). What is of interest here is that the variance decreases as the population size N increases. In other words, as the population size gets bigger, so it becomes more and more likely that the actual next population is very close to the expected next population. This has led to the function $\mathcal{G}$ being referred to as the *infinite population model*, since in the limit as $N \rightarrow \infty$ the next generation actually is $\mathcal{G}$. Moreover, this is true for any *finite* number of time steps. It is, of course, not true that the random process follows the deterministic sequence $p, \mathcal{G}(p), \mathcal{G}^2(p), \ldots$ forever, since if the Markov chain is irreducible, then all populations will be visited infinitely often, whereas the deterministic sequence may converge to some fixed-point of $\mathcal{G}$.

Actually, it turns out that the fixed-points of $\mathcal{G}$ do have something to say about the long-term behaviour of the Markov process, as long as $\mathcal{G}$ satisfies certain technical conditions (which are usually satisfied in the case of the simple genetic algorithm). Put simply, the genetic algorithm likes to spend its time in populations which (when considered as vectors) are near fixed-points of $\mathcal{G}$. The long-term behaviour of the system is characterised by relatively stable periods near such points (sometimes referred to as *metastable states*) followed by rapid transitions to other such states. This gives rise to a picture of *punctuated equilibria*, which is described in more technical detail in chapters 13 and 14 of Vose's book. In terms of the Markov chain, we would say that states near fixed-points tend to have higher probability in the stationary distribution.

These results apply, technically speaking, in the case when the population size is sufficiently large. However, there is experimental evidence which suggests that populations stay near fixed-points even for small population sizes. To see why this might be the case, consider the following. Call the distance $|\mathcal{G}(p) - p|$ the *force* of $\mathcal{G}$ at the point p. We would expect that if the force were large, then the GA is more likely to jump to a population which is very different to the current one. Obviously, at a fixed-point the force is zero. And since $\mathcal{G}$ is continuous (for genetic algorithms, anyway), then the force near a fixed-point is small. Moreover, when the population size is small, the population vectors representing the possible populations are rather spread out in the simplex. That is, they are a long distance apart. This seems to imply that, if a population is near a fixed-point, the probability of making the jump to the next nearest population should be rather small. The genetic algorithm will therefore tend to "stall" in such states. This would explain the punctuated equilibria effect, even for small populations. So far, there has been no formal proof of this idea, but it seems that finding the fixed-points of $\mathcal{G}$ is an important tool for describing the behaviour of the genetic algorithm. We will see some examples of this later.

3   The simple genetic algorithm

We define the *simple genetic algorithm* to be a generational GA, comprising selection (usually fitness proportional selection), mutation and crossover. The search space is usually the set of binary strings of length ℓ. We identify this search space with the set $\Omega = \{0, 1, \ldots, n-1\}$ by setting $n = 2^\ell$ and interpreting each binary string as being an integer written in base 2 notation. Given a current population of size N, we create the new one as follows.

1. Select first parent from population (with probability proportional to fitness).
2. Select second parent from population (with probability proportional to fitness).
3. Cross the two parents to form an offspring.
4. Apply mutation to the offspring.
5. Add the result to the next generation.
6. Repeat until next generation contains N individuals.

As we described above, this process is equivalent to the Random Heuristic Search, as long as we define the operator $\mathcal{G}$ in such a way as to accurately reflect the effects of the three operators. In this section, we will describe how this can be done, and give some examples of a fixed-point analysis.

We start by describing fitness proportional selection, although it is quite possible to describe other kinds of selection within the same mathematical framework (for example, tournament selection, rank-based selection). We denote the heuristic operator corresponding to proportional selection by $\mathcal{F} : \Lambda \to \Lambda$. Recall that $\mathcal{F}(p)_k$ should be the probability of selecting individual $k \in \Omega$ from a population described by the vector p. It is simple to check that the following definition achieves this:

$$\mathcal{F}(p)_k = \frac{f_k p_k}{\widehat{f}(p)}$$

where f_k is the fitness of item k and $\widehat{f}(p)$ represents the average fitness of the population p. We get a slightly simpler formulation by placing the elements of the vector $f = (f_0, f_1, \ldots, f_{n-1})$ along the diagonal of a diagonal matrix, which we will denote $\mathrm{diag}(f)$. We also note that the average fitness may be found by calculating the inner product $f^T p$, giving

$$\mathcal{F}(p) = \frac{\mathrm{diag}(f)p}{f^T p}$$

It is worth remarking that, if you run a genetic algorithm with just selection, then the system is a Markov chain with absorbing states. The absorbing states are the *uniform* populations: that is, populations containing only copies of a single individual. The corresponding population vectors are the standard basis vectors, containing a one in the position of the individual that is in the population and zeros elsewhere. These vectors are the corners of the simplex. We denote them $\mathbf{e}_j$ where j is the individual in the population. That is:

$$(\mathbf{e}_j)_k = [j = k]$$

The square bracket notation $[expr]$ denotes 1 if the expression $expr$ is true, and 0 if it is false.

It is easy to check that these vectors are fixed-points of $\mathcal{F}$ as follows:

$$\mathcal{F}(\mathbf{e}_j)_k = \frac{f_k(\mathbf{e}_j)_k}{\boldsymbol{f}^T \mathbf{e}_j}$$
$$= \frac{f_k[j = k]}{f_j}$$
$$= [j = k]$$

Starting with a population $\boldsymbol{p}(0)$, write $\boldsymbol{p}(1) = \mathcal{F}(\boldsymbol{p}(0))$, $\boldsymbol{p}(2) = \mathcal{F}(\boldsymbol{p}(1))$ and so on. The dynamics can be determined as:

$$\boldsymbol{p}(t)_k = \frac{(f_k)^t \boldsymbol{p}(0)_k}{\sum_j (f_j)^t \boldsymbol{p}(0)_j}$$

(see [15] for a proof).

The effects of mutation will be given by an operator $\mathcal{U} : \Lambda \to \Lambda$. The quantity $\mathcal{U}(\boldsymbol{p})_i$ should be the probability of creating item $i \in \Omega$ from population $\boldsymbol{p}$ using mutation. If we denote the probability that item j mutates to item i by $U_{i,j}$ then

$$\mathcal{U}(\boldsymbol{p})_i = \sum_j U_{i,j} p_j$$

and so the effects of mutation are given by a matrix U with entries $U_{i,j}$ so that

$$\mathcal{U}(\boldsymbol{p}) = U\boldsymbol{p}$$

In the usual case in which mutation is implemented by flipping each bit of a string independently, with probability u, we find an explicit formula for U

$$U_{i,j} = u^{d(i,j)}(1 - u)^{\ell - d(i,j)}$$

where $d(i, j)$ is the Hamming distance between the strings i and j.

The set of binary strings of length ℓ can be given a natural algebraic structure. We define the *sum* of two strings i and j, which we will denote $i \oplus j$, by combining them together using bitwise exclusive-or (or, equivalently, bitwise addition modulo-2). In fact, Ω forms a *group* under this definition, with identity element 0. Every element of Ω is its own inverse: $i \oplus i = 0$ for all $i \in \Omega$. Mutation, as defined above, has the interesting property that it *commutes* with the group. That is:

$$U_{k \oplus i, k \oplus j} = U_{i,j}$$

for all $i, j, k \in \Omega$. Many other types of mutation also have this property. In general, define mutation *by mask* by assigning a probability distribution $\mu \in \Lambda$ to the set Ω. That is, the probability of picking $k \in \Omega$ is μ_k. Then we mutate an element j in the population by picking k according to this distribution and forming $j \oplus k$. Bitwise mutation by a rate is just a special case of this with

$$\mu_k = u^{d(k,0)} (1 - u)^{\ell - d(k,0)}$$

We can use the fact that mutation commutes with the group to prove that the centre of the simplex, $v = (1/n, 1/n, \ldots, 1/n)$ is a fixed-point of such a mutation:

$$(Uv)_i = \sum_j U_{i,j} v_j = \frac{1}{n} \sum_j U_{i,j} = \frac{1}{n} \sum_j U_{j \oplus i, 0} = \frac{1}{n} \sum_k U_{k,0} = \frac{1}{n} = v_i$$

The Perron-Frobenius theorem tells us that, if $U_{i,j} > 0$ for all i, j (for example, if mutation is bitwise by a rate $u > 0$), then this is in fact the only fixed-point in the simplex for $\mathcal{U}$.

Finally, we will deal with crossover. This will be represented as an operator $\mathcal{C} : \Lambda \to \Lambda$. Naturally, we want $\mathcal{C}(p)_k$ to be the probability of producing item $k \in \Omega$ from the population p. Let us denote the probability that parents i and j combine to form k by $r(i, j, k)$. Then we want

$$\mathcal{C}(p)_k = \sum_{i,j} p_i p_j r(i, j, k)$$

This suggests that, for each item k, we put the probabilities $r(i, j, k)$ into a matrix M_k, so that

$$\mathcal{C}(p)_k = p^T M_k p$$

which is a *quadratic form*. $\mathcal{C}$ is called a *quadratic operator* [18]. The entries of M_k are

$$(M_k)_{i,j} = \frac{r(i, j, k) + r(j, i, k)}{2}$$

and so M_k is symmetric.[2]

To go further, we need the details of $r(i, j, k)$. For fixed-length binary strings, the most common crossovers are done by *masks*. Let b be a binary string of length ℓ. Then define

---

[2] We get exactly the same quadratic form if we set $M_k = r(i, j, k)$, so one may as well assume that the matrices M_k are symmetric.

$i \otimes b$ to be the bitwise AND operator. We define a probability distribution χ over the set of masks. That is, χ_b is the probability of picking b. Then

$$r(i, j, k) = [(i \otimes b) \oplus (j \otimes \bar{b})]\chi_b$$

(where $\bar{b}$ is the complement of b). For example, if $i = 10110011$ and $j = 01100110$, then mask $b = 11110000$ produces offspring

$$(10110011 \otimes 11110000) \oplus (01100110 \otimes 00001111) = (10110000 \oplus 00000110)$$
$$= 10110110$$

which corresponds to a one-point crossover, with the cut point in the middle of the string. We get different forms of crossover by choosing different distributions χ. For example, uniform crossover is given by assigning all masks equal probabilities. If crossover is performed with probability less than 1, then this corresponds to assigning higher probability to the mask $b = 0$, which has the effect of cloning the first parent.

Crossover by masks shares with mutation by masks the nice property of commuting with the group. That is

$$r(i, j, k) = r(a \oplus i, a \oplus j, a \oplus k)$$

for all $i, j, k, a \in \Omega$. This property means that we can simplify the definition of the crossover operator as follows. For each $k \in \Omega$ define an $n \times n$ matrix σ_k by

$$(\sigma_k)_{i,j} = [i = k \oplus j]$$

Then it is easy to check that, for all $k \in \Omega$

$$M_k = \sigma_k M_0 \sigma_k^T$$

In other words, the probabilities found in the matrix M_k are identical to those found in M_0, but have been moved around (by the permutation matrix σ_k). This means that, once the group is defined, all the information necessary to describe crossover is in the matrix M_0. This matrix is referred to as the *mixing matrix* of crossover. In fact, this happens whenever the search space has a group structure, or indeed, has a group acting transitively upon it — see [20, 21] for details.

If we just run crossover over and over again, starting from some initial population, we again get to a fixed-point. The fixed-points of crossover are described by Geiringer's Theorem [5], a well-known theorem from population genetics. Notice that applying crossover to a population cannot change the number of ones and zeros in any bit position. All that happens is that these bits get shuffled around. This means that the fixed-point we get to depends on the initial population. It is characterised by the fact that, if we draw a string at random from the fixed-point population, the probability that we see a one or zero in any position is *independent* of the bit values at the other positions. Moreover, that probability is given simply by the frequency of ones and zeros in that position in the starting population. In other words, crossover tends to *de-correlate* the bit values, whilst not introducing any new genetic material. By contrast, selection tends to drive evolution towards uniform (high fitness) populations, losing genetic material. Mutation tends to randomise the population (move it towards the centre of the simplex), creating new individuals. Of course, the behaviour that we see in an actual genetic algorithm will be a balance of these three effects working in combination.

Let's consider what happens when we combine selection and mutation. In other words, we set the crossover probability to zero. This is equivalent to setting the probability of mask $b = 0$ to 1. This has the effect of always cloning the first parent selected, and ignoring the second parent. A single generation of the genetic algorithm therefore reduces to:

1. Select a parent (with probability proportional to fitness).
2. Apply mutation to the parent to create an offspring.
3. Add offspring to the next generation.
4. Repeat until next generation contains N individuals.

The heuristic function for selection plus mutation is found by composing the heuristic for the two operators:

$$\mathcal{G} = \mathcal{U} \circ \mathcal{F}$$

Putting in the definitions of these operators we get

$$\mathcal{G}(p) = \frac{U \operatorname{diag}(f) p}{f^T p}$$

The fixed-points of $\mathcal{G}$ satisfy the equation

$$U \operatorname{diag}(f) p = (f^T p) p$$

In other words, the fixed-point population is an eigenvector of the matrix $U \operatorname{diag}(f)$. The corresponding eigenvalue is the average fitness of this population. Again, the Perron-Frobenius Theorem tells us that there is exactly one fixed-point in the simplex, and that this corresponds to the eigenvalue of largest magnitude. It is also interesting to note that every point in the interior of the simplex corresponds to a fixed-point for some fitness function [17]. Choosing

$$f = (U \operatorname{diag}(p))^{-1} p$$

(or a scalar multiple) does the trick, assuming that U is invertible (which it usually is).

Finding the eigenvectors of $U \operatorname{diag}(f)$ becomes computationally intractable for large search spaces. However, in the case of *functions of unitation* (where the fitness of a string depends only on the number of ones it contains), the search space can be collapsed onto one of size $\ell + 1$. Each element of the new search space simply counts the number of ones in the strings of the old search space. One defines the mutation matrix U by

$$U_{i,j} = \sum_{k=0}^{\ell-j} \sum_{l=0}^{j} [i = j + k - l] \binom{\ell - j}{k} \binom{j}{l} u^{k+l} (1 - u)^{\ell - k - l}$$

(see [14] for details). Note that i and j are now unitation classes, taking values from $\{0, 1, \ldots, \ell\}$.

As an example, consider the OneMax function $f_k = d(k, 0)$. This is obviously a function of unitation. Working with the reduced search space $\Omega = \{0, 1, \ldots, \ell\}$, we can calculate the fixed-points of $\mathcal{G}$ for quite large problems. For example, for $\ell = 50$ and $u = 0.05$, the fixed-point distribution is illustrated in figure 1, with a corresponding average fitness of 28.65. As hinted at earlier, the effects of other fixed-points, even those outside the simplex, may be important in explaining where the genetic algorithm will spend most of its time. A number of examples are worked out in [17].

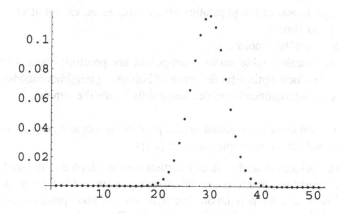

Fig. 1. The fixed-point population, shown as a distribution over unitation classes, for the OneMax function with $\ell = 50$, $u = 0.05$ and no crossover. The corresponding average fitness is 28.65.

Now let us consider the case when there is crossover, but no mutation (that is, set $u = 0$ and $\chi_0 < 1$). A moment's thought tells us that, as with selection-only, the uniform populations are again absorbing states. This is because neither selection nor crossover can introduce new genetic material. It is also easy to check that such populations are also fixed-points of $\mathcal{G} = \mathcal{C} \circ \mathcal{F}$. However, it turns out that some of these points are *asymptotically stable*, while others are not. An asymptotically stable fixed-point is one for which, if the initial population is chosen close enough, then the evolutionary trajectory is guaranteed to converge to the fixed-point. A theorem by Michael Vose and Alden Wright proves the following interesting property. Each vertex of the simplex corresponds to a uniform population, and therefore also corresponds to a single point of the search space (with N copies). If the vertex is an asymptotically stable fixed-point of $\mathcal{G}$ then the corresponding element of the search space must be a *local optimum* with regards the usual Hamming neighbourhood. That is, it must have a fitness that is greater than all strings which differ from it in exactly one bit position [26]. It may happen that there are also some fixed-points that are not vertices. However, it is believed that any such points would not be asymptotically stable. This has not be proved, however, and remains an open question.

Slightly more is known about a different form of crossover — one that is not a quadratic operator. The so-called *genepool* crossover works on the whole population simultaneously. We generate each new string of the next generation one bit at a time, by choosing its value according to the frequency of ones and zeros in the corresponding position in the current population. The effect of this is to immediately de-correlate the bit values. In effect, the algorithm goes in one step to the Geiringer limit. If we use genepool crossover with proportional selection, the algorithm is known as UMDA (Univariate Marginal Distribution Algorithm), and is related to the class of *estimation of distribution* algorithms. This algorithm has been analysed thoroughly by Heinz Mühlenbein [13] who shows, amongst other things:

- The average fitness of the population always increases, except at vertices (that is, uniform populations).
- The vertices are fixed-points.
- The asymptotically stable vertex fixed-points are precisely those vertices which correspond to local optima (in the sense of Hamming neighbourhoods).
- There are no asymptotically stable fixed points inside the simplex.

The situation when mutation is added to genepool crossover and proportional selection has been analysed for a few simple case (see [29]).

Some of the techniques described in this section can be adapted to the study of *steady-state* genetic algorithms, in which a single offspring is created in a given time-step, and it is inserted back into the population. The dynamics of this approach can be approximated by considering time to be continuous [28]. One of the crucial implementation decisions that has to be made is how to decide which element of the population is to be replaced by the newly created offspring. Two popular choices are:

1. Replace the worst item in the population.
2. Replace a random item.

This choice has a strong influence on the fixed-points of the infinite population model. In the first case (replace worst), it can be shown that the fixed-points are uniform populations provided some technical conditions are met (which they are, for example, if there is a positive mutation rate). In the second case, however, (replace random) the fixed-points turn out to be identical to the fixed-points of the generational GA, using the same selection and mixing heuristic.

4   Search spaces with variable-sized objects

The simple genetic algorithm has, as its canonical search space, the finite set of binary strings of some fixed length. Aspects of the theory have been generalised to other finite search spaces, for example, for combinatorial problems such as the Travelling Salesman Problem [20]. However, there are a number of application areas in which the search space is not necessarily finite, due to the fact that the objects in the search space have variable size. One example of this occurs when the elements of the search space are rules, made up of IF and THEN parts, each of which may contain a variable number of clauses (conditions which have to be true, in the case of the IF-part, and things that become true in the case of THEN-parts). A related example is the problem of trying to evolve rules for a grammar, for example, if the representation is a developmental one, like an L-system. In the case of Learning Classifier Systems, variable-length structures have been used in conjunction with XCS [11]. They have also been used to grow hidden layers in neural classifiers [4].

Perhaps the most common example of variable-length structures comes from the field of Genetic Programming, in which the individuals are (representations of) computer programs. Programs can be represented as sequences of instructions (for example, an assembly language program) or as a parse-tree of a lisp-like language. While a lot of work has gone into developing these representations and the operators that act upon them, as well as empirical work justifying their use in practice, there is nothing like a comparable amount of theoretical analysis or understanding. Most of what is known has been developed by Riccardo Poli, William Langdon and colleagues [10]. Much of this work concerns the generalisation of ideas from genetic algorithm theory to the case of trees. We will look briefly at the rather simpler case of variable-length linear structures (that is, strings), where there is a more obvious extension of the situation for genetic algorithms.

There are many ways in which mutations and crossovers can be defined for variable-length strings. We will give as examples three of the most common (which are, in fact, counterparts of operators defined for trees).

Crossover Two strings are truncated at random places along their length. The left part of one of the strings is appended to the right part of the other string.

Grow mutation A string is truncated at some random point along its length. New symbols are then added according to a sequence of Bernoulli trials. That is, there is some probability q that a new symbol (chosen arbitrarily from the available alphabet) will be appended, and a probability of $1 - q$ that the growth process will stop. Growth continues until this latter event happens.

Full mutation A string is truncated at some random point along its length. A new random string of length D (which is a parameter to be set by the user) is appended.

Some of the effects of applying these operators have been investigated [19], if we just consider what is happening to the *average length* of strings in the population. In the absence of selection, this is the only significant factor.

For crossover, it should first be noted that the average length of items in the population does not change. The fixed-point therefore depends on the average length of the initial population. It can be shown that populations of the form

$$p_k = (1 - a)^2 k a^{k-1}$$

are fixed-points of this kind of crossover, where a is a parameter related to the average length, m, of the initial (and subsequent) populations by

$$a = \frac{m-1}{m+1}$$

This was originally proved in [12]. Some examples of this distribution are illustrated in figure 2.

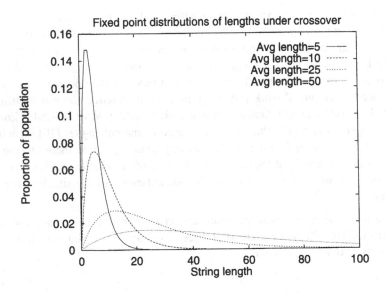

Fig. 2. Fixed-point populations for crossover, acting on variable length strings, for different average lengths. (Graph taken from [19].)

Remarkably, for grow mutation, the fixed-points have exactly the same form as for crossover, with the parameter a set to the value q. That is, the fixed-point of grow mutation is:

$$p_k = (1-q)^2 k q^{k-1}$$

(see [19]).

For full mutation, the result obviously depends on the value of the parameter D. For $D > 2$ there is no known closed form for the fixed-point. However, for $D = 2$ we have:

$$p_k = \frac{[k > 1]}{e(k-2)!}$$

[19] which is illustrated in figure 3. Notice that the height of the "plateau" is exactly $1/e$.

There has been very little work done on studying the combination of selection and crossover or mutation for variable-length structures. This is partly because the subject

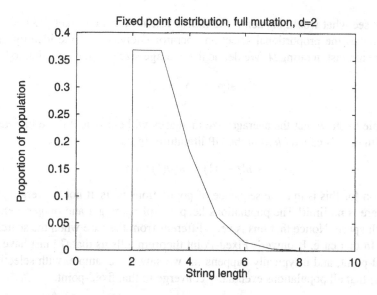

Fig. 3. The fixed-point population for full mutation with $D = 2$. (Graph taken from [19].)

is rather difficult, but mostly because there are very few researchers working on these problems.[3] We give a simple example to illustrate some of the technicalities.[4]

Let us, then, consider the simplest possible case of an infinite search space, namely taking Ω to be the natural numbers $\{1, 2, 3, 4, \ldots\}$. You can think of each integer k as representing the *size* of a program or string. Populations are then associated with probability distributions over this set:

$$p = (p_1, p_2, p_3, \ldots)$$

where as usual p_k indicates the proportion of the population which is made up of copies of item k. The set of all such probability distributions is:

$$\Lambda = \left\{ (p_1, p_2, p_3, \ldots) | \sum_k p_k = 1 \text{ and } p_k \geq 0 \text{ for all } k \right\}$$

We will use fitness proportional selection (which is well-defined as long as the fitness function is bounded) and the following mutation operator (for some rate $0 < u < 1$):

$$\mathcal{U}(p)_k = \begin{cases} (1-u)p_1 & \text{if } k = 1 \\ (1-u)p_k + up_{k-1} & \text{if } k > 1 \end{cases}$$

The effect of this mutation is that item k is mutated into item $k + 1$ with a probability u. This mutation was christened the *super-bloater* by Riccardo Poli, for reasons which we will see.

---

[3] I counted just six researchers at the time of writing!

[4] These ideas were originally presented at a Dagstuhl Seminar in January 2002 by the author.

First let's see what would happen if the fitness function is a constant $f(k) = c$ for all k. This makes the proportional selection operator equivalent to the identity. In other words, we are just iterating $\mathcal{U}$. We define the average size of the population to be

$$s(\boldsymbol{p}) = \sum_k k p_k$$

It is simple to show that the average size increases without bound if we just iterate $\mathcal{U}$. Such a situation is called *bloat* in the GP literature. In fact

$$s(p(t+1)) = s(p(t)) + u$$

The reason for this is that the sequence of populations is itself not converging to anything: there is no limit! The populations keep sampling bigger and bigger elements of the search space. Notice that this is very different from the case when the search space is finite. In that case, Brouwer's Fixed-Point theorem tells us that $\mathcal{G}$ must have at least one fixed-point, and it typically happens (as we saw, for example, with selection plus mutation) that all populations eventually converge to that fixed-point.

We could try to change this by having a non-constant fitness function, that favours smaller individuals. The resulting behaviour should be a balance of these two forces. Let's suppose we have a fitness function such that there is a fixed-point which is the population v where

$$v_k = \frac{6}{(\pi k)^2}$$

This is a valid population since

$$\sum_k v_k = \sum_k \frac{6}{(\pi k)^2} = \frac{6}{\pi^2} \sum_k \frac{1}{k^2} = 1$$

But the average size of this population is:

$$s(\boldsymbol{v}) = \sum_k k v_k = \sum_k \frac{6}{k \pi^2} = \frac{6}{\pi^2} \sum_k \frac{1}{k} = \infty$$

So it's possible that a sequence of populations might converge to this fixed-point, but nevertheless, the average size would still grow to infinity! The only question is: does there exist a fitness function such that v is the fixed-point? It turns out that the argument given above for finite search spaces, showing that any point in the simplex is a fixed-point for some fitness function, can be adapted to the infinite case, and the answer is, yes there is such a fitness function. The details are rather technical and are omitted here, but if you run an evolutionary algorithm with that fitness and the given mutation operator, you indeed see two things happening simultaneously: the population converges to the fixed-point; and the average length increases without bound (that is, the system bloats, even though it is converging to a limit).

Clearly, much more theoretical investigation is required if we are going to understand how evolution can take place in these infinite search spaces. Our example is highly simplistic, yet it already highlights some difficulties, as well as interesting phenomena that do not occur in the finite case.

5 Dynamic fitness functions

We now return to finite search spaces, and the simple genetic algorithm, but consider the situation when the fitness function changes over time. There are a number of situations when this might happen, for example:

- We are using the genetic algorithm to try to find an optimal control strategy for a manufacturing process. The correct control settings will vary depending on parameters governing that process.
- We are trying to evolve solutions to a problem set by a user, whose requirements change with time.
- We are trying to model an ecosystem, in which the external environment is changing.

In this section we will consider the simpler case when the fitness varies in a known way with time. When the fitness also depends on the state of the population, we have a *co-evolutionary* algorithm, which we address in the following section. One option, if the fitness function changes, is simply to run the genetic algorithm over again from scratch. However, we will assume that the changes are rapid enough that this method would not be able to keep track of them, and we will assume that the changes are smooth enough that continuing the evolution from the current population will give viable results.

Let us suppose we have a genetic algorithm with proportional selection and mutation, but no crossover. We consider the case when the fitness function is varying in a known periodic manner, with period τ. That is, there is a sequence of fitness functions, $f(0), f(1), \ldots, f(\tau - 1)$, and the function that applies in generation t is $f(t \bmod \tau)$. We define a corresponding set of selection heuristics:

$$\mathcal{F}_i(p) = \frac{\mathrm{diag}\,(f(i))p}{f(i)^T p}$$

Of course, we no longer expect the evolutionary dynamics to have a fixed-point, but rather a periodic cycle, of order τ. For each $i \in \{0, 1, \ldots, \tau - 1\}$, define an operator $\mathcal{G}_i = \mathcal{U} \circ \mathcal{F}_i$, and then define operators $\mathcal{H}_i$ as follows:

$$\mathcal{H}_0 = \mathcal{G}_{\tau-1} \circ \mathcal{G}_{\tau-2} \circ \cdots \circ \mathcal{G}_0$$
$$\mathcal{H}_1 = \mathcal{G}_{\tau-2} \circ \mathcal{G}_{\tau-3} \circ \cdots \circ \mathcal{G}_0 \circ \mathcal{G}_{\tau-1}$$
$$\mathcal{H}_2 = \mathcal{G}_{\tau-3} \circ \mathcal{G}_{\tau-4} \circ \cdots \circ \mathcal{G}_0 \circ \mathcal{G}_{\tau-1} \circ \mathcal{G}_{\tau-2}$$
$$\vdots$$
$$\mathcal{H}_{\tau-1} = \mathcal{G}_0 \circ \mathcal{G}_{\tau-1} \circ \cdots \circ \mathcal{G}_2 \circ \mathcal{G}_1$$

Then the periodic attractor for the system is $v(0), v(1), \ldots, v(\tau - 1)$ where

$$\mathcal{H}_i(v(i)) = v(i)$$

for each $i \in \{0, 1, \ldots, \tau - 1\}$. We can solve for these vectors using a method similar to the case of a static fitness function. For example, the vector $v(0)$ is an eigenvector of the matrix

$$U \operatorname{diag}(\boldsymbol{f}(\tau - 1))U \operatorname{diag}(\boldsymbol{f}(\tau - 2))\ldots U \operatorname{diag}(\boldsymbol{f}(0))$$

The corresponding eigenvalue is not the average fitness, but rather

$$\prod_{i=0}^{\tau-1} \boldsymbol{f}(i)^T \boldsymbol{v}(i)$$

As a simple example, lets consider the following pair of functions of unitation:

$$f_0(x) = [d(x,0) = \ell/2 - 5] + 1, \quad f_1(x) = [d(x,0) = \ell/2 + 5] + 1$$

That is, f_0 scores 2 for strings that contain exactly $\ell/2 - 5$ ones, and 1 otherwise and f_2 scores 2 for strings with $\ell/2 + 5$ ones, and 1 otherwise. Using a string length $\ell = 50$ and a mutation rate $u = 0.001$, we calculate the fixed-points of $\mathcal{H}_0$ and $\mathcal{H}_1$ as described above to find the periodic attractor, shown in figure 4. We can see that at each generation, the majority of the population is clustered around the optimal point. However, a large minority of individuals are placed at the optimal point for the other fitness function. In this way, the population manages to keep track of the changing optimum over time.[5]

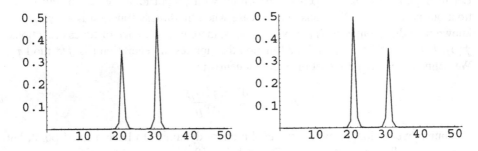

Fig. 4. The periodic attractor corresponding to f_0 and f_1 described in the text.)

6   Co-evolutionary dynamics

In this final section, we will continue to consider dynamic fitness functions, but now we assume that the fitness changes in a way that depends on the current population. That is, the fitness of an individual will depend on which other individuals are sharing the same population. This situation is sometimes referred to as *co-evolution*. One of the

---

[5] For an alternative, but equivalent, method, see [16]

most common examples of this phenomenon in evolutionary algorithms, is in *classifier systems*. Here, the aim is to evolve a population of rules which will control the activities of an agent in some environment. The success of a rule depends not only on the environment, but also on what other rules are in the population, which it can work with to form chains of activations.

In theoretical biology, the situation is described using evolutionary game theory [6]. It is assumed that each member of the population interacts with all the other members (or perhaps randomly encounters a sample) during one time period. As a result of each interaction, the individual either gains or loses some fitness. Its final fitness is a measure of its ability to reproduce. In order to represent what is happening, we need to store all of the possible interactions. We do this in a matrix, called the *payoff matrix*, which we will denote P. The entry $P_{i,j}$ gives us the increment in fitness to individual i as a result of an encounter with individual j. Negative entries correspond to a decrease in fitness. We assume that each individual i has some baseline fitness c_i, which is how fit it would be if there were no encounters. If the current population is $p \in \Lambda$, then we can work out the fitness of individual i in this population as

$$f_i = c_i + \sum_j P_{i,j} p_j$$

In other words, the fitness function (considered as a vector) can be written as a function of the population as follows:

$$f(p) = Pp + c$$

We will assume that the time period under consideration is one generation, and also that individuals reproduce in proportion to their fitness. In this case, we get a discrete-time *replicator equation*:

$$p(t+1) = \frac{\mathrm{diag}\,(f(p))p(t)}{f(p)^T p(t)}$$

which is the same as we had for proportional selection for the simple genetic algorithm, except now the fitness varies as a function of p.

The assumption that each individual interacts with all the others in the population (with equal probability), and that the results of these interactions are independent in their effects, results in the fitness being a *linear* function of the population vector. It is possible, in what follows, to relax this assumption, and consider fitness to depend in any (possibly non-linear) way on p. However, to keep things simple, we will stick with the assumption of linearity.

We can solve for the fixed-points of the system as follows. First, let's suppose that there is a fixed-point v inside the simplex (so all its entries are non-zero). Let the average fitness of this population be λ. Then v must satisfy

$$v = \frac{\mathrm{diag}\,(f)v}{\lambda} = \frac{\mathrm{diag}\,(v)f}{\lambda} = \frac{\mathrm{diag}\,(v)(Pv + c)}{\lambda}$$

and so, rearranging,

$$v = \lambda P^{-1}\mathbf{1} - P^{-1}c$$

where $\mathbf{1}$ is the vector containing all ones, and we have assumed that P is invertible. The value of λ can be found from the fact that $\sum_k v_k = 1$, giving

$$\lambda = \frac{1 + \sum_k \left(P^{-1}c\right)_k}{\sum_k \left(P^{-1}\mathbf{1}\right)_k}$$

The assumption that all the entries of v were non-zero was important, because during this calculation, we need the inverse of the matrix $\mathrm{diag}(v)$ to exist. So what happens if this is not the case, and $v_k = 0$ for some $k \in \Omega$? If this happens, then the type k is extinct — we only have selection, so it can never come back. Since it is no longer a player in the game, we simply remove the corresponding row and column from P to get a reduced payoff matrix. We also delete c_k from c. Then we can proceed as before (assuming the new P is invertible) to find a new fixed-point. There are, therefore, potentially 2^n possible fixed-points of the system. An algorithm for finding them is:

1. For each subset $A \subset \Omega$ repeat the following:
2. Remove rows and columns corresponding to elements of A from the payoff matrix.
3. Remove the corresponding elements from c.
4. If the resulting payoff matrix is invertible, find the fixed-point (as above).

There is no point taking $A = \Omega$ as this corresponds to the case where everything is extinct. Also, the cases where there is just one species in the game give rise to the trivial fixed-points e_j at the corners of the simplex.

Let's work through an example with three strategies. We expect to get a maximum of $2^3 - 1 = 7$ fixed-points (ignoring 0). We assume that $c = 0$ and consider the payoff matrix:

$$P = \begin{pmatrix} 4 & 2 & 1 \\ 3 & 3 & 1 \\ 2 & 1 & 2 \end{pmatrix}$$

This matrix is invertible and

$$P^{-1} = \begin{pmatrix} 0.5556 & -0.3333 & -0.1111 \\ -0.4444 & 0.6667 & -0.1111 \\ -0.3333 & 0 & 0.6667 \end{pmatrix}$$

Multiplying P^{-1} by $(1,1,1)$ and normalising gives the fixed-point $(0.2, 0.2, 0.6)$ corresponding to $\lambda = 1.8$, which is the average fitness value at that fixed-point.

Now we must consider what happens when $p_0 = 0$. The projection of P is

$$\begin{pmatrix} 3 & 1 \\ 1 & 2 \end{pmatrix}$$

which is also invertible. Multiplying the inverse by $(1, 1)$ and normalising gives us the fixed-point $(0, 1/3, 2/3)$ corresponding to fitness value $\lambda = 5/3$. Note that this is the fitness value for strategies 1 and 2. The fitness for the extinct strategy 0 is not defined (and, of course, irrelevant).

When $p_1 = 0$ a similar process gives us the fixed-point $(1/3, 0, 2/3)$ corresponding to $\lambda = 2$. And when $p_2 = 0$ we get the fixed-point $(0.5, 0.5, 0)$ with $\lambda = 3$.

You can also check that $(1, 0, 0)$, $(0, 1, 0)$ and $(0, 0, 1)$ are fixed-points with corresponding average fitness 4, 3 and 2 respectively. We therefore have seven fixed-points for this system, all of which could, in fact, correspond to actual finite populations. Only one of the fixed-points, however, contains copies of all three strategies.

The fact that there are potentially so many fixed-points can have a large influence on the dynamics of the system, when run with a finite population. Suppose the initial population contains members of all elements of Ω, and there is a fixed-point in the simplex. As the evolution unfolds, sampling effects due to the finite size of the population will create fluctuations. This might lead to an accidental extinction, if there aren't enough representatives of a particular individual in the population. Once this happens, the game changes, and there is a new fixed-point. As the evolutionary trajectory heads in that direction, it is possible that further extinctions might take place. Such events are often observed in runs of co-evolutionary genetic algorithms, with the end result being a population with many elements of the search space missing. One extinction event can trigger others in a cascade. For an investigation into the dynamics of extinctions, see [27].

Some co-evolutionary algorithms are implemented with a number of isolated populations. The system is co-evolutionary in the sense that the fitness function that is applied to one population depends on the contents of the other populations. An example of this is *co-operative* co-evolution, in which different parts of a problem are assigned to different populations. The fitness of an element from one population depends on which members of the other populations it is put together with in order to be evaluated. We can extend our analysis to this situation in a straightforward manner. Let us suppose, for simplicity, that there are two populations, p_1 and p_2. The fitness function to be applied to the first population is a (linear) function of the second, and vice versa:

$$f_1 = P_1 p_2, \; f_2 = P_2 p_1$$

In this example, we take the baseline fitness to be zero. Again, by assuming that there are fixed-points for each population v_1 and v_2 in the interior of the simplex, we find:

$$v_1 = \lambda_1 (P_2)^{-1} 1$$

and

$$v_2 = \lambda_2 (P_1)^{-1} 1$$

where the constants λ_1 and λ_2 can be found by normalising. The same considerations as before apply if certain elements become extinct.

7 Conclusions

Learning classifier systems adapt their behaviour on two different time-scales. In the short-term, the existing population of classifiers responds to the changing environment. In the long-term, the underlying genetic algorithm seeks to evolve better populations. As such, the theory of genetic algorithms provides a starting point for building a framework of LCS population dynamics. However, the description of the Simple Genetic Algorithm must be extended in important ways, both structurally and dynamically. Structurally, because it is possible for classifiers to take on a number of syntactic forms, rather than simply being fixed-length strings (for example, S-expressions and neural classifiers have recently been proposed [11, 4]). Dynamically, because the fitness of an individual classifier depends critically on the context of the population in which it resides. This makes the long-term process a *co-evolutionary* one, based on performance data from the short-term success of the system. Moreover, the environment faced by the system may itself change in time. Consequently, these extensions to standard GA theory, which have only recently begun, are essential to the development of a fuller understanding of learning classifier systems.

References

1. A.Prügel-Bennett and A.Rogers. Modelling genetic algorithm dynamics. In L. Kallel, B. Naudts, and A. Rogers, editors, *Theoretical aspects of evolutionary computation*, pages 59–86. Springer, 2001.
2. H.-G. Beyer. *Theory of evolution strategies*. Springer, 2001.
3. J. Branke. *Evolutionary optimization in dynamic environments*. Kluwer Academic Publishers, 2001.
4. L. Bull. On using constructivism in neural classifier systems. In J. Merelo, P. Adamidis, H.-G. Beyer, J.-L. Fernandez-Villacanas, and H.-P. Schwefel, editors, *Parallel Problem Solving from Nature — PPSN VII.*, pages 558–567. Springer Verlag, 2002.
5. H. Geiringer. On the probability theory of linkage in mendelian heredity. *Annals of Mathematical Statstics*, 15(1):25–57, 1944.
6. J. Hofbauer and K. Sigmund. *Evolutionary games and population dynamics*. Cambridge University Press, 1998.
7. D. L. Isaacson and R. W. Madsen. *Markov chains: theory and applications*. John Wiley & Sons, 1976.
8. T. Jansen and I. Wegener. Real royal road functions — where crossover provably is essential. In L. Spector, E. D. Goodman, A. Wu, W. B. Langdon, H.-M. Voigt, M. Gen, S. Sen, M. Dorigo, S. Pezeshk, M. H. Garzon, and E. Burke, editors, *Proceedings of the Genetic and Evolutionary Computation Conference (GECCO 2001)*, pages 1034–1041. Morgan Kaufmann, 2001.
9. L. Kallel, B. Naudts, and A. Rogers, editors. *Theoretical aspects of evolutionary computation*. Springer, 2001.
10. W. B. Langdon and R. Poli. *Foundations of genetic programming*. Springer, 2002.
11. P.-L. Lanzi. Extending the representation of classifier conditions part II: from messy coding to S-expressions. In W. Banzhaf, J. Daida, A. E. Eiben, M. H. Garzon, V. Honovar, M. Jakiela, and R. E. Smith, editors, *Proceedings of the Genetic and Evolutionary Computation Conference (GECCO 1999)*, pages 11–18. Morgan Kaufmann, 1999.

12. N. F. McPhee, R. Poli, and J. E. Rowe. A schema theory analysis of mutation size biases in genetic programming with linear representations. In *Proceedings of the 2001 Congress on Evolutionary Computation CEC 2001*, pages 1078–1085, Seoul, Korea, May 2001.

13. H. Mülenbein and T. Mahnig. Convergence theory and applications of the factorized distribution algorithm. *Journal of Computing and Information Technology*, 7:19–32, 1999.

14. E. Van Nimwegen, J. P. Crutchfield, and M. Mitchell. Finite populations induce metastability in evolutionary search. *Physics Letters A*, 229(2):144–150, 1997.

15. C. R. Reeves and J. E. Rowe. *Genetic algorithms: principles and perspectives*. Kluwer Academic Publishers, 2002.

16. J. E. Rowe. Finding attractors for periodic fitness functions. In W. Banzhaf, J. Daida, A. E. Eiben, M. H. Garzon, V. Honovar, M. Jakiela, and R. E. Smith, editors, *Proceedings of the Genetic and Evolutionary Computation Conference (GECCO 1999)*, pages 557–563. Morgan Kaufmann, 1999.

17. J. E. Rowe. Population fixed-points for functions of unitation. In W. Banzhaf and C. R. Reeves, editors, *Foundations of Genetic Algorithms*, volume 5, pages 69–84. Morgan Kaufmann, 1999.

18. J. E. Rowe. A normed space of genetic operators with applications to scalability issues. *Evolutionary Computation*, 9(1):25–42, 2001.

19. J. E. Rowe and N. F. McPhee. The effects of crossover and mutation operators on variable length linear structures. In L. Spector, E. D. Goodman, A. Wu, W. B. Langdon, H.-M. Voigt, M. Gen, S. Sen, M. Dorigo, S. Pezeshk, M. H. Garzon, and E. Burke, editors, *Proceedings of the Genetic and Evolutionary Computation Conference (GECCO 2001)*, pages 535–542. Morgan Kaufmann, 2001.

20. J. E. Rowe, M. D. Vose, and A. H. Wright. Group properties of crossover and mutation. *Evolutionary Computation*, 10(2):151–184, 2002.

21. J. E. Rowe, M. D. Vose, and A. H. Wright. Structural search spaces and genetic operators. *Evolutionary Computation*, 12(4), 2004.

22. G. Rudolph. *Convergence properties of evolutionary algorithms*. Kovacs, 1997.

23. J. Shapiro. Statistical mechanics theory of genetic algorithms. In L. Kallel, B. Naudts, and A. Rogers, editors, *Theoretical aspects of evolutionary computation*, pages 87–108. Springer, 2001.

24. C. R. Stephens and H.Waelbroeck. Schemata evolution and building blocks. *Evolutionary Computation*, 7(2):109–124, 1999.

25. M. D. Vose. *The simple genetic algorithm*. MIT Press, 1999.

26. M. D. Vose and A. H. Wright. Stability of vertex fixed points and applications. In L. D. Whitley and M. D. Vose, editors, *Foundations of Genetic Algorithms*, volume 3, pages 103–114. Morgan Kaufmann, 1995.

27. C. O. Wilke. *Evolutionary dynamics in time-dependent environments*. Shaker Verlag, 1999.

28. A. H. Wright and J. E. Rowe. Continuous dynamical system models of steady-state genetic algorithms. In W. N. Martin and W. M. Spears, editors, *Foundations of Genetic Algorithms*, volume 6, pages 209–226. Morgan Kaufmann, 2001.

29. A. H. Wright, J. E. Rowe, R. Poli, and C. R. Stephens. Bistability in a gene pool ga with mutation. In K. De jong, R. Poli, and J. E. Rowe, editors, *Foundations of Genetic Algorithms*, volume 7, pages 63–80. Morgan Kaufmann, 2003.

Approximating Value Functions in Classifier Systems

Lashon B. Booker

The MITRE Corporation
7515 Colshire Drive
McLean, VA 22102-7508, USA
booker@mitre.org

1 Introduction

While there has been some attention given recently to the issues of function approximation using learning classifier systems (e.g. [13, 3]), few studies have looked at the quality of the value function approximation computed by a learning classifier system when it solves a reinforcement learning problem [1, 8]. By contrast, considerable attention has been paid to this issue in the reinforcement learning literature [12]. One of the fundamental assumptions underlying algorithms for solving reinforcement learning problems is that states and state-action pairs have well-defined values that can be computed and used to help determine an optimal policy. The quality of those approximations is a critical factor in determining the success of many algorithms in solving reinforcement learning problems.

In most classifier systems, the information about the value function is stored and computed by individual rules. Each rule maintains an independent estimate of the value of taking its designated action in the states that match its condition. From this standpoint, each rule is treated as a separate function approximator. The quality of the approximations that can be achieved by simple estimates like this is not very good. Even when those estimates are pooled together to compute a more reliable collective estimate, it is still questionable how good the overall approximation will be. It is also not clear what the best way is to improve the quality of those approximations.

One approach to improving approximation quality is to increase the computational abilities of individual rules so that they become more capable function approximators [13]. Another idea is to look back to the original concepts underlying the classifier system framework and seek to take advantage of the properties of distributed representations in classifier systems [2]. This paper follows the latter approach. We describe a new way to tap the distributed representational power present in a collection of rules to improve the quality of value function approximations. The basic idea is to treat rules as features that collectively specify a linear gradient-descent function approximator.

The paper begins with a brief overview of the role of value functions and approximations in reinforcement learning. Then we examine the corresponding issues in classifier systems and make an empirical comparison with a widely

L.B. Booker: *Approximating Value Functions in Classifier Systems*, StudFuzz **183**, 45–61 (2005)
www.springerlink.com

used technique from the reinforcement learning community. This comparison points out some weaknesses in the typical classifier system methods. Finally, a new approach to value function approximation — called hyperplane coding — is introduced along with empirical results showing how effective it is.

2   Value Function Approximation and Reinforcement Learning

We begin with a formal characterization of value functions in the context of reinforcement learning problems. Assume that the problem environment can be characterized as a discrete time, stochastic dynamic system with a finite set of states. This setting is well studied in the theory of reinforcement learning, as it provides the starting point for the analysis of finite Markov decision processes. A Markov decision process satisfies the Markov property and therefore can be characterized by the one-step dynamics of the environment. This means that in addition to specifying all possible states and actions, a problem definition includes two other things: transition probabilities $p_{ij}(u)$, which give the probability that the next state is j if action u is taken in the current state i; and, scalar rewards $r_i(u)$ which indicate the immediate feedback available after applying action u in state i. An agent trying to solve such a problem uses a decision policy π to specify which action is selected as a function of the current observed state. A decision policy is a function of states and actions that computes the probability of taking action u when in state i. Given a fixed policy π, the value function is a mapping that computes, for each state, the long term expected reward the agent will accrue by using the policy π to make decisions. If a discounted reward criterion is used to compute the long term reward, the formal recursive definition of the value function is given by

$$V_\pi(i) = \sum_u \pi(i, u) \sum_j p_{ij}(u)[r_i(u) + \gamma V_\pi(j)]$$

where $\gamma \in [0, 1]$ is a discount factor that determines the influence of future rewards on current decisions.

Most approaches to solving reinforcement learning problems explicitly compute and store some representation of V_π. For very simple problems, a lookup table is an adequate way to represent the value function. In most cases of interest, however, the input space is too large to represent V_π exhaustively in tabular form so the function must be represented more compactly. Efficient storage is not the only important issue though. In a large state space the learning agent will only directly experience a relatively small number of inputs. The agent nevertheless needs to leverage that experience to determine how to behave when it encounters inputs that have not been seen before. This implies that generalization is a key issue for reinforcement learning problems with large state spaces. The most common approach to addressing these issues is to use function approximation techniques to compute a compact representation of V_π that generalizes well.

The approach to approximating V_π used in learning classifier systems belongs to a class of techniques known as soft state aggregation [10]. In the simplest forms of state aggregation, the states are partitioned into a set of disjoint groups or clusters. A reinforcement learning problem can be solved at the cluster level to compute a value function for the clusters. The value of a cluster is then used as the value for each of the states in that cluster. Soft state aggregation techniques allow a single state to belong to more than one cluster, providing for cluster overlap. This is accomplished by defining cluster probabilities $P(x|i)$ that specify the degree to which state i is associated with cluster x. The value for a state is given by a weighted average of the values of the clusters the state is associated with; that is,

$$V_\pi(i) = \sum_x P(x|i)V_\pi(x)$$

Rule input conditions designate the clusters of states used by learning classifier systems. Each condition represents a set of states whose value is summarized in various ways by the rule's utility measure. In XCS, for example, a cluster's value is represented by the prediction parameter of the corresponding rule. The cluster probabilities are given by the rule's fitness divided by the sum of the fitnesses of all the rules matching state i.

While state aggregation approaches to function approximation can be useful in some settings, they are known to have serious shortcomings [12]. First, they tend to scale poorly as the number of dimensions of the state space increases. Second, large numbers of clusters may be needed to represent smooth functions accurately[1]. The most widely used approaches to function approximation for reinforcement learning avoid these problems by relying on linear gradient-descent methods.

The remainder of this paper takes a brief look at linear gradient-descent methods and one important special case that uses binary features. We then propose a new approach to using linear gradient descent in a classifier system setting and present empirical results showing that the idea has merit.

3   Linear Approximations and Coarse Coded Features

Linear gradient-descent methods for value function approximation begin with a linearly parameterized representation of the value function given by

$$V(x_t) = \sum_i w_i(t)\phi_i(x_t)$$

where the ϕ_i are features defined on the state space and the w_i are real-valued adjustable weight parameters. The weights are adjusted to try to reduce the error

---

[1] This limitation will become more important to the classifier system community as classifier systems are applied to function approximation problems [13].

on the observed sample points x, and to generalize from that data to provide good approximations for other points that have not yet been seen.

Gradient-descent methods try to minimize error by adjusting the weights on each step in the direction that reduces error the most. In the linear case, the gradient descent update for adjusting the weights is given by

$$w_i(t+1) = w_i(t) + \alpha[v(t) - V(x_t)]\nabla_{w_i} V(x_t)$$

where $\nabla_{w_i} V(x_t) = \phi_i(x_t)$ is the gradient of the linear function with respect to weight parameter w_i and $v(t)$ is the true function value for x_t.

Linear gradient-descent methods are simple and they are particularly well-suited to reinforcement learning [12]. A key aspect determining how well these methods work in practice, though, is the quality of the features they use. The features must represent whatever task-relevant qualities of the state may be needed to discriminate one state from another, as well as any associated feature interactions that may be important.

3.1   Tile coding

Coarse coding [7] is a general approach to defining a set of adequate features. In this form of representation, each feature corresponds to some subset of the state space (the feature's "receptive field"). For a given state, a feature is said to be activated if the state belongs to that receptive field. The representation of state is coarse coded in the sense that the receptive fields overlap to produce a distributed representation whose acuity is proportional to the number of features activated in a given state. One general purpose way to define receptive fields suitable for efficient on-line learning is called tile coding [12].

Tile coding is a particular form of coarse coding in which the receptive fields for all features are organized into exhaustive partitions of the input space. The features are assumed to be binary, the receptive fields are called *tiles*, and each partition is called a *tiling*. The tilings are offset from each other in order to achieve the overlap needed for local generalizations. For a single input dimension, the offsets typically used in tile coding are given by $i(w/n)$ where i is the index of the tiling, w is the tile width, and n is the number of tilings ($0 \leq i < n$). There are several advantages to organizing the receptive fields in this way. Every point in the input space activates the same number of tiles, so there is strict control over the density of tiles and the resulting precision of the approximation. It is also easy to set the learning rate for a linear gradient-descent function approximator based on tile coding. Since the number of features active for each point is equal to the number of tilings m, the learning rate can be expressed intuitively as a fraction of the rate $1/m$ which gives exact one-trial learning. The weight update for activated features is given by

$$w_i(t+1) = w_i(t) + \frac{\alpha}{m}[v(t) - V(x_t)]$$

where α is the desired fraction.

Tile coding has been been used extensively for reinforcement learning, and the overall coarse coding approach is known to be capable of computing high quality approximations [9]. It is not clear how well classifier system methods compare to these approaches from the standpoint of function approximation. We try to answer that question with an empirical comparison of tile coding with the widely used classifier system mechanisms in XCS [4] for predicting expected payoff.

3.2   Comparing tile coding with XCS predictions

The effectiveness of classifier system methods for function approximation can be assessed by using function values as rewards [13] and allowing the system to generate outputs in the usual way. In order to test the XCS prediction mechanism, a skeletal classifier system was implemented. This skeletal system has traditional ternary rules with no actions and no rule discovery mechanisms. On every step the system is presented with a data point x, and the reward received is the function value $f(x)$. The system forms a match set and proceeds to update the basic XCS parameters: experience, prediction, prediction error, and fitness. The system prediction is calculated in the usual way and that prediction becomes the system's estimate for the value of x. The parameter settings were consistent with those used for XCS in the literature [13]: learning rate 0.2, error threshold 0.2, fitness power 5.0, and fitness scale (i.e., α) 0.1. See Butz and Wilson [4] for details about these parameters and computations.

The test function suite was taken from a set of functions proposed by Donoho and Johnstone [5] that has been widely used in the literature on statistical estimation and reconstruction of signals from data. We use four one-dimensional functions — Blocks, Bumps, Doppler, and HeaviSine — that provide a good variety of spatial variability and smoothness (see definitions in the Appendix). The training data for each function was drawn from a set of 2048 equally spaced sample points. A separate distinct set of 2000 equally spaced sample points was set aside to use as a test set. The quality of an approximation is measured in terms of the average squared error at those sample points. More specifically, the performance measure is

$$R = n^{-1} \sum_{i=0}^{n-1} (\hat{f}(x_i) - f(x_i))^2$$

where $\hat{f}$ is the approximation and f is the true function. In all of the experiments reported here, learning proceeded over 100 trials with 10,000 steps per trial, and with a random data point selected from the training set on each step. This gave the function approximators ample time to converge to their most accurate output. Results were averaged over 10 replications, and statistical significance was assessed using a Student's t-test with significance level 0.05.

The goal of this comparison is to assess how well each approach makes use of a fixed allocation of approximation resources. For tile coding this means that the

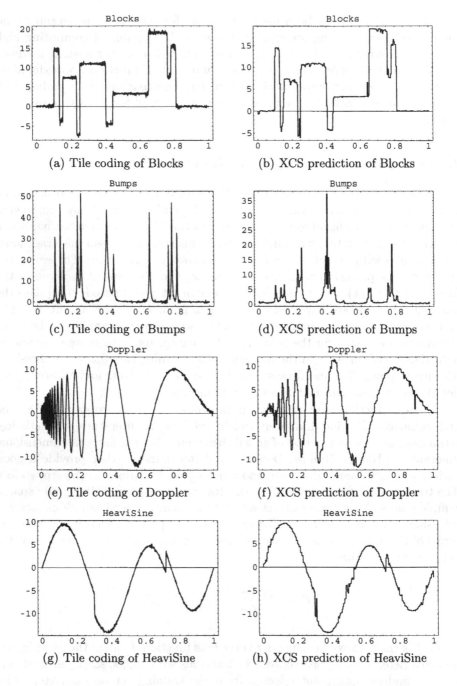

Fig. 1. Reconstructions computed by tile coding and XCS prediction

Algorithm	Approximation Error							
	Blocks		Bumps		Doppler		HeaviSine	
	Train	Test	Train	Test	Train	Test	Train	Test
Tile coding	0.06988	1.7535	0.16809	0.93068	0.03579	0.08922	0.02327	0.06458
XCS prediction	3.4697	3.2368	25.111	25.977	2.1472	2.1360	0.08345	0.08865

Table 1. Average square errors for tile coding and XCS prediction

number of tiles and the way they are organized is fixed. On these test functions, we use 2048 grid-like tiles each having width 1/256. The tiles are organized into 8 tilings that are offset as described previously. The learning rate is specified by the assignment $\alpha = 0.2$. For the XCS prediction mechanism, the population of classifiers is fixed at 2048 rules generated randomly using a probability of 1/3 for placing the # symbol at any given position in a rule condition. Each classifier condition is 8 bits, giving every classifier the same input resolution as one of the grid-like tiles.

The results on the suite of test functions are summarized in Table 1. All of the differences in performance between the tile coding approximation and the XCS prediction are statistically significant. Tile coding is substantially more effective than the XCS prediction on these functions. Tile coding shows an impressive ability to reconstruct functions with respect to the training data. Its performance on the four test functions compares favorably with results on the same data achieved by more sophisticated approximation techniques like a discrete wavelet transform [5]. The reconstructions shown in Figure 1 show that the tile coding representation has enough precision to pinpoint the location of abrupt changes in function values. Moreover, tile coding also has sufficient local generalization properties that the approximations are fairly smooth.

The XCS prediction, on the other hand, does poorly from the standpoint of both precision and smoothness. There is a sense in which this is not surprising, since the mechanisms were intended to be used in combination with rule discovery to compute a good approximation of the value function. There is a dilemma with that arrangement, however. Rule discovery depends on guidance from the prediction computations in order to know what type of rules to generate. If that guidance is poor, then rule discovery will have to thrash around somewhat randomly until it discovers something that improves the approximation.

It should be possible to take the information in a population of classifiers, even if that population is random, and reliably compute good approximations that provide useful guidance for rule discovery. What aspects of the tile coding approach can be leveraged to improve the value function approximations computed in classifier systems? One straightforward approach would be to restrict our attention somehow to hyperplane features that define an exhaustive partition. This would allow the tile coding computational mechanisms to be used directly, but would be overly restrictive from the standpoint of typical classifier system operating principles. For many reasons, the heterogeneity of the classifier population is a feature not a bug. An alternative approach is to use that heterogeneity to our advantage by devising a variation of tile coding that relies more on the strengths of distributed representations. The next section introduces a new alternative based on this idea called hyperplane coding.

3.3 Hyperplane coding

Hyperplane coding is a closely related variation of tile coding in which classifier rule conditions fill the role of tiles, and there are few restrictions on the way those "tiles" are organized. The hypothesis behind this idea is that classifier rules can be more effective as function approximators if they collectively implement a distributed representation of the value function. The distributed representation is realized by treating individual rules as features rather than as independent function approximators whose estimates are pooled to compute an overall result.

In a random population, the classifier conditions serving as tiles do not cover the space like an exhaustive partition. Nevertheless, a population of classifiers does richly cover the space with a collection of overlapping coordinate hyperplanes. Each point in the input space is covered by an expected number of tiles (or matching conditions) k given by

$$k = N \left(\frac{p_\# + 1}{2} \right)^l$$

where N is the population size, $p_\#$ is the probability of the # symbol appearing at any position in the condition, and l is the length of the condition. For the population sizes typically used in classifier system applications, this expected value is much larger than the fixed number of tilings most often used for tile coding. This bodes well for the resolution of approximations based on hyperplane coding, since greater tile density usually means higher precision.

The coarse coding idea requires the ability to represent patterns of contiguous inputs (the tiles arranged in a tiling) that can be offset from each other by arbitrary amounts. This requirement is trivial to fulfill in tile coding. The tiles are fixed sized intervals in each dimension, and the interval endpoints can be adjusted as needed. The standard syntax for the input condition of a classifier rule does not provide this kind of flexibility. It is not clear how to adjust that syntax to represent hyperplanes offset by arbitrary amounts in input space, while preserving the simple matching operations between rules and messages. For example, it is easy to represent the lower half of the input range [0, 1] with the condition 0#...#, which corresponds to the interval [0, 0.5] using the standard binary encoding. How do we represent the hyperplane corresponding to the offset interval $[0 + \epsilon, 0.5 + \epsilon]$?

One obvious way to manage this issue is to apply the offset to the input space, then define hyperplanes on that transformed space in the usual way. Looking at the offset interval $[0 + \epsilon, 0.5 + \epsilon]$ again, we can determine if some input value x belongs to that interval by checking if a message encoding the translated value $x - \epsilon$ matches the condition 0#...#. This leads to the following ideas for the way a population of classifiers is organized to implement coarse coding. Each classifier is assigned to a specific tiling[2], just like each tile belongs to a specific tiling under tile coding. In this case, though, there is no specific organization imposed on the tiling. Continuing with the analogy, we do associate a fixed offset with each tiling. The classifier system operating principles are also adjusted somewhat. Instead of having a single message matched against all rules on each cycle, we generate a separate message for each tiling. Each message is computed from the raw input by applying the offset associated with the tiling in question.

The only remaining details needing attention have to do with tile width and offsets. Since hyperplanes in general do not correspond to simple contiguous regions of the input space, some thought must be given to the issue of how to define tile width. There are several possibilities and we choose one of the simplest. The smallest possible contiguous region defined by a hyperplane is one that corresponds to a single binary value. The width of this region is simply the resolution size used to discretize the raw input. The width of every contiguous region matched completely by some hyperplane is a multiple of this resolution size. Consequently, the resolution size can be used as the tile width for all hyperplanes[3].

As noted previously, the offsets typically used in tile coding are given by $i(w/n)$ where i is the index of the tiling, w is the tile width, and n is the number of tilings ($0 \leq i < n$). Under this arrangement, every point has at least one tile in common with all points that are within a tile width away in each direction. This translation scheme uses only positive offsets that translate tiles to the right. A point gets grouped with its neighboring points on the left when the adjacent tile on the left (that does not originally contain all the points) gets translated to cover those points. This scheme does not work well in the classifier system setting, however. An unmodified input message matches classifiers representing the base tiles (i.e., tiling $i = 0$) covering a point x. If we adhere to the usual concept of a match set, the only way that x will be grouped into a tile with any other point is if the match set contains a classifier that matches both points. Offsets can change the groupings by excluding some points, but there is no way to include points that are not covered by the base match set. This makes it important to group the matched points in as many ways as possible. Accordingly, we use a more symmetric set of offsets given by $i(w/n)$ with $-n/2 \leq i < n/2$ so that points get grouped in both directions.

---

[2] We will call each major grouping of classifiers a tiling, even though the set does not partition the input space (i.e., the elements are not disjoint, and they may not span the entire space).

[3] Each hyperplane has its own smallest width determined by the position of the lowest order specific bit in the classifier condition. Giving each condition its own tile width and offset would lead to a potentially unmanageable number of messages on each cycle, though, so that option is not considered here.

| Algorithm | Approximation Error | |
Variant	Train	Test
Baseline	0.24656	0.34647
Gray code	0.22560	0.32736
Salience	0.19654	0.30969
Better offsets	0.13746	0.30148

Table 2. Average square errors for variations of hyperplane coding on the Blocks function

4 Experiments With Hyperplane Coding

In order to evaluate this idea, the skeletal classifier system described previously for the experiment with XCS prediction mechanisms was modified to implement the hyperplane coding algorithm described above. This section briefly describes that implementation, empirically evaluates its performance, then describes a series of modifications that improve performance.

4.1 Baseline implementation

The initial implementation of linear function approximation based on hyperplane coding starts with the algorithm described above and uses parameters taken from the previous experiments with tile coding. We use a population of 2048 random classifiers organized into 8 tilings of 256 classifiers each. The classifier conditions were 8 bits long to provide the same resolution for discretizing the input as the tile coding approach. Each classifier has a weight parameter w that is adjusted by gradient descent just as in tile coding. The learning parameter α for gradient decent was set to 0.2, again in agreement with the tile coding experiments. These choices give the linear approximator based on hyperplane coding roughly the same amount of approximation resources to work with as the tile coding version had.

We begin by focusing our attention on how the algorithm performs on the Blocks function. Performance on Blocks is summarized in Table 2. The average square error on the training data was 0.24656, which is a statistically significant drop in performance from the error of 0.06988 for tile coding in Table 1. Interestingly, the roles were reversed on the testing data. The average square error for hyperplane coding was 0.34647, a statistically significant improvement over tile coding's value of 1.7535. The relatively large number of hyperplanes covering each point apparently gives the hyperplane coding scheme a huge advantage in generalization. Note that the performance advantage of hyperplane coding over the XCS prediction is statistically significant on both the test and the training data.

4.2 Gray coded inputs

One of the important properties of approximation techniques like tile coding is that the generalizations they compute are localized. Points that are sufficiently close in input space will produce output values that are close. Moreover, values in widely spaced regions can be learned with relatively little interference. This property is compromised somewhat with hyperplane coding since hyperplanes are not restricted to contain localized collections of points. Some of the approximation error we observe in the results so far can probably be attributed to this effect.

If this is true, then a representation that provides more localized collections of points should boost performance. The Gray code is known to be such a representation for bit strings [6]. In order to see why, consider the classifier condition ##10. The bit strings matching that condition are 0010, 0110, 1010, and 1110. None of these points are contiguous under a binary coding. A binary reflected Gray code, however, groups these points into two clusters of consecutive points: (0010, 0110) and (1110, 1010). This example is illustrative of a more general phenomenon. A Gray code will never group bit strings matching some condition into more clusters of consecutive points than a binary code does. Furthermore, for some conditions, the Gray code will organize the points into fifty percent fewer clusters than the binary code (as in our simple example). See Faloutsos [6] for more details.

This analysis suggests that significant improvements should be obtained by using a binary reflected Gray code to encode the inputs for our function approximator. Table 2 shows that those improvements do indeed occur. The performance improvement is statistically significant for both the test data and the training data.

4.3 Feature salience

Under tile coding, every point belongs to exactly one tile in every tiling. As noted previously, a point belongs to many tiles in each tiling under hyperplane coding. Because the hyperplanes in a tiling are so diverse, they may not all be equally useful for approximating the function. It might matter that some are more specific than others, some may correspond more closely to key regularities in the function, and so on. This presents the approximation algorithm with a feature selection problem that does not occur with tile coding. The problem is important because irrelevant features are a source of noise that can slow down learning of relevant features.

One way to address this feature selection problem is to use dynamically adjusted learning rates to identify which features are most relevant to the task at hand. The idea is to give small learning rates to weights for irrelevant features and large learning rates to weights for relevant features. The individual learning rates thereby become a source of bias that make learning and generalization more efficient. Sutton [11] describes an algorithm — called the Incremental Delte-Bar-Delta (IDBD) method — that uses experience to incrementally adjust learning rates in a linear learning system. That algorithm is well suited to this setting and was incorporated into our hyperplane-based function approximator.

The intuition behind the IDBD algorithm is to adjust rates based on the correlation between successive weight changes: if the weight changes have all been in the same direction, the rate was too small; if the weight changes have been in opposite directions, the rate was too large. The algorithm has one free parameter, the meta learning rate θ. It also uses two parameters for each feature ξ: a learning rate parameter β_ξ and a memory parameter h_ξ that stores a trace of the cumulative sum of recent errors. Each h_ξ is initialized to zero. Given a match set with binary features ξ, weights w_ξ, and approximation error $\delta(t)$, the version of IDBD used here performs the following updates in the order indicated:

1. $\beta_\xi(t+1) = \beta_\xi(t) + \theta\delta(t)h_\xi(t)$
2. $\alpha_\xi(t+1) = e^{\beta_\xi(t+1)}$
3. $w_\xi(t+1) = w_\xi(t) + \alpha_\xi(t+1)\delta(t)$
4. $h_\xi(t+1) = h_\xi(t)[1 - \alpha\xi(t+1)]^+ + \alpha_\xi(t+1)\delta(t)$ where the notation $[x]^+$ indicates a quantity equal to x if $x > 0$ and 0 otherwise.

See Sutton [11] for more details about this algorithm and the reasons why it works.

The linear function approximator based on hyperplane coding was augmented with the IDBD algorithm using meta parameter $\theta = 0.01$. Following Sutton's advice about implementation details, bounds were enforced on each β_ξ to prevent arithmetic underflow. The lower bound was $\ln(\alpha)$ so that the adjusted rates never fell below the global rate α specified for the function approximator. We also enforced an upper bound of 1.0 and limited the change in β_ξ on any one step to ± 1 to help ensure that the weight adjustments remain stable. The results in Table 2 show that these changes had the anticipated effect. Statistically significant performance improvements were seen on both the test data and the training data.

4.4   Improved feature offsets

The large number of overlapping hyperplanes in a match set help to provide a strong generalization capability for linear function approximation based on hyperplane coding. Since the performance on the training data still lags far behind the levels achieved by tile coding, there appears to be room for improvement in the acuity of this approximation. Some performance improvements might be achieved if the available features could be reorganized to cut through the input space in a larger variety of ways. Two changes were implemented to test this hypothesis. First, we change the number of tilings. The number of tilings was set to 8 in the baseline implementation simply to be consistent with the parameters used in tile coding. On closer examination, though, this setting does not achieve the desired effect. In the tile coding implementation, every point has at least one tile in common with all points that are within a tile width away in each direction. The hyperplane coding using the symmetric offsets in the match set is

Algorithm	Approximation Error							
	Blocks		Bumps		Doppler		HeaviSine	
	Train	Test	Train	Test	Train	Test	Train	Test
Tiles	0.06988	1.7535	0.16809	0.93068	0.03579	0.08922	0.02327	0.06458
Hyperplanes	0.13746	0.30148	0.35245	0.98615	0.04709	0.08056	0.01444	0.02288

Table 3. Average square errors for tile coding and hyperplane coding

more limited. Every point can potentially be grouped only with points that are within half a tile width away in each direction. This deficiency is easily remedied by doubling the number of tilings to 16. The size of the population of classifiers remains the same, so the algorithm still has the same amount of approximation resources to work with.

Second, while the choice of the resolution size as the tile width for all tilings was convenient, it does not take full advantage of the possibilities for tile offsets. The resolution size is the simplest tile width that makes sense for classifier conditions with a specific bit at the lowest order bit position, and those classifiers occupy a large fraction of a random population. Larger tile widths are possible for the remaining classifiers, however, and the increased overlap could improve the acuity of the overall approximation. This possibility can be easily tested by organizing the classifiers into two types of features, coarse and fine, stored in two separate groups of tilings. The fine features are those classifiers with a specific bit at the lowest order bit position. These classifiers use the resolution size as the tile width. The remaining classifiers are all treated as coarse features, which use a tile width equal to twice the resolution size. As a consequence of these changes, on each cycle the system generates up to $4n$ potentially distinct messages where n is the number of tilings used in a comparable tile coding scheme.

The results in Table 2 show that these changes had the desired effect. There was a statistically significant and large improvement on the training data, with no significant change in performance on the test data. The overall performance of this version of hyperplane coding is summarized in Table 3. Tile coding still has a statistically significant performance advantage on the training data for all of the test functions except HeaviSine, where hyperplane coding is far superior. Hyperplane coding has a statistically significant performance advantage on the test data for all of the test functions except Bumps, where tile coding has a small but significant advantage. The reconstructions computed by each type of coding for all of the test functions are shown in Figure 2. The precision and smoothness properties of the two representations are remarkably similar. It appears that hyperplane coding offers an alternative for linear approximations that is comparable in performance to what can be achieved with a more conventional approach like tile coding.

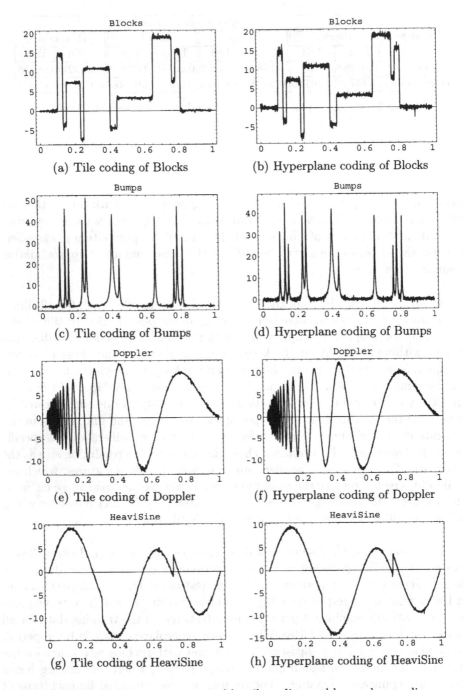

Fig. 2. Reconstructions computed by tile coding and hyperplane coding

5 Conclusions

This paper has shown that by carefully using the resources available in a random population of classifiers, continuous value functions can be approximated with a high degree of accuracy. The results demonstrate that hyperplane coding can achieve levels of performance comparable to those achieved by more well-known approaches such as tile coding. Hyperplane coding treats classifier system rules as features that contribute to a distributed representation of the value function. This approach computes much better approximations than more conventional classifier system methods in which individual rules compute approximations independently. High quality value function approximations that provide both data recovery and generalization are a critically important component of most approaches to solving reinforcement learning problems. Because these results substantially improve the quality of the approximations that can be computed by a classifier system using relative small populations of classifiers, this work provides the foundation for significant improvements in classifier system performance.

Conventional approaches such as linear gradient-descent function approximation based on tile coding are faster, but the hyperplane coding approach seems to offer more opportunities for increasing precision without incurring significantly greater computational costs. The density of tiles in hyperplane coding is naturally higher than the density in tile coding. This contributes to more resolution in the final approximation. The precision of the approximation can also be increased by increasing the length of the classifier input conditions instead of by adding more tiles. Moreover, the hyperplane coding scheme makes it possible to adapt the collection of tiles to achieve more precision. The obvious next step in this research is to use the approximation resources available in a random population as a starting point for a more refined approach to approximation that reallocates resources adaptively to gain greater precision in those regions of the input space where it is needed.

Finally, we note that in hyperplane coding the classifier conditions serve the role of value-based generalizations [14] in the way they organize inputs according to similar function values. While this clearly allows for the specification of decision policies for solving reinforcement learning problems, it ignores the attribute-based generalizations that have been a key feature of the rule-based policies produced by learning classifier systems. Future work will show how attribute-based rule conditions can be learned along with value-based generalizations in a tightly coupled fashion.

† Acknowledgments

This work is based on research originally funded by the MITRE Sponsored Research program. That support is gratefully acknowledged. The author's affiliation with The MITRE Corporation is provided for identification purposes only, and is not intended to convey or imply MITRE's concurrence with, or support for, the positions, opinions or viewpoints expressed by the author.

References

1. Lashon B. Booker. Viewing Classifier Systems as an Integrated Architecture. In *Collected Abstracts for the First International Workshop on Learning Classifier System (IWLCS-92)*, 1992. October 6–8, NASA Johnson Space Center, Houston, Texas.
2. Lashon B. Booker, David E. Goldberg, and John H. Holland. Classifier Systems and Genetic Algorithms. *Artificial Intelligence*, 40:235–282, 1989.
3. Larry Bull and Toby O'Hara. Accuracy-based neuro and neuro-fuzzy classifier systems. In W. B. Langdon, E. Cantú-Paz, K. Mathias, R. Roy, D. Davis, R. Poli, K. Balakrishnan, V. Honavar, G. Rudolph, J. Wegener, L. Bull, M. A. Potter, A. C. Schultz, J. F. Miller, E. Burke, and N. Jonoska, editors, *GECCO 2002: Proceedings of the Genetic and Evolutionary Computation Conference*, pages 905–911. Morgan Kaufmann Publishers, 9-13 July 2002.
4. Martin V. Butz and Stewart W. Wilson. An Algorithmic Description of XCS. In Pier Luca Lanzi, Wolfgang Stolzmann, and Stewart W. Wilson, editors, *Advances in Learning Classifier Systems*, volume 1996 of *LNAI*, pages 253–272. Springer-Verlag, Berlin, 2001.
5. David L. Donoho and Iain M. Johnstone. Ideal spatial adaptation by wavelet shrinkage. *Biometrika*, 81:425–455, 1994.
6. Christos Faloutsos. Gray codes for partial match and range queries. *IEEE Transactions on Software Engineering*, 14(10):1381–1393, October 1988.
7. Geoffrey E. Hinton, James L. McClelland, and David E. Rumelhart. Distributed representations. In David E. Rumelhart, James L. McClelland, and CORPORATE PDP Research Group, editors, *Parallel distributed processing: explorations in the microstructure of cognition, vol. 1: foundations*, pages 77–109. MIT Press, 1986.
8. David LeRoux and Michael Littman. Reinforcement learning using lcs in continuous state space. Seventh International Workshop on Learning Classifier Systems (IWLCS-2004), Extended Abstract, 2004.
9. W. Thomas Miller, Filson H. Glanz, and L. Gordon Kraft. CMAC: An associative neural network alternative to backpropagation. *Proceedings of the IEEE*, 78(10):1561–1567, October 1990.
10. Satinder P. Singh, Tommi Jaakkola, and Michael I. Jordan. Reinforcement learning with soft state aggregation. In G. Tesauro, D. Touretzky, and T. Leen, editors, *Advances in Neural Information Processing Systems*, volume 7, pages 361–368. The MIT Press, 1995.
11. Richard S. Sutton. Adapting bias by gradient descent: An incremental version of delta-bar-delta. In *AAAI*, pages 171–176, 1992.
12. Richard S. Sutton and Andrew G. Barto. *Introduction to Reinforcement Learning*. MIT Press, Cambridge,MA, 1998.
13. Stewart W. Wilson. Classifiers that approximate functions. *Natural Computing*, 1(2-3):211–234, 2002.
14. Richard C. Yee. Abstraction in control learning. Technical Report COINS Technical Report 92-16, Department of Computer and Information Science. University of Massachusetts, Amherst, MA 10003, 1992. A dissertation proposal.

Appendix - Test Functions

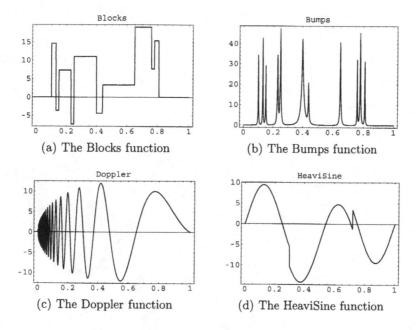

(a) The Blocks function   (b) The Bumps function

(c) The Doppler function   (d) The HeaviSine function

Fig. 3. The Four Donoho Test Functions

Blocks.

$$f(t) = 3.65948 * \sum h_j K(t - t_j) \text{ where } K(t) = (1 + \text{sgn(t)})/2$$

$(t_j) = (0.1, 0.13, 0.15, 0.23, 0.25, 0.4, 0.44, 0.65, 0.76, 0.78, 0.81)$
$(h_j) = (4, -5, 3, -4, 5, -4.2, 2.1, 4.3, -3.1, 2.1, -4.2)$

Bumps.

$$f(t) = 10.5174 * \sum h_j K((t - t_j)/w_j) \text{ ,where } K(t) = (1 + |t|)^{-4}$$

$(t_j) = (0.1, 0.13, 0.15, 0.23, 0.25, 0.4, 0.44, 0.65, 0.76, 0.78, 0.81)$
$(h_j) = (4, 5, 3, 4, 5, 4.2, 2.1, 4.3, 3.1, 5.1, 4.2)$
$(w_j) = (0.005, 0.005, 0.006, 0.01, 0.01, 0.03, 0.01, 0.01, 0.005, 0.008, 0.005)$

Doppler.

$$f(t) = 24.2158 * \sin(2\pi(1 + \epsilon)/(t + \epsilon))\sqrt{t(1 - t)} \text{ , where } \epsilon = 0.05$$

HeaviSine.

$$f(t) = 2.3564 * [4\sin(4\pi t) - \text{sgn(t} - 0.3) - \text{sgn}(0.72 - t)]$$

Two Simple Learning Classifier Systems

Larry Bull

Faculty of Computing, Engineering & Mathematical Sciences
University of the West of England, Bristol, BS16 1QY, U.K.
larry.bull@uwe.ac.uk

1. Introduction

Since its introduction Holland's Learning Classifier System (LCS) [Holland, 1976] has inspired much research into 'genetics-based' machine learning [Goldberg, 1989]. Given the complexity of the developed system [Holland, 1986], simplified versions have previously been presented (e.g., [Goldberg, 1989][Wilson, 1994]) to improve both performance and understanding. It has recently been shown that Wilson's simpler 'zeroth-level' system (ZCS) [Wilson, 1994] can perform optimally [Bull & Hurst, 2002] but "it would appear that the interaction between the rate of rule updates and the fitness sharing process is critical" [ibid.]. In this chapter, a simplified version of ZCS is explored - termed a 'minimal' classifier system, MCS.

Most current research has made a shift away from Holland's formalism, moving LCS much closer to the field of reinforcement learning [Sutton & Barto, 1998], after Wilson introduced XCS [Wilson, 1995]. XCS uses the accuracy of rules' predictions of expected payoff as their fitness. In this way a full map of the problem space is created, rather than the traditional search for only high payoff rules, with (potentially) maximally accurate generalizations over the state-action space [ibid.]. That is, XCS uses a genetic algorithm (GA)[Holland, 1975] to evolve generalizations over the space of possible state-action pairs with the aim of easing the use of such approaches in large problems. XCS can also avoid problematic 'overgeneral' rules which receive a high optimal payoff for some inputs but are sub-optimal for other, lower payoff, inputs. Since their average payoff is higher than that for the optimal rules in the latter case the overgenerals tend to displace them, leaving the LCS sub-optimal. However, the payoffs received by overgeneral rules typically have high variance (they are inaccurate predictors) and so have low fitness in XCS. Holland's LCS was shown to suffer due to such rules emerging [e.g., Dorigo, 1993]. XCS has been shown to perform well in a number of domains [e.g., Bull, 2004] but exactly how this is achieved is not well-understood, although considerable progress has recently been made [e.g., Butz et al., 2003]. In this chapter, a simple accuracy-based LCS, which keeps much of Wilson's framework but simplifies it to increase understandability, is presented and explored - termed YCS (as something of a bridge between ZCS and XCS).

In the following sections the simple examples of these general forms of LCS are presented and modelled before being implemented. Initial findings suggest that, with the parameters used, accuracy-based fitness is more effective for the tasks considered. The issue of generalization is then discussed in more detail with two approaches examined. It is found that the performance of the two LCS can be equivalent but that the underlying fitness pressure is different for each system.

L. Bull: *Two Simple Learning Classifier Systems*, StudFuzz **183**, 63–89 (2005)
www.springerlink.com

2. A Simple Accuracy-based Learning Classifier System

2.1 YCS

YCS is a Learning Classifier System without internal memory, where the rulebase consists of a number (N) of condition/action rules in which the condition is a string of characters from the usual ternary alphabet {0,1,#} and the action is represented by a binary string. Associated with each rule is a predicted payoff value p, a scalar which indicates the error (ε) in the rule's predicted payoff and an estimate of the average size of the niches (action sets - see below) in which that rule participates (σ). The initial random population have these initialized to 10.

On receipt of an input message, the rulebase is scanned, and any rule whose condition matches the message at each position is tagged as a member of the current match set [M]. An action is then chosen from those proposed by the members of the match set and all rules proposing the selected action form an action set [A]. A variety of action selection schemes are possible but a version of XCS's explore/exploit scheme will be used here. That is, on one cycle an action is chosen at random and on the following the action with the highest average payoff is chosen deterministically.

In this paper the simplest case of immediate reward (payoff P) is considered. Reinforcement in YCS consists of updating the error, the niche size estimate and then the payoff estimate of each member of the current [A] (after XCS [Butz & Wilson, 2001]) using the Widrow-Hoff delta rule with learning rate β:

$$\varepsilon_j \leftarrow \varepsilon_j + \beta(|P - p_j| - \varepsilon_j) \qquad (1)$$
$$\sigma_j \leftarrow \sigma_j + \beta(\|[A]\| - \sigma_j) \qquad (2)$$
$$p_j \leftarrow p_j + \beta(P - p_j) \qquad (3)$$

YCS employs two discovery mechanisms, a panmictic GA and a covering operator. On each time-step there is a probability g of GA invocation. The GA uses roulette wheel selection to determine two parent rules based on the inverse of their error:

$$\text{fitness, } f_j = 1/(\varepsilon_j + 1) \qquad (4)$$

Offspring are produced via mutation (probability μ, turned into a wildcard at rate $p_\#$) and crossover (single point with probability χ), inheriting the parents' parameter values or their average if crossover is invoked. Replacement of existing members of the rulebase uses roulette wheel selection based on estimated niche size. If no rules match on a given time step, then a covering operator is used which creates a rule with the message as its condition (augmented with wildcards at the rate $p_\#$ above) and a random action, which then replaces an existing member of the rulebase in the usual way. The GA is not invoked on exploit trials.

Thus YCS represents a simple accuracy-based LCS which captures many of the key features of XCS: "[E]ach classifier maintains a prediction of expected payoff, but the classifier's fitness is *not* given by the prediction. Instead, the fitness is a separate number based on an inverse function of the classifier's average prediction error" [Wilson, 1995] and "[a] classifier's deletion probability is set proportional to the [niche] size estimate, which tends to make all [niches] have about the same size, so that classifier resources are allocated more or less equally to all niches" [ibid.].

The main difference between YCS and XCS is that the former, unlike the latter, has no mechanisms by which to form a maximally general mapping of the state-action space since it does not use a triggered niche GA nor any form of subsumption (see [Butz & Wilson, 2001] for a detailed description of XCS). However, those mechanisms, along with the others found in XCS, need not be considered as pre-requisites for the accuracy-based fitness approach within the LCS paradigm. Indeed, Holland's initial implementation of an LCS was accuracy-based [Holland & Reitman, 1978]. Other early systems included payoff-accuracy hybrids (e.g., [Booker, 1989] - see [Wilson, 1995] for discussions). As will be shown, the simpler system can be effective - the term effective being taken to mean able to solve problems of low complexity whilst remaining open to close modelling. The canonical GA may be defined in much the same way. The mechanisms of YCS are now modelled, in keeping with its philosophy, in a simple way.

2.2 A Simple Model of YCS

The evolutionary algorithm in YCS is a steady-state GA. A simple steady-state GA without genetic operators can be expressed in the form (after [DeJong & Sarma, 1993]):

$$n_j \leftarrow n_j + n_j R_j - n_j D_j \qquad (5)$$

where n_j refers to the number of individuals of type j in the population, R_j refers to their probability of reproductive selection and D_j to their probability of deletion. Roulette-wheel selection is used in YCS:

$$R_j = f_j / f_{[P]} \qquad (6)$$

where f_j is the fitness of individuals of type j (Equation 4) and $f_{[P]}$ is the total population ([P]) fitness. Replacement is proportional to estimated action set size, i.e.:

$$D_j = \sigma_j / \sigma_{[P]}$$

Table 1 shows the payoffs for the single-step task with a single-bit condition and single-bit action considered here (after [Kovacs, 2000]). The last two entries in Table 1 show the expected payoff for the general rules, i.e., the predicted payoff of a general rule is the average of the payoffs it receives (assuming equal probability). It can be seen that under this scheme for input '1' the general rule #:0 has a higher predicted payoff than the correct rule 1:1; #:0 is an overgeneral rule which would cause sub-optimal performance (it is a 'strong' overgeneral [Kovacs, 2001]). The progress of all six rules is examined here, with rulebase size $N=400$.

Table 1. Reward payoffs for the single-step task considered.

Input	Action	Payoff
1	1	1000
1	0	800
0	1	1000
0	0	3000
#	0	1900
#	1	1000

After [Bull & Hurst, 2002], using equations of the general form shown in Equation 5 the expected proportions of each rule type in the next generation can be determined; by specifying the initial proportions of each rule in the population ($N/6$), it is possible to generate the trajectory of their proportions over succeeding generations. Note partial individuals are allowed and hence it is in effect an infinite population model. The trajectory of their related parameters can also be generated. In the following it is assumed that both inputs are presented with equal frequency, that both actions are chosen with equal frequency and that the GA fires once every four cycles (i.e., always explore trials and $g=0.25$). The rules' parameters are updated according to Equations 1 to 3 on each cycle with $\beta=0.2$.

Figure 1 shows the behaviour of the modelled YCS on the single-step task defined in Table 1. It can be seen that the overgeneral rule #:0 is rapidly squeezed out of the population (Figure 1(a)) since it has a high error (Figure 1(b)). The two accurate rules with action '0' gain a larger fraction of the rulebase than those with action '1' but the system is very roughly balanced. That is, the rule replacement scheme based on action set size appears to work effectively here. However, the accurate generalization #:1, whilst slightly more numerous than the specific rules 0:1 and 1:1, does not displace them as there is no explicit generalization pressure within the system. Figure 1(b) shows the errors of the rules over the first 100 GA events (400 system cycles). It can be seen that, apart from the overgeneral rule #:0, errors rapidly drop to zero after initial adjustments. The error for #:1 is adjusted more quickly than those of 0:1 and 1:1, due to its participating in more action sets than either of the specific rules, which explains how it achieves a slightly higher numerosity (see [Wilson, 1987] for a related discussion). Therefore the simple accuracy-based fitness scheme results in a rulebase capable of optimal performance under the exploit action selection scheme described above.

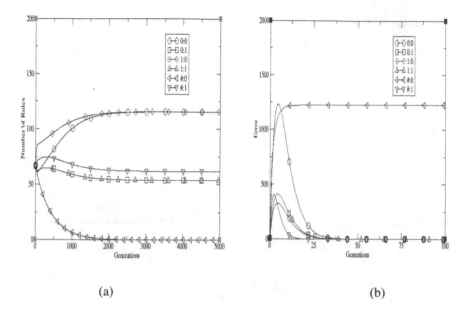

(a)   (b)

Fig. 1: Behaviour of model YCS on the task in Table 1: numerosities (a) and errors (b).

2.3 YCS on the Multiplexer Problem

YCS has been implemented and investigated using versions of the well-known multiplexer task. These Boolean functions are defined for binary strings of length $l = k + 2^k$ under which the first k bits index into the remaining 2^k bits, returning the value of the indexed bit. A correct classification results in a payoff of 1000, otherwise 0.

Figure 2(a) shows the performance of YCS, as described in Section 2.1, on the 6-bit multiplexer problem using the same parameters as in Section 2.2, with $p_\#=0.6$, $\chi=0.5$ and $\mu=0.01$. After [Wilson, 1995], performance from exploit trials only is recorded (fraction of correct responses are shown), using a 50-point running average, averaged over ten runs. It can be seen that YCS is capable of optimal performance and that the average error (shown divided by the payoff range) of the rules drops below 10% of the payoff range. Figure 2(b) shows the performance of the same system on the 11-bit multiplexer with $N=2000$. Again, it can be seen that YCS achieves optimal performance.

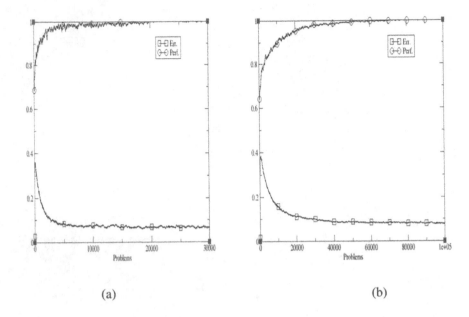

Fig. 2: Performance of YCS on the multiplexer task, 6-bit (a) and 11-bit (b) versions.

Figure 3(a) shows the performance of YCS on the 20-bit multiplexer problem using the same parameters as for the 11-bit problem. It can be seen that the simple system is only able to achieve around 80% performance in the time allowed and the average error of rules reflects this inability to solve the task, being comparatively higher than those shown in Figure 2. Longer runs, e.g., 200,000 problems, find the system is just about capable of optimal performance with these parameters (not shown). Figure 3(b) shows the average specificity (fraction of non-# bits in a condition) for YCS on the three tasks as described. That is, the amount of generalization produced by the simple system is shown. The maximally general solution to the 6-bit multiplexer has specificity 3/6 = 0.5, for the 11-bit problem it is 4/11 = 0.36 and for the 20-bit it is 5/20 = 0.25. It can be seen that, for the problems YCS is able to solve, the degree of specificity is approximately 10% higher than it could be and that it is considerably higher than 0.25 for the 20-bit problem (also true for longer runs, not shown). Hence, YCS does appear to exploit the generalizations available to it, as noted above, but there is no pressure for maximal generality.

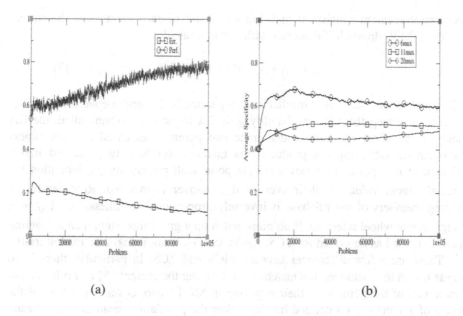

Fig. 3: Performance of YCS on the 20-bit multiplexer task (a) and specificities for all tasks (b).

3. A Simple Payoff-based Learning Classifier System

3.1 MCS

MCS is a minimal system without internal memory, where the rulebase consists of a number (N) of condition/action rules in which the condition is a string of characters from the usual ternary alphabet $\{0,1,\#\}$ and the action is represented by a binary string. Associated with each rule is a fitness scalar (f) and the initial random population have the parameter initialized to 10 (f_0) here.

The matching procedure and formation of action sets is as described for YCS. Again, a variety of action selection schemes are possible but a version of XCS's explore/exploit scheme will be used here. That is, on one cycle an action is chosen at random and on the following the action with the highest total payoff is chosen deterministically.

Although the use of fitness sharing for externally received payoff had been suggested before [Holland, 1985], it was not until Wilson introduced the action set-based scheme in ZCS that simple but effective fitness sharing in LCS became possible [Bull & Hurst, 2002]. MCS uses the fitness sharing mechanism of ZCS, i.e., within action sets. The simplest case of immediate reward (payoff P) is again considered and hence reinforcement consists of updating the fitness of each member of the current [A] using the Widrow-Hoff delta rule with learning rate β:

$$f_j \leftarrow f_j + \beta \left((P / |[A]|) - f_j \right) \qquad (7)$$

MCS employs two discovery mechanisms, a panmictic GA and a covering operator. On each time-step there is a probability g of GA invocation. When called, the GA uses roulette wheel selection to determine two parent rules based on their fitness (Equation 6). Offspring are produced via mutation (probability μ, turned into a wildcard at rate $p_\#$) and crossover (single point with probability χ), inheriting the parents' fitness values or their average if crossover is invoked. Replacement of existing members of the rulebase is inversely proportional to fitness, i.e., $1/(f_j +1)$, using roulette wheel selection. If no rules match on a given time step, then a covering operator is used as described for YCS. Again, the GA is not invoked on exploit trials.

There are a few differences between MCS and ZCS. In particular, there is no fitness tax on the members of a matchset not forming the current [A] and rules do not donate half of their fitness to their offspring in MCS. Also, cover is not fired if the fitness of a matchset is a defined fraction below the population mean in MCS. Again, those mechanisms, as will be shown, need not be considered as pre-requisites for the use of a payoff-based fitness scheme. MCS is now modelled in the same way as YCS.

3.2 A Simple Model of MCS

Equations of the general form shown in Equation 5 can be used to determine the expected proportions of each rule type in the next generation for MCS, after specifying the initial proportions of each rule in the population ($N/6$). Again, it is assumed that both inputs are presented with equal frequency, that both actions are chosen with equal frequency and that the GA fires once every four cycles (i.e., always explore trials and $g=0.25$). The rules' fitnesses are updated according to Equation 7 on each cycle.

Figure 4 shows the behaviour of the modelled MCS on the single-step task defined in Table 1 with $\beta=0.2$. From Figure 4(a) it can be seen that the overgeneral rule #:0 maintains a greater proportion of the population than any rule for the action '1'. Since, as shown in Figure 4(b), under fitness sharing the fitness of all rules goes to the same value (see [Wilson, 1987] for discussions), this means that the LCS will provide a sub-optimal response for input '1'. Figure 5 shows the behaviour of the modelled LCS on the same task with $\beta=1.0$. It can be seen that the overgeneral rule #:0 has the lowest numerosity along with the lowest payoff rule (1:0) and that there are an equal number of #:1, 0:1 and 1:1 rules. That is, numerosities reflect rule payoffs more appropriately with the instantaneous update but the more general solution (#:1) does not win out as there is no explicit generalization pressure. Also, given that payoff-based LCS using fitness sharing hold their estimation of utility in *rule numerosity*, the instantaneous fitness update means a rule's fitness can immediately consider the current numerosity, something which is constantly changed by the actions of the GA; it appears that a high learning rate allows the LCS to approximate rule utility more efficiently. This results contrasts with that reported in [Bull & Hurst, 2002] who showed an instantaneous fitness update failing to solve this task (their Figure 2). However, their model assumed that the fitness of general rules at any time was the average of the corresponding specific rules at that time. The assumption is not made here and hence explains the difference. Experiments with higher learning rates in the modelled YCS have not produced significant differences to those reported in Section 2 (not shown). It can be noted that XCS has been shown to be sensitive to the learning rate, although this can be reduced using a rank-based selection scheme [Butz et al., 2003b] - a scheme which is problematic for fitness sharing systems (e.g., [Deb, 2001]).

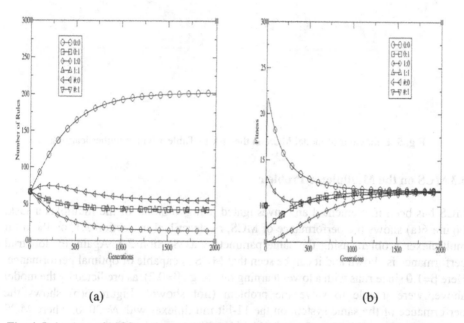

(a)   (b)

Fig. 4: Behaviour of MCS on the task in Table 1, showing numerosities (a) and fitnesses (b).

Despite the fact that both YCS and MCS use niche (action set) size to enforce diversity in the rulebase, they do it in different ways, as described above, which will usually result in different solutions. YCS uses niche size to equally balance rulebase resources. In contrast, MCS apportions resources based on relative payoff (as highlighted in [Bull & Hurst, 2002]). Since action '0' receives almost twice as much payoff as action '1' in the task presented in Table 1, MCS converges on a rulebase containing almost twice as many rules for the former action than the latter (Figure 5(a)). The effects, if any, on performance from this difference remain open to further investigation.

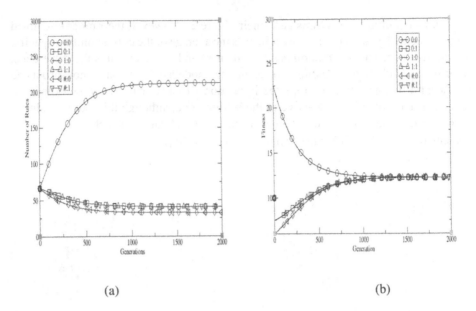

(a)   (b)

Fig. 5: Behaviour of model MCS on the task in Table 1 with a higher learning rate.

3.3 MCS on the Multiplexer Problem

MCS has been implemented and investigated using versions of the multiplexer task. Figure 6(a) shows the performance of MCS, as described in Section 3.1, on the 6-bit multiplexer problem using the same parameters as Section 2.3. Again, exploit trial performance is shown and it can be seen that MCS is capable of optimal performance. Here $\beta=1.0$ since runs with a lower learning rate (e.g., $\beta=0.2$), as predicted by the model above, were unable to solve the problem (not shown). Figure 6(b) shows the performance of the same system on the 11-bit multiplexer with $N=2000$, where MCS appears just about able to perform the 11-bit multiplexer task with the parameters used.

Figure 7(a) shows the performance of MCS on the 20-bit multiplexer problem using the same parameters as for the 11-bit problem. It can be seen that the simple system is only able to achieve around 50% (random) performance. Figure 7(b) shows the average specificity for MCS on the three tasks as described, i.e., the amount of generalization produced by the simple system is shown. It can be seen that, for the problems MCS is able to solve, the degree of specificity is approximately 30% higher than optimal and the values indicate that non-critical bits have been determined solely by the mutation rate, i.e., randomly. That is, there appears to be no pressure for generalization in the system. Hence the simple accuracy-based YCS appears better able to solve these problems with the parameters used both in terms of performance and the degree of generalization.

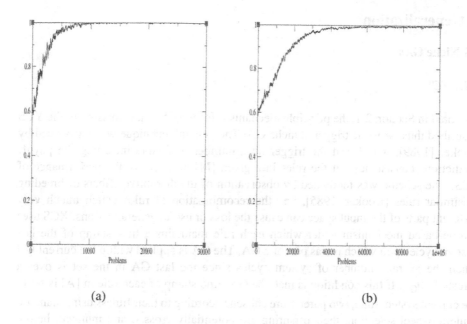

Fig. 6: Performance of MCS on the multiplexer task, 6-bit (a) and 11-bit (b) versions.

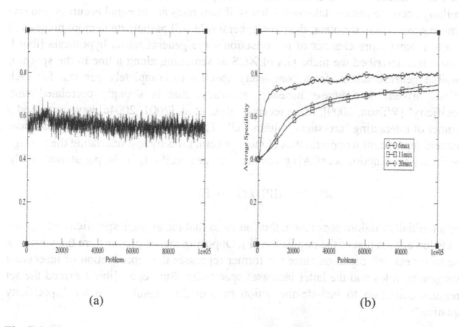

Fig. 7: Performance of MCS on the 20-bit multiplexer task (a) and specificities for all (b).

4. Generalization

4.1 Niche GAs

4.1.1 YCS

As noted in Section 2.1, the principle mechanism from XCS that is missing in the YCS described thus far is the triggered niche GA. The general technique was introduced by Booker [1989], who based the trigger on a number of factors including the payoff prediction "consistency" of the rules in a given [M], to improve the performance of LCS. The scheme was motivated by observation of the disruptive effects of breeding dissimilar rules [Booker, 1985], i.e., the recombination of rules which match very different parts of the input space can cause the loss of useful generalizations. XCS uses a time-based mechanism under which each rule maintains a time-stamp of the last system cycle upon which it was part of a GA. The GA is applied within the current [A] when the average number of system cycles since the last GA in the set is over a threshold θ_{GA}. If this condition is met, the GA time-stamp of each rule in [A] is set to the current system time, two parents are chosen according to their fitness using standard roulette-wheel selection, their offspring are potentially crossed and mutated, before being inserted into the rulebase.

When introducing XCS, Wilson [1995] highlighted how the triggered niche GA leads to a tendency for accurate rules which participate in more niches than other similarly accurate rules to takeover. That is, if two rules are of equal accuracy and one is more general than the other, the more general rule will participate in more niches and therefore have more chances of reproduction - the generalization hypothesis [ibid.]. Wilson has described the niche GA of XCS as searching along a line in the space of possible generalizations, from completely specific to completely general, for each action "driven by a fitness measure, accuracy, that is strongly correlated with specificity" [Wilson, 2000]. More recently, Butz et al. [2001; 2004a] have proposed a number of interacting "pressures" within XCS. Their "set pressure" considers the more frequent reproduction opportunities of more general rules by approximating the average specificity of an action set s([A]) given the average specificity in the population s([P]):

$$s([A]) = s([P]) / (2 - s([P]))\qquad(8)$$

For an initially random population, this indicates that the average specificity of a given [A] is lower than that of the population [P]. Opposing the set pressure are the pressures due to fitness and mutation since the former represses the reproduction of inaccurate overgeneral rules and the latter increases specificity. Butz et al. [ibid.] extend the set pressure definition to include the action of mutation, resulting in the "specificity equation":

$$s([P(t+1)]) = s([P(t)]) + f_{ga} \frac{2 (s([A]) + \delta_{mut} - s([P(t)])}{N} \qquad (9)$$

where δ_{mut} is the average change in specificity between a parent of specificity $s(cl)$ and its offspring under mutation, defined as $0.5\mu(2 - 3s(cl))$, and f_{ga} is the frequency of GA application per cycle. It is shown that, for a number of simple scenarios such as a random Boolean function, Equation 9 is a good predictor of resulting specificity and they note this "represent[s] the first theoretical confirmation of Wilson's generalization hypothesis" [ibid.].

Bull [2002] presented a simple Markov model of the GA working within niches, examining the difference in fitness pressure between two rule types - accurate and inaccurate or general and specific. The reproductive bias inherent in more general rules was approximated very simply for such executable models through increasing the fitness of the more general rule by a given factor. A similar approach can be used within the simple difference equation model presented here by altering Equation 4:

$$f_j = \pi_j (1 / (\epsilon_j+1)) \qquad (10)$$

where π_j is the proportion of all possible action sets in which the rule participates. Thus for the problem in Table 1, the two rules containing generalization have $\pi_j = 2/4$, whereas for the specific rules $\pi_j = 1/4$. Figure 8 shows the effects of this approximation of the niche GA within the YCS modelled in Figure 1. As can be seen, the accurate general rule #:1 is the most numerous, with an equal number of the accurate specific rules 1:0 and 0:0. That is, the system has converged upon the maximally general solution to the problem (compare with Figure 1).

Figure 9 shows the performance of the implemented YCS of Section 2.3 with XCS's triggered niche GA incorporated (θ_{GA}=25, after [Butz et al., 2001]). All other aspects of selection and the replacement procedure remain the same as before. Figure 9(a) shows how YCS produces a solution with the optimal average specificity for the 11-bit multiplexer. However Figure 9(b) shows how the same system performs much worse than the panmictic GA version on the 20-bit multiplexer problem, becoming swamped with overgeneral rules. There are signs of learning around 50,000 problems as the error starts to drop but longer runs, e.g., 200,000 problems, display little improvement in performance (not shown). That is, in the more difficult 20-bit problem, the set pressure would appear to be significantly stronger, with the same parameters, than the fitness and mutation pressures. That the fitness pressure is less in the niche GA LCS is predicted by the slight delay in rule numerosities reaching their equilibrium values in Figure 8(a) when compared with the panmicitc system in Figure 1(a).

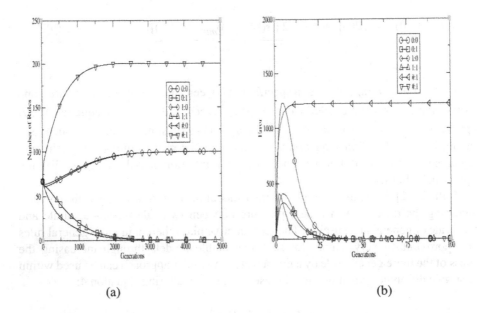

(a)   (b)

Fig. 8: Behaviour of model YCS with approximated niche GA, numerosities (a) and errors (b).

A distinguishing feature of XCS is its accuracy function which is a negative power of error. The function is controlled by three variables α, ε_0 and υ, with typical values of 0.1, 10 and 5 respectively, which make it a very harsh, almost step-like, function with a flat top and slight curve at the bottom (e.g., see [Butz et al., 2001]): $f_j = \alpha (\varepsilon_j / \varepsilon_0)^{-\upsilon}$ unless $\varepsilon_j < \varepsilon_0$ where $f_j = 1$. Hence slight differences in error are greatly magnified using this function. In YCS, as described in Section 2.1, fitness is simply inversely proportional to error. As noted above, Butz et al. [2001; 2004a] suggest that the set pressure is balanced by the fitness and mutation pressures. One way to control the fitness pressure under roulette wheel selection is to vary the degree of separation between fitnesses and in XCS this can be achieved by altering υ, although this has not previously been explored. A similar mechanism can be added to YCS by extending Equation 4 to include a power term υ such that:

$$f_j = 1/(\varepsilon_j^{\upsilon} + 1) \qquad (11)$$

Figure 10(a) shows how, with $\upsilon=10$, the system is able to solve the 20-bit problem in around 200,000 trials. Smaller values of υ, e.g., $\upsilon=5$, show very little signs of improved learning in the same timespan (not shown). With $\upsilon=20$ no further significant benefits are found (not shown). Figure 10(b) shows how increasing the mutation rate to $\mu=0.09$

also produces optimal behaviour. Smaller values, e.g., μ=0.04 (as used in [Butz et al., 2001; 2004a]), give improved learning but not optimality in the same timespan (not shown). As predicted by Equation 9, a slight increase in specificity is seen with higher mutation rates. Thus these results confirm expectations that the detrimental effects of the set pressure can be reduced by increasing the fitness pressure or mutation pressure.

Figure 11(a) shows how by increasing both the mutation rate (μ=0.04, i.e., lower than 0.09 to reduce the rise in average specificity) and the fitness pressure (υ=10) at the same time, YCS using a niche GA is able to solve the 20-bit multiplexer problem and create solutions with the average specificity of a maximally general solution in a shorter number of problems. Figure 11(b) shows the performance of the same system on the 37-bit multiplexer, with the same parameters except N=5000. Optimal performance and optimal average specificity are again obtained and this performance is comparable to that of XCS on the same task, using the same parameters (where relevant) as reported in [Butz et al., 2001; 2004a].

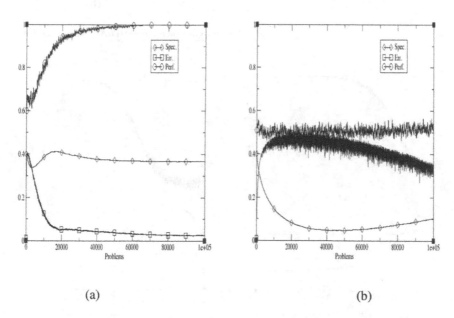

(a)   (b)

Fig. 9: Performance of YCS with a triggered niche GA on the 11-bit (a) and 20-bit tasks (b).

According to Equation 9 the rate of change in specificity is inversely proportional to the population size N and this has also been found to be a mechanism by which to counteract the set pressure. For example, the system in Figure 9(b) with N=20,000 takes around 350,000 trials to solve the 20-bit problem optimally (not shown - see [Butz et al., 2003] for discussions on population sizing).

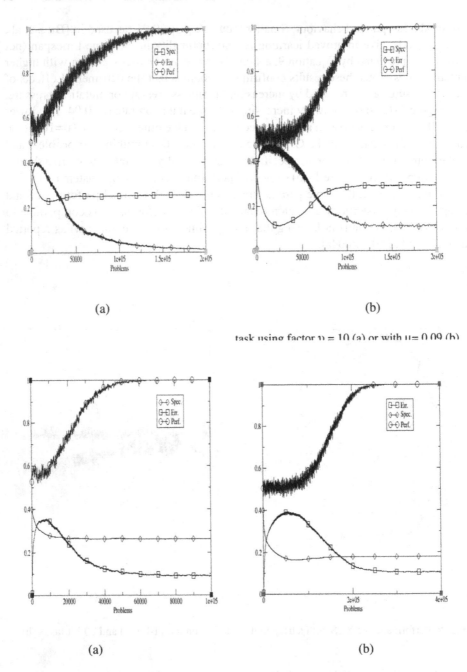

task using factor υ = 10 (a) or with υ= 0.09 (b)

Fig. 11: Improved performance on the 20-bit multiplexer task (a) and on the 37-bit task (b).

4.1.2 MCS

After [Booker, 1985] a number of studies have used a niche GA within payoff-based LCS such as ZCS (e.g, [Bull, 1998][Tomlinson & Bull, 1999]). The effects of an action set GA can be modelled in MCS in the same way as YCS by simply factoring fitness by the proportion of actions sets in which a rule participates, i.e., $f_j' \leftarrow \pi_j f_j$, where f_j' is used in all subsequent calculations for Equation 5.

Figure 12 shows the behaviour of the system modelled in Figure 5 with the approximated niche GA. As can be seen, the system will produce an optimal solution since the appropriate rules #:1 and 0:0 are the most numerous. However, the fitness sharing process has been disrupted by the actions of the niche GA with the rule 0:0 maintaining a higher fitness than all other rules (compare with Figure 5). That is, despite the global deletion scheme, the fitness balancing is adversely affected by localised selection (see [Deb, 2001] for related discussions).

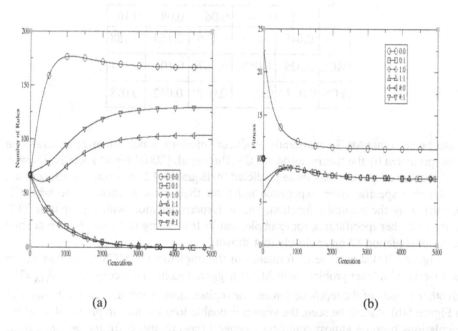

(a)   (b)

Fig. 12: Behaviour of model MCS on the task in Table 1 with an approximated niche GA.

Significantly, the analysis of the niche GA in XCS presented by Butz et al. (e.g., [Butz et al. 2001; 2004a]) described above would appear to apply equally well to payoff-based LCS since it does not explicitly consider XCS's accuracy-based fitness scheme; it is derived via the consideration of rule specificity. That is, their analysis makes no assumptions beyond the GA selecting from within [A] and using a global deletion scheme. Butz et al. [2003] note how Equation 9 can be used to predict the specificity that a population will converge upon for a given mutation rate (see [ibid.] for the derivation):

$$s([P]) = \frac{1 + 2.5\mu - (6.25\mu^2 - 3\mu + 1)^{0.5}}{2} \tag{12}$$

Table 2 shows the predicted and actual specificity of YCS and MCS with the niche GA on a 20-bit random Boolean function (randomly returns 0 or 1000) as used in [Butz et al., 2001; 2004a] for a number of mutation rates. All other parameters used are the same as in Figure 9(b) but with θ_{GA}=1.0 (after [Butz et al., 2001; 2004a]). The value shown is taken at the end of 200,000 problems, averaged over ten runs. The average difference between the empirical and predicted value is 0.035 for MCS and 0.066 for YCS. Butz et al. [2003] present tabulated results for the same function with XCS where the average error is 0.053 over the same values of μ, i.e., similar to the values obtained here.

Table 2. Converged specificities for random Boolean problem.

μ	0.02	0.04	0.06	0.08	0.10
(12)	0.040	0.078	0.116	0.153	0.188
MCS	0.029	0.059	0.087	0.101	0.124
YCS	0.017	0.038	0.047	0.062	0.078

It can be seen that MCS consistently produces solutions which are more general than those predicted by the theory, as does YCS. Butz et al. [2003] report specificities with XCS that are all greater than those predicted by Equation 12. It is noted in [Butz et al., 2004a] that specific rules experience noise on their error estimates here which is magnified by the accuracy function, i.e., v. Experimentation with higher v in YCS produces higher specificities, for example, results from using v=10 match the predicted values of Equation 12 more closely (not shown).

Figure 13(a) shows the performance of the implemented MCS of Section 3.3 on the 11-bit multiplexer problem with XCS's triggered niche GA incorporated (θ_{GA}=25). All other aspects of the reproduction and the replacement procedure remain the same as in Figure 6(b). As can be seen, the system is unable to solve the simple problem as the populations become almost completely general (specificity -> 0); the set pressure is dominant. Use of β=0.2 appears to have no effect (not shown). The same was found to be true for the 6-bit problem (not shown). Given that YCS was able to solve the 11-bit multiplexer within 100,000 problems (Figure 9(a)) before alterations needed to be made to balance the set pressure, it can be concluded that the fitness pressure within MCS is significantly less with these parameters. This is despite MCS selecting and replacing based on fitness, the latter being the reciprocal of fitness, whereas YCS selects based on the reciprocal of error but replaces on niche size, i.e., not fitness related. Again, results in the models show a comparative increase in the time taken to equilibrium.

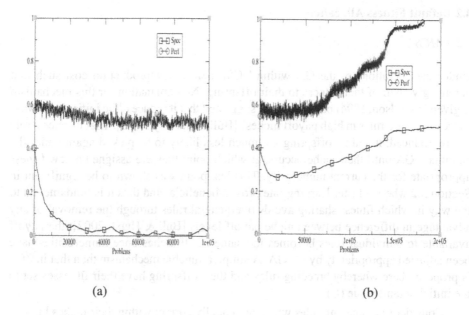

Fig. 13: MCS with niche GA on 11-bit problem (a) and with altered parameters (b).

Figure 13(b) shows the effects of making the same alterations to MCS as were made for YCS. That is, by adjusting the fitness pressure through raising a rule's fitness value for selection (reproduction and deletion) via a power term v, i.e. f_j^v, and increasing the mutation rate, optimal performance can be obtained in the 11-bit multiplexer with the same parameters but where $v=10$ and $\mu=0.30$. The specificity is high but represents an improvement over results reported in Section 3.3 where the same problem was only just solved by MCS with a specificity of around 0.7.

The results with the model suggest one of the difficulties for MCS is that the niche GA disrupts the fitness sharing process. These multiplexer results indicate that the disruption may be too severe for MCS to work effectively with a niche GA on the harder multiplexer problems. Of course, it may also be that the appropriate parameters to use with MCS to increase the fitness pressure sufficiently are very different from those of YCS (XCS) used here. Certainly, experimentation (not shown) has found it difficult to solve either the 20-bit or 37-bit multiplexer problems with the same population sizes as those used for YCS above, over similar timespans. Despite these difficulties, the predictions of Equation 12, with respect to the effects of altering the mutation and fitness pressures, apply as well here as they did for YCS; Equation 12 is a general predictor of specificity behaviour for such niche GAs.

However it is possible to improve the generalization capabilities and performance of MCS with a simple heuristic, as discussed in the following section.

4.2 Default Fitness Allocation

4.2.1 MCS

Under the operations of the GA within ZCS, there is a reproduction cost such that parents give half of their fitness to their offspring. No explanation for this mechanism is given in [Wilson, 1994] but it has been suggested that it reduces "the initial 'runaway' success of those rules in high payoff niches" [Bull & Studley, 2002]. That is, once a rule has reproduced, it and its offspring are much less likely to be picked again under the panmictic GA until their niche occurs, at which point they are assigned a new fitness appropriate for the current numerosity. This last point was shown to be significant in Section 3.2 where a faster learning rate proved beneficial and thus it is fundamental to the way in which fitness sharing avoids overgeneral rules through the removal of any advantage in difference between niche payoff levels [Bull & Hurst, 2002]; the payoff available to individual rules becomes the same in all niches once numerosities have been adjusted appropriately by the GA. A simpler, tunable mechanism than that in ZCS is proposed here whereby breeding rules and their offspring have their fitnesses set to the initial default value (f_0).

Consider two breeding rules which are equally correct within their niches but one is more general than the other. Once their fitnesses have been set to the default value they must wait until one of their niches occurs before their fitnesses will be reset. On average, this will occur more quickly for the more general rule meaning it will tend to be selected again for reproduction more quickly than the less general rule; a more general correct rule will increase in numerosity more rapidly than a less general rule. The effects of the scheme can be modelled by alteration to fitnesses such that:

$$f_j' \leftarrow S_j (((f_j d_{fa}) (1 - \pi_j)) + (f_j \pi_j)) + (1 - S_j) f_j \qquad (13)$$

where d_{fa} is the percentage change to the fitness of the given rule type under the default fitness allocation, S_j is the probability of selecting a rule of type j for reproduction, i.e., $n_j R_j$ as defined in Equations 5 and 6, and f_j' is used for all subsequent calculations.

Figure 14 shows the behaviour of the model MCS of Figure 5 with fitnesses altered according to Equation 13 and $d_{fa} = 0.7$. Here the more general rule #:1 has a higher numerosity than 0:1 and 1:1, where all three rules were of equal numerosity without the heuristic; the degree of generalization within the solution is improved. Figure 14(b) shows how there has been some disruption to the fitness sharing process but it is less than that experienced by MCS under the niche GA. Decreasing d_{fa} improves the numerosity of the rule #:1 but increases the disruption to the fitness sharing (not shown).

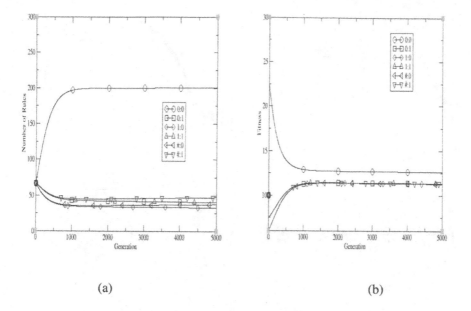

(a)   (b)

Fig. 14: Behaviour of model MCS on the task in Table 1 with default fitness allocation.

Figure 15(a) shows the effect of assigning the initial default fitness value to both the parents and the offspring in the MCS of Section 3.3. Further, the GA is fired on every explore trial to increase the rate at which rules are apportioned to niches. It can be seen that the system now solves the 20-bit multiplexer, with all other parameters as in Figure 7(a), and produces solutions in which the average specificity is close to that of the maximally general solution(+5%). It was found that increasing the GA rate gave increased benefit to the default fitness allocation technique (faster learning and better generalization - not shown). Here the GA is constantly working on assigning an appropriate numerosity to each niche, whilst new and existing rules have their fitnesses set according to its latest actions through a maximum learning rate. Figure 15(b) shows the performance of the same system in the 37-bit multiplexer with all parameters as before, except N=5000. It can be seen that optimal performance is obtained around 250,000 exploit trials - matching the performance of YCS with the triggered niche GA. It can also be seen that the average specificity of the solutions produced by the simple LCS is very close to that of the maximally general solution.

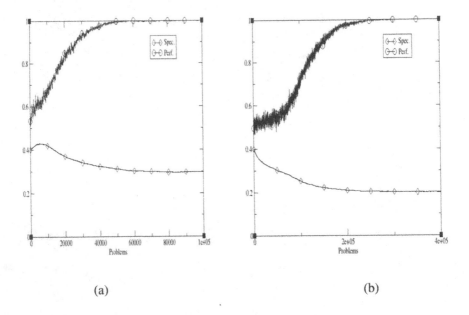

(a)   (b)

Fig. 15: Performance of MCS with the fitness heuristic on the 20-bit (a) and 37-bit (b) tasks.

As suggested by the model above, the default value controls the degree of generalization but this must be balanced against disruption of the fitness sharing process. Figure 16 shows how using the same systems in Figure 15 but with $f_0 = 1.0$ produce solutions with optimal average specificity but they are slightly slower, presumably due to the increased disruption.

4.2.2 YCS

Despite the fact that the triggered niche GA is known to work well with accuracy-based fitness, the default fitness allocation scheme can also be used within such systems. Again, this can be modelled via an adjustment to the error according to:

$$\varepsilon_j' \leftarrow S_j (((\varepsilon_j \, d_{fa})(1 - \pi_j)) + (\varepsilon_j \, \pi_j)) + (1 - S_j) \varepsilon_j \qquad (14)$$

In this case, since YCS minimises error, $d_{fa} > 1.0$, i.e., parents and offspring have their errors increased. Figure 17 shows the behaviour of the modelled YCS in Figure 1 but with $d_{fa} = 1.3$. As can be seen, there is a slight increase in the number of #:1 rules, i.e., an improvement in the degree of generalization but to nothing like the degree seen in Figure 8 under the niche GA. Increasing d_{fa} causes an increase in the maximum error experienced, particularly by the more specific but accurate rules, as expected (not shown).

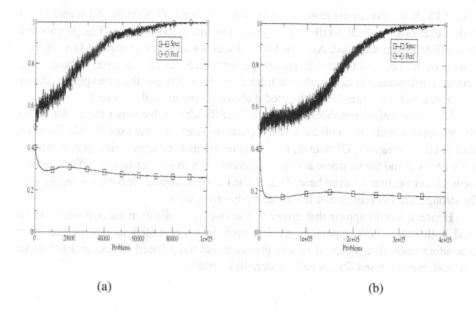

(a)   (b)

Fig. 16: Performance of MCS with lower default fitness on the 20-bit (a) and 37-bit (b) tasks.

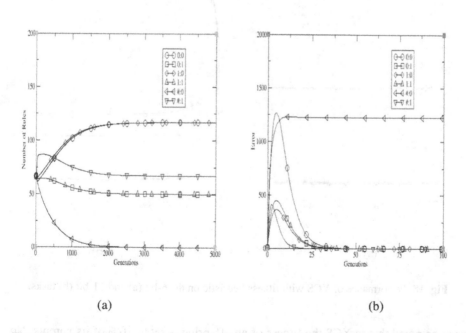

(a)   (b)

Fig. 17: Behaviour of model YCS on the task in Table 1 with default fitness allocation.

Figure 18 shows the results from using the heuristic with YCS on the 6-bit and 11-bit multiplexer tasks with all details as in Figure 2 but the fitnesses of breeding parents and their offspring are set to 10. As with MCS, it was found that running the GA on each system cycle, i.e., $g=1.0$, gave improved performance. As can be seen, in both cases optimal performance is achieved in similar times to before but the average specificity is now optimal, as suggested by the model above (compare with Figure 2).

Experimentation (not shown) has found it difficult to solve either the 20-bit or 37-bit multiplexer problems with the same population sizes as those used for MCS above, over similar timespans. Of course, it may again be that the appropriate parameters to use with YCS and the heuristic are very different from those best suited to YCS and the niche GA of Section 4.1 used here. That is, the heuristic appears less easy to apply with the stronger fitness pressure of the accuracy-based system.

Hence it would appear that either of the two generalization mechanisms can be used with either LCS approach but that each is better suited to one or the other depending upon their inherent fitness pressure: accuracy-based fitness and the niche GA; and, payoff-based fitness and the default heuristic.

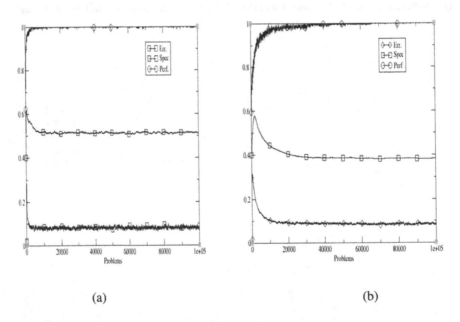

(a)   (b)

Fig. 18: Performance of YCS with fitness heuristic on the 6-bit (a) and 11-bit (b) tasks.

It can be noted that in XCS the fitness of an offspring is set to 10% of its parents. No explanation for this is given in [Butz & Wilson, 2001] but the above findings suggest the adjustment adds to the generalization pressure, even if only relatively slightly.

5. Conclusions

This paper has presented simple examples of the two main forms of Learning Classifier System, where each contains only the fundamental mechanisms of their type: a simple accuracy-based system has been presented which uses rules' prediction errors for reproduction and niche size for replacement; and, a simple pay-off based system has been presented which uses fitness attributed under a niche-based sharing scheme for reproduction and its inverse for replacement. The purpose of this work is not to suggest that these LCS should replace the more sophisticated systems from which they are derived - XCS [Wilson, 1995] and ZCS [Wilson, 1994] respectively. Rather, it is to present canonical examples of each form thereby enabling a greater understanding of exactly how such LCS work and the relative benefits of each. Simple executable models which capture the basic features of each have also been presented.

The aforementioned systems currently being used each contain a number of heuristics and mechanisms which allow them to solve relatively complex problems, inevitably more complex than the basic systems presented here are capable of. By introducing these minimalist (but effective) versions of each type of LCS, the role of such heuristics can be examined in isolation and in combination on tasks of varying complexity/type. Towards this aim, two mechanisms for improving the generalization capabilities of LCS have been investigated here. A default fitness allocation scheme has been examined and found better suited to the payoff-based LCS. It has been shown that previously presented formal analysis of the niche GA in XCS (e.g., [Butz et al., 2001; 2004a]) can be more generally applied. However *experimentation found the fitness pressure within payoff-based LCS to not be as strong as in accuracy-based LCS*, particularly with a niche GA. This result was predicted by the models wherein the time taken for fitnesses and numerosities to reach their equilibrium was always longer for the payoff-based LCS, particularly fitnesses. Presumably this difference is due to the sharing process relying upon rule numerosity *and* fitness to indicate utility. Future work must determine whether this is also the case in delayed reward, i.e., multi-step, environments and other more complex cases such as noisy and non-stationary problems. Future work should also examine whether the related formal analyses of XCS (e.g., [Butz et al., 2003; 2004b; 2004c]) are more generally applicable and whether parameters/mechanisms exist which aid the use of a niche GA in payoff-based LCS.

Current work is using other approaches from evolutionary computing theory (e.g., [Vose, 1999], after [Horn et al., 1994][Bull, 2002]) to further formal understanding of these systems and to ascertain whether fitness pressure is the underlying difference between the two general approaches, as suggested here.

Acknowledgements

I am indebted to the members of the Learning Classifier Systems Group at UWE for many useful discussions about this work. I would also like to thank the anonymous reviewers of a previous version of this paper for their comments.

References

Booker, L.B. (1985) Improving the Performance of Genetic Algorithms in Classifier Systems. In J.J. Grefenstette (ed) *Proceedings of the First International Conference on Genetic Algorithms and their Applications*. Lawrence Erlbaum Associates, pp80-92.

Booker, L.B. (1989) Triggered Rule Discovery in Classifier Systems. In J.D. Schaffer (ed) *Proceedings of the Third International Conference on Genetic Algorithms*. Morgan Kaufmann, pp265-274.

Bull, L. (1998) On ZCS in Multi-Agent Environments. In A.E. Eiben, T. Baeck, M. Schoenauer & H-P. Schwefel (eds) *Parallel Problem Solving from Nature - PPSN V*. Springer, pp471-480.

Bull, L. (2002) On Accuracy-based Fitness. *Soft Computing* 6(3-4): 154-161.

Bull, L. (2004)(ed) *Applications of Learning Classifier Systems*. Springer.

Bull, L. & Hurst, J. (2002) ZCS Redux. *Evolutionary Computation* 10(2): 185-205.

Bull, L. & Studley, M. (2002) Consideration of Multiple Objectives in Neural Learning Classifier Systems. In J. Merelo, P. Adamidis, H-G. Beyer, J-L. FernandezVillicanas & H-P. Schwefel (eds) *Parallel Problem Solving from Nature - PPSN VII*. Springer, pp558-567.

Butz, M. & Wilson, S.W. (2001) An Algorithmic Description of XCS. In P-L. Lanzi, W. Stolzmann & S.W. Wilson (eds) *Advances in Learning Classifier Systems: IWLCS 2000*. Springer, pp253-272.

Butz, M., Kovacs, T., Lanzi, P-L & Wilson, S.W. (2001) How XCS Evolves Accurate Classifiers. In *Proceedings of the 2001 Genetic and Evolutionary Computation Conference - Gecco 2001*. Morgan Kaufmann, pp927-934.

Butz, M., Goldberg, D.E. & Tharakunnel, K. (2003) Analysis and Improvement of Fitness Exploitation in XCS: Bounding Models, Tournament Selection, and Bilateral Accuracy. *Evolutionary Computation* 11(3): 239-278.

Butz, M., Kovacs, T., Lanzi, P-L & Wilson, S.W. (2004a) Toward a Theory of Generalization and Learning in XCS. *IEEE Transactions on Evolutionary Computation* 8(1): 28-46

Butz, M., Goldberg, D.E. & Lanzi, P-L. (2004b) Bounding Learning Time in XCS. In *Proceedings of the 2004 Genetic and Evolutionary Computation Conference - Gecco 2004*. Springer, pp927-934.

Butz, M., Goldberg, D.E. & Lanzi, P-L. & Sastry, K. (2004c) Bounding the Population Size to Ensure Niche Support in XCS. *IlliGAL Technical Report 2004033*. Available from http://www-illigal.ge.uiuc.edu/

Deb, K. (2001) *Evolutionary Multiobjective Optimization Algorithms*. Wiley.

DeJong, K. & Sarma, J. (1993) Generation Gaps Revisited. In D. Whitley (ed) *Foundations of Genetic Algorithms 2*. Morgan Kaufmann, pp19-28.

Dorigo, M. (1993) Genetic and Non-Genetic Operators in ALECSYS. *Evolutionary Computation* 1(2):151-164.

Goldberg, D.E. (1989) *Genetic Algorithms in Search, Optimization and Machine Learning*. Addison Wesley.

Holland, J.H. (1975) *Adaptation in Natural and Artificial Systems*. University of Michigan Press.

Holland, J.H. (1976) Adaptation. In R. Rosen & F.M. Snell (eds) *Progress in Theoretical Biology, 4*. Academic Press, pp313-329.

Holland, J.H. (1985) Properties of the Bucket Brigade. In J.J. Grefenstette (ed) *Proceedings of the First International Conference on Genetic Algorithms and their Applications*. Lawrence Erlbaum Associates, pp1-7.

Holland, J.H. (1986) Escaping Brittleness. In R.S. Michalski, J.G. Carbonell & T.M. Mitchell (eds) *Machine Learning: An Artificial Intelligence Approach, 2*. Morgan Kauffman, pp48-78.

Holland, J.H. & Reitman, J.S. (1978) Cognitive Systems based on Adaptive Algorithms. In D.A. Waterman & F. Hayes-Roth (eds) *Pattern Directed Inference Systems*. Academic Press, pp313-329.

Horn, F., Goldberg, D.E. & Deb, K. (1994) Implicit Niching in a Learning Classifier System: Nature's Way. *Evolutionary Computation* 2(1):37-66.

Kovacs, T. (2000) Strength or Accuracy? A Comparison of Two Approaches to Fitness Calculation in Learning Classifier Systems. In P-L. Lanzi, W. Stolzmann & S.W. Wilson (eds) *Learning Classifier Systems: From Foundations to Applications,* Springer, pp194-208.

Kovacs, T. (2001) Toward a Theory of Strong Overgeneral Classifiers. In W. Martin & W. Spears (eds) *Foundations of Genetic Algorithms 6*. Morgan Kaufmann, pp165-184.

Sutton, R.S. & Barto, A.G. (1998) *Reinforcement Learning*. MIT Press.

Tomlinson, A. & Bull, L. (1999) On Corporate Classifier Systems: Improving the use of Rule-Linkage. In *Proceedings of the 1999 Genetic and Evolutionary Computation Conference - Gecco 1999*. Morgan Kaufmann, pp649-656.

Vose, M. (1999) *The Simple Genetic Algorithm*. MIT Press.

Wilson, S.W. (1987) Classifier Systems and the Animat Problem. *Machine Learning* 2:199-228.

Wilson, S.W. (1994) ZCS: A Zeroth-level Classifier System. *Evolutionary Computation* 2(1):1-18.

Wilson, S.W. (1995) Classifier Fitness Based on Accuracy. *Evolutionary Computation* 3(2):149-177.

Wilson, S.W.(2000) State of XCS Classifier System Research. In P-L. Lanzi, W. Stolzmann & S.W. Wilson (eds) *Learning Classifier Systems: From Foundations to Applications*. Spinger, pp63-81.

Computational Complexity of the XCS Classifier System

Martin V. Butz, David E. Goldberg, and Pier Luca Lanzi

Illinois Genetic Algorithms Laboratory (IlliGAL)
University of Illinois at Urbana-Champaign
Urbana, IL, 61801
{butz,deg,lanzi}@illigal.ge.uiuc.edu

1 Introduction

Learning classifier systems (LCSs) are online-generalizing rule-based learning systems that use evolutionary computation techniques to evolve an optimal set of rules, that is, a *population of classifiers* (1; 2). LCSs tackle both *single-step* classification problems and *multi-step* reinforcement learning (RL) problems. Although the LCS proposal dates back over twenty years ago, there has been hardly any theory regarding convergence, computational effort, problem instances, etc. Successful applications seemed to rather rely on a "black art" of correct parameter settings, supported by powerful computers, than on actual insight.

XCS (3) can be viewed as a mile-stone of learning classifier system research. The XCS system combines an accuracy-based fitness approach with a niched genetic algorithm (4; 5). Recent results show that XCS can solve typical data-mining problems in a machine-learning competitive way, providing classification accuracy that is comparable to that of well-known machine learning algorithms, such as C4.5, nearest neighbor, Naive Bayes, and support vector machines (6; 7; 8).

This chapter connects LCSs, and specifically the XCS classifier system, to important elements of computational learning theory. We focus on the most fundamental class of concept learning problems, the learning of Boolean functions. Based on previous facetwise analyses resulting in several bounding models (9; 10; 11; 12), we show that k-DNF problems that satisfy few additional properties are PAC-learnable (13; 14) by XCS. That is, XCS scales polynomially in time and space complexity learning with high probability an approximately correct solution. The proof also confirms Wilson's previous conjecture on XCS's scalability (15).

The analysis essentially proves that XCS is an effective machine learning system that learns complex machine learning problems with a computational effort that scales similarly to other machine learning systems. Moreover, since XCS is an evolutionary learning system with very generally applicable learning mechanisms, the analysis actually confirms the general learning competence of the XCS system, which searches effectively for accuracy structures in any provided problem search space.

Due to its general applicability and its online learning capability, XCS is certainly not the most effective learning algorithm to solve k-DNF problems (see

M.V. Butz et al.: *Computational Complexity of the XCS Classifier System*, StudFuzz **183**,
91–125 (2005)
www.springerlink.com

e.g. (16)). However, XCS is not particularly targeted to solve k-DNF problems. Rather, XCS is a much more general, noise-robust problem solver. The aim of this paper is to show that LCSs, and XCS in particular, can be shown to be PAC-learning algorithms. The advantage of XCS compared to more specialized PAC-learning algorithms is its flexibility and generality in their applicability to different problem types, problem representations, and problem domains as well as to online learning problems.

To prevent complete problem dependence, our analysis does not model the system behavior exactly, as could be done for example with a Markov-chain model. Only general problem features are required to predict system behavior using our approach. Consequently, the derived problem bounds are flexible, modifiable, and less problem-dependent than an exact analysis. Additionally, the facet-wise analysis improves the overall understanding of XCS providing a general theory of how XCS evolves solutions that are maximally accurate and maximally general.

The chapter first gives a short introduction to XCS. Next, we provide a short evolutionary pressure overview that shows how XCS works. Section 4 derives the computational complexity of XCS. Summary and conclusions discuss the impact and extendibility of the presented research.

2   The XCS Classifier System

XCS learns in typical RL settings interacting online with an unknown environment that provides problem instances and reward feedback. Hereby, it learns to predict future reward values accurately.

XCS (as all other LCSs) represents the problem solution by a *population* of *classifiers*. At each time step, XCS receives a problem instance. Based on its current knowledge, XCS proposes a solution for the instance. Depending on the problem instance and the solution proposed, XCS receives numerical reward characterizing the goodness of the proposed solution. In essence, XCS is designed to evolve a complete, maximally accurate, and maximally general representation of the optimal problem solution. This section gives an overview over the basic structure and mechanisms in XCS, further details can be found in (17).

For the purpose of our analysis, we introduce XCS as a pure classification system in which no reward propagation is necessary. However, the reader should keep in mind that XCS is a much more general learning system that is able to learn in more general, multistep RL problems, in which (back-)propagation of reward is necessary in order to learn an optimal problem solution.

2.1   Problem Definition

We define a classification problem as a set of problem instances $X = \{0,1\}^l$ with length l. Each problem instance $S \in X$ is consequently characterized by l binary features. The target concept assigns each problem instance a corresponding class $A \in \{1, 2, ..., n\}$. Instances are generated at random from X underlying a certain probability distribution D. If not stated differently, we assume a uniform distribution over all 2^l possible problem instances. Problem instances are iteratively presented to XCS. In response to the resulting classification, the problem provides scalar reinforcement r reflecting the correctness of the classification. In the simplest case, reward 0 indicates an incorrect classification and a non null constant reward (e.g., 1000) indicates a correct classification.

2.2 Knowledge Representation

As mentioned above, knowledge is represented by a *population* $[P]$ (that is, a set) of *classifiers* (that is, classification rules). Each classifier is characterized by five major attributes: (1) the condition part C specifies when the classifier matches; (2) the action part A specifies the action (or classification); (3) the reward prediction R estimates the average reward received given conditions C executing action A; (4) prediction error ε estimates the mean absolute deviation of the reward prediction; (5) fitness F estimates the average relative accuracy of the classifier. In the problem setting considered here, conditions are strings of l symbols in the ternary alphabet $\{0, 1, \#\}$ ($C \in \{0, 1, \#\}^l$) where the symbol $\#$ (called *don't care*) matches both zero and one. Effectively, a problem solution is represented by a disjunctive normal form in which each accurate classifier specifies one conjunctive clause.

2.3 Classifier Evaluation

Given the current problem instance S, XCS forms a *match set* $[M]$ consisting of all classifiers in $[P]$ whose conditions match S. The match set $[M]$ essentially represents the knowledge about the current problem instance. $[M]$ is used to decide on the classification on the current problem forming fitness-weighted reward predictions of each possible classification. After the execution of the chosen classification A, and the resulting reward R, an *action set* $[A]$ is formed consisting of all classifiers in $[M]$ that specify the chosen action A. Parameters R, ε, and F of all classifiers in $[A]$ are updated according to the following equations:

$$R \leftarrow R + \beta(r - R), \tag{1}$$

$$\varepsilon \leftarrow \varepsilon + \beta(|r - R| - \varepsilon), \tag{2}$$

$$\kappa = \begin{cases} 1 & \text{if } \varepsilon < \varepsilon_0 \\ \alpha(\varepsilon/\varepsilon_0)^{-\nu} & \text{otherwise} \end{cases}, \quad \kappa' = \frac{\kappa}{\sum\limits_{x \in [A]} \kappa_x}, \tag{3}$$

$$F \leftarrow F + \beta(\kappa' - F) \tag{4}$$

Parameter β denotes the learning rate, ε_0 denotes the error tolerance, α and ν are additional constants that scale the fitness evaluation. In XCS, classifier prediction R (Equation 1) corresponds to the Q-values (formal details in (18)); classifier fitness F (equations 2–4) essentially estimates the scaled, average, relative accuracy of the classifier derived from the current reward prediction error ε with respect to the competing classifiers.

2.4 Rule Evolution

Initially, the population [P] is empty. When a problem instance is presented and no classifier in [P] matches, XCS applies a covering mechanism that generates a classifier for each possible classification. Covering classifiers match the problem instance and have an average predefined *specificity* $(1 - P_\#)$. If $P_\#$ is close to one, XCS basically starts from very general hypotheses and basically pursues a *general-to-specific* search. If $P_\#$ is small, XCS starts from very specific hypotheses and the search follows a *specific-to-general* approach.

The genetic algorithm (GA) is the main rule learning component. Because classifier fitness estimates the accuracy of the reward prediction, the GA favors the evolution of classifiers which provide an accurate prediction of the expected payoffs. The genetic algorithm used is a steady-state niched genetic algorithm (19). Each problem iteration, the GA may be applied. The GA reproduces two classifiers selecting from the current action set [A] maximizing fitness. The introduction of tournament selection with a tournament size proportionate to the current action set size strongly increased the noise-robustness of the system (9). Offspring classifiers are crossed with probability χ, mutated with probability μ, and inserted in the population. To keep the population size constant, two classifiers are deleted from the population. Additionally, a subsumption-deletion mechanism is applied which favors more general, accurate previous classifiers over more-specialized offspring classifiers.

3 Evolutionary Pressures in XCS

XCS is designed to evolve a complete, accurate, and maximally general representation of the reward function of the problem. More accurate classifiers are favored due to the accuracy-based selection. Since classifiers are favorably selected if they are more accurate, the GA in XCS pushes toward a population of maximally accurate classifiers. More general classifiers are favored since more general classifiers match and therefore reproduce more often. Since more general classifiers match and therefore reproduce more often, the overall effect of the GA in XCS is an evolutionary pressure toward a solution consisting of maximally accurate and maximally general classifiers. This simple principle, firstly stated by (3), has been theoretically analyzed in (10), in which a basic idea of XCS functioning is provided. The analysis characterizes the different evolutionary biases as *evolutionary pressures*, which together guide XCS in the evolution of the desired complete, maximally accurate, and maximally general problem solution representation. The four most important pressures are (1) *set pressure*, (2) *mutation pressure*, (3) *deletion pressure*, and (4) *fitness pressure*.

Set pressure refers to the generalization pressure in XCS resulting from niche reproduction but population-wide deletion. Mutation pressure generally can be regarded as a diversification pressure that results in a local search in the neighborhood of the selected classifier. Mutation essentially pushes the population towards an equal distribution of symbols in classifier conditions. In terms of specificity in the ternary case, mutation pushes towards an equal distribution of zeroes, ones, and don't care symbols and thus towards a specificity of $2/3$. Deletion pressure additionally prefers the deletion of inaccurate classifiers and classifiers that populate over-populated niches. In general, though, deletion pressure in XCS is weak and can be approximated by random deletion from the population.

Combining set pressure and mutation pressure assuming random deletion in $[P]$, we can derive a general *specificity equation* that expects the specificity change in the population.

$$\sigma[P'] = \sigma[P] + f_{ga}\frac{2(\sigma[A] + \Delta_\mu(\sigma[A]) - \sigma[P])}{N} \qquad (5)$$

where $\sigma[X]$ denotes the average specificity of all classifier conditions in set $[X]$, f_{ga} denotes the frequency of GA application, Δ_μ denotes the specificity change due to mutation, and N denotes the population size. (10) evaluate the accuracy of the specificity equation in detail confirming that given no further fitness influence, the specificity in the population behaves in the derived way and converges to the predicted value.

Fitness pressure is the main pressure towards higher accuracy. In essence, since more accurate classifiers are selected for reproduction, classifiers with higher accuracy and thus classifiers with lower error are propagated. In general, the degree of fitness pressure is strongly problem dependent. In (9) it was shown that tournament selection results in a much more reliable fitness pressure towards higher accuracy.

In sum, XCS's evolutionary pressures propagate the evolution of accurate, maximally general classifiers in the population. Imagining a specificity axis along which the classifier population evolves, the pressures can be visualized as shown in Figure 1. While fitness pressure prevents over-generalization, set pressure prevents over-specialization. Mutation serves as the diversification mechanism that enables the fitness pressure to evolve more accurate classifiers. However, it also has a general specialization tendency. Crossover is not visualized since it has no immediate impact on specificity. However, the mechanism can be very important in the effective propagation of accurate sub-solutions dependent on the problem (19; 9).

4   Towards Computational Complexity

With the knowledge of XCS's learning biases at hand, we can now analyze XCS's learning complexity. Essentially, we now assume that the learning pressures are set correctly and that XCS will evolve a complete problem solution as long as enough computational resources are available. Thus, we do not investigate the chosen fitness approach, the type of offspring selection, or improvements in the applied search operators in further detail. The interested reader is referred to the available literature (20; 21; 22). Rather, we investigate the computational effort necessary to assure that the evolutionary pressures can apply.

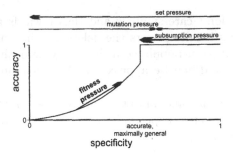

Fig. 1. The visualized evolutionary pressures give an intuitive idea how XCS evolves the intended complete, accurate, and maximally general problem representation.

Each learning iteration, all classifiers need to be monitored as matching candidates so that the computational effort in each learning iteration is bounded by the population size N. We may denote the number of iterations until an optimal solution is found by the learning time t^* so that the overall computational effort grows in Nt^*.

Several constraints need to be satisfied to ensure that the evolutionary pressures can apply, that enough time is allocated to evolve a complete solution, and that the evolved solution is sustained. In essence, the following four facets need to be satisfied.

1. *Population initialization* needs to ensure time for classifier evaluation and successful GA application.
2. *Schema supply* needs to be ensured to have better classifiers available.
3. *Schema growth* needs to be ensured to grow those better classifiers evolving a complete problem solution.
4. *Solution sustenance* needs to be ensured to sustain a complete problem solution.

We address these issues in the subsequent sections.

First, we derive the *covering bound* to ensure proper initialization. Second, we derive the *schema bound* to ensure the availability of better schema representatives for reproduction. We use the schema bound to derive initial settings for population specificity and population size. Also the time it takes to generate a better classifier at random is considered.

Next, we derive the *reproductive opportunity bound* to ensure schema growth by ensuring that better classifiers can be detected and reproduced successfully. We show that the schema bound and the reproductive opportunity bound somewhat interact since, intuitively, too much supply implies too large specificity consequently disabling reproduction.

While the reproductive opportunity bound assures that better classifiers grow, we are also interested in how long it takes them to grow. This is expressed in the learning time bound that assures solution growth.

Once enough learning time is available to make better classifiers grow until the final solution is found, we finally need to assure *sustenance* of the grown solution. XCS applies niching in that it reproduces in problem subspaces and deletes from the whole population. Larger niches are preferred for deletion. The analysis results in a final population size bound ensuring the sustenance of a complete problem solution as long as there are no severe problem solution overlaps.

Putting the results together, we are able to derive a positive computational learning theory result for an evolutionary-based learning system with respect to k-DNF functions. We show that XCS is able to PAC-learn k-DNF functions with few restrictions. However, the reader should keep in mind that XCS is a system that is much more broadly applicable and actually an online generalizing RL system. XCS's capability of PAC-learning k-DNF functions confirms its general learning scalability and potential widespread applicability.

4.1 Proper Population Initialization: The Covering Bound

Several issues need to be considered when intending to make time for classifier evaluation and thus the identification of better classifiers. The first bound is derived from the rather straight-forward requirement that the evolutionary algorithm in XCS needs to apply. That is, genetic reproduction in action sets and deletion in the population needs to take place.[1]

Reproduction will not occur if XCS gets stuck in an infinite covering-deletion cycle. This can happen if the initial population is filled with over-specialized classifiers. In this case, inputs may continuously not be covered since covering will continue to generate over-specialized classifiers and delete random other overspecialized classifiers. In this case, the GA will never take place and no evolutionary pressures apply. This scenario can only happen if the maximal population size N is set too small and initial specificity, controlled by the *don't care parameter* $P_{\#}$, is set too high.

As specified above, XCS triggers the creation of a covering classifier if the current problem instance is not matched by at least one classifier for each possible classification. If the population is already filled up with classifiers, other classifiers are deleted to make space for the new covering classifiers. Given the population is filled up with overspecialized classifiers that did not undergo any evaluation so far, and since all covering classifiers have identical fitness F and action set size estimate as values, classifiers are effectively selected uniformly randomly for deletion. Thus, in this case, a *covering-random deletion* cycle may continue for a long time.

Assuming a uniform problem instance distribution over the whole problem space $\mathcal{S} = \{0, 1\}^l$, we can determine the probability that a given problem instance is covered by at least one classifier in a randomly generated population:

$$P(\text{cover}) = 1 - \left(1 - \left(\frac{2 - \sigma[P]}{2}\right)^l\right)^N , \tag{6}$$

---

[1] Related publications of parts of this and the following section can be found elsewhere (23; 9; 10).

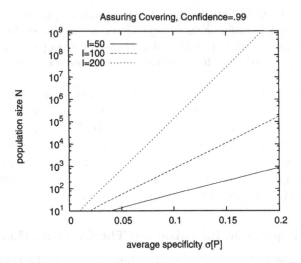

Fig. 2. To ensure that better classifiers can be identified, population size needs to be set high enough and specificity low enough to satisfy the covering bound.

where $\sigma[P]$ may be equated with $1 - P_{\#}$ in the beginning of a run. Similarly, we can derive the actual necessary maximal specificity given a certain population size using the inequality $1 - \exp^x < x$ setting $(1 - \text{cover})$ to $1/\exp$:

$$\sigma[P] < 2(1 - (\frac{1}{N})^{1/l}) < 2 - 2(1 - (1 - \text{cover})^{1/N})^{1/l}, \tag{7}$$

showing that increasing N results in an increase in maximal specificity polynomial in $1/l$ deriving an effective rule of thumb of how low the don't care probability $P_{\#}$ may be set to assure an appropriate specificity $\sigma[P]$. Figure 2 shows the resulting boundary conditions on population size and specificity for different problem lengths requiring a confidence level of 0.99.

To automatically avoid the covering bound, XCS could be enhanced to detect infinite covering and consequently increase the $P_{\#}$ value. However, we did not experiment with such an enhancement so far since usually the covering bound can be easily circumvented by setting the parameter $P_{\#}$ large enough.

Given that the problem instances are not sampled uniformly randomly in the problem space, the covering bound can be used as an upper bound. Since the covering mechanism generates classifiers that cover actual instances and the genetic algorithm mainly focuses on generating offspring classifiers that apply in the current problem niche, the smaller the set of sampled problem instances, the larger the probability that an instance is covered. In most RL problems as well as in datamining problems the number of distinct problem instances is usually much smaller than the whole problem space so that the covering bound becomes less important in these problems.

4.2   Ensuring Supply: The Schema Bound

The supply question relates to the schema supply in GAs. Similar to GAs, LCSs process *building blocks* (BBs)—small dependency structures in the overall problem structure that result in an increase in fitness. Since the fitness structure of one BB may point towards the mediocre solution (that is, a local optimum), the optimal solution to a BB as a whole needs to be processed. Thus, disregarding mutation effects, individuals that represent the optimal BB structure need to be available from the beginning to be able to propagate the BB.

The same observation applies in XCS albeit in slightly different form. The question arises, what is a BB in the XCS classifier system? We know that fitness is based on accuracy. Thus, a BB in XCS is a substructure in the problem that increases classification accuracy. As there are BBs for GAs, there are minimal substructures in classification problems that result in higher accuracy (and thus fitness).

To establish a general notion of BBs in XCS, we use the notion of a *schema* as suggested elsewhere (24; 4). A schema for an input of length l is defined as a string that specifies some of the positions and ignores others. The number of specified positions is termed the *order* k of the schema. A schema is said to be *represented* by a classifier if the classifier correctly specifies *at least* all positions that are specified in the schema. Thus, a representative of a schema of order k has a specificity of at least $\sigma(.) = k/l$. For example, a classifier with condition C =##10#0 is a representative of schema **10*0, but also of schemata **10**, **1**0, ***0*0, **1***, ***0**, *****0, and ******.

Let's consider now a specific problem in which a minimal order of at least k_m bits needs to be specified to reach higher accuracy. We call such a problem a problem that has a *minimal order of difficulty* k_m. That is, if less than k_m bits are specified in a classifier, the class distribution in the specified problem subspace is equal to the overall class distribution. In other words, the *entropy* of the class distribution decreases only if at least some k_m bits are specified. Since XCS's fitness is derived from accuracy, representatives of the schema of order k_m need to be present in the population.

Population Size Bound To assure the supply of the representatives of a schema of order k_m, the population needs to be specific enough and large enough. It is possible to determine the probability that a randomly chosen classifier from the current population is a schema representative by:

$$P(\texttt{representative}) = \frac{1}{n} \left(\frac{\sigma[P]}{2} \right)^{k_m}, \tag{8}$$

where n denotes the number of possible actions and $\sigma[P]$ denotes the specificity in the population, as before. From Equation 8 we can derive the probability of the existence of a representative of a specific schema in the current population

$$P(\texttt{representative exists}) = 1 - \left(1 - \frac{1}{n} \left(\frac{\sigma[P]}{2} \right)^{k_m} \right)^N, \tag{9}$$

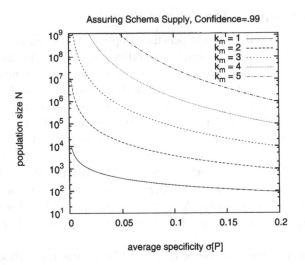

Fig. 3. The schema bound requires population size to be set sufficiently high with respect to a given specificity. The larger the population size, the lower the necessary specificity.

basically deriving the probability that at least on schema representative of order k_m exists in $[P]$.

As shown in previous publications (25; 10; 9), in a problem in which no current fitness pressure applies, specificity can be approximated by twice the mutation probability μ, that is, $\sigma[P] \approx 2\mu$. Additionally, the population may be initialized to a desired specificity value by choosing parameter $P_\#$ appropriately. It should be kept in mind, though, that albeit $P_\#$ might bias specificity further early in the run, without any fitness influence, specificity converges to a value that can be approximated by 2μ. Thus, mutation determines specificity on the long term. Well-chosen $P_\#$ values may boost initial XCS performance.

Requiring a high probability for the existence of a representative, we can derive the following population size bound using the inequality $x < -ln(1-x)$:

$$N > -n\left(\frac{2}{\sigma[P]}\right)^{k_m} ln(1 - P(\texttt{rep.exists})) > \frac{\log\left(1 - P(\texttt{rep.exists})\right)}{\log\left(1 - \frac{1}{n}\left(\frac{\sigma[P]}{2}\right)^{k_m}\right)}, \quad (10)$$

which shows that N needs to grow logarithmically in the probability of error and exponentially in the minimal order of problem difficulty k_m given a certain specificity. Enlarging the specificity, we can satisfy the schema bound. However, schema growth may be violated if specificity is chosen too high as shown in the subsequent sections.

Figure 3 shows the schema bound plotting the required population size with respect to a given specificity and several minimal orders k_m (left-hand side). Population size is plotted in log scale due to its exponential dependence on the order k_m.

Specificity Bound Similar to the bound on population size, given a certain problem of minimal order of problem difficulty k_m, we can derive a minimal specificity bound from Equation 9 assuming a fixed population size:

$$\sigma[P] > 2n^{1/k_m}(1-(1-P(\texttt{rep.exists}))^{1/N})^{1/k_m} \tag{11}$$

Setting $(1 - P(\texttt{rep.exists}))$ to $1/\exp$ we can use the inequality $1 - \exp^{-x} < x$ to derive that:

$$\sigma[P] > 2\left(\frac{n}{N}\right)^{1/k_m} \tag{12}$$

Note that an identical derivation is possible determining the expected number of schema representatives in $[P]$ given specificity $\sigma[P]$:

$$E(representative) = \frac{N}{n}\left(\frac{\sigma[P]}{2}\right)^{k_m} \tag{13}$$

Requiring that at least one representative can be expected in the current population

$$E(representative) > 1, \tag{14}$$

also yields Equation 12.

We may rewrite Equation 12 using the O-notation. Given a population size of N and the necessary representation of an unknown schema of order k_m, the necessary specificity $\sigma[P]$ can be bounded by

$$\sigma[P]:\ O\left(\left(\frac{n}{N}\right)^{1/k_m}\right), \tag{15}$$

showing that the required specificity decreases polynomially with increasing population size N and increases exponentially with increasing problem complexity k_m. Since we have shown that population size N also needs to increase exponentially in k_m but necessary specificity decreases polynomially in N, the two effects cancel each other. Thus, it is possible to leave specificity and thus mutation constant and to focus only on a proper population size to assure effective schema supply.

Extension in Time Given we start with a completely general or highly general initial classifier population (that is, $P_\#$ is close to 1.0), the schema bound also extends in time. In this case, it is the responsibility of mutation to push the population towards the intended specificity generating initial supply.

Given a mutation probability μ, the probability can be approximated that a classifier is generated that has all k_m relevant positions specified given a current specificity $\sigma[P]$:

$$P(\texttt{generation of representative}) = (1-\mu)^{\sigma[P]k_m}\cdot\mu^{(1-\sigma[P])k_m} \tag{16}$$

With this probability, we can determine the expected number of steps until at least one classifier may have the desired attributes specified. Since this is a geometric distribution:

$$E(t(\text{generation of representative})) =$$
$$1/P(\text{generation of representative}) =$$
$$\left(\frac{\mu}{1-\mu}\right)^{\sigma[P]k_m} \mu^{-k_m} \qquad (17)$$

Given a current specificity of zero, the expected number of steps until the generation of a representative consequently equals to μ^{-k_m}. Thus, given we start with a completely general population, the expected time until the generation of a first representative is less than μ^{-k_m} (since $\sigma[P]$ increases over time). Requiring that the expected time until a classifier is generated is smaller than some threshold Θ, we can generate a lower bound on the mutation μ:

$$\mu^{-k_m} < \Theta$$
$$\mu > \Theta^{\frac{1}{-k_m}} \qquad (18)$$

The extension in time is directly correlated with the specificity bound in Equation 12. Setting Θ to N/n we get the same bound (since σ can be approximated by 2μ).

As mentioned before, although supply may be assured easily by setting the specificity σ and thus $P_\#$ and more importantly the mutation rate μ sufficiently high, we yet have to assure that the supplied representatives can grow. This is the concern of the following section in which the the reproductive opportunity bound is derived.

4.3   Making Time for Growth: The Reproductive Opportunity Bound

To ensure the growth of better classifiers, we need to ensure that *better* classifiers get reproductive opportunities. So far, the covering bound only assures that reproduction and evaluation are taking place. This is a requirement for ensuring growth but not sufficient for it. This section derives and evaluates the *reproductive opportunity bound* that provides a population size and specificity bound that assures the growth of better classifiers.[2]

The idea is to assure that more accurate classifiers need to be ensured to undergo reproductive opportunities before being deleted. To do this, we minimize the probability of a classifier being deleted before being reproduced. The constraint effectively results in another population and specificity bound since only a larger population size and a sufficiently small specificity can assure reproduction before deletion.

---

[2] Related publications of parts of this section can be found elsewhere (26; 9).

General Population Size Bound To derive the bound, we first determine the expected number of steps until deletion. Assuming neither any fitness estimate influence nor any action set size estimate influence, the probability of deletion is essentially random. Thus, the probability of deleting a particular classifier in a learning iteration equals:

$$P(\texttt{deletion}) = \frac{2}{N},\qquad(19)$$

since two classifiers are deleted per iteration. A reproductive opportunity takes place if the classifier is part of an action set. As shown elsewhere (22) the introduced action-set size proportionate tournament selection bounds the probability of reproduction of the best classifier from below by a constant. Thus, the probability of being part of an action set directly determines the probability of reproduction:

$$P(\texttt{reproduction}) = \frac{1}{n}2^{-l\sigma(cl)}\qquad(20)$$

Note that as before this derivation assumes binary input strings and further uniform random sampling from the problem space. Combining Equation 19 with Equation 20, we can determine that neither reproduction nor deletion occurs at a specific point in time:

$$P(\texttt{no rep., no del.}) = (1 - P(\texttt{del.}))(1 - P(\texttt{rep.})) =$$

$$= (1 - \frac{2}{N})(1 - \frac{2^{-l\sigma[P]}}{n}) = 1 - \frac{2}{N} - \frac{2^{-l\sigma[P]}}{n}(1 - \frac{2}{N})\qquad(21)$$

Together with equations 19 and 20, we can now derive the probability that a certain classifier is part of an action set before it is deleted:

$$P(\texttt{rep.before del.}) = P(\texttt{rep.})(1 - P(\texttt{del.}))\sum_{i=0}^{\infty} P(\texttt{no rep., no del.})^i =$$

$$P(\texttt{rep.})(1 - P(\texttt{del.}))\frac{1}{1 - P(\texttt{no rep., no del.})} =$$

$$\frac{\frac{1}{n}2^{-l\sigma[P]}(1 - \frac{2}{N})}{\frac{2}{N} + \frac{1}{n}2^{-l\sigma[P]}(1 - \frac{2}{N})} = \frac{\frac{1}{n}2^{-l\sigma[P]}}{\frac{2}{N-2} + \frac{1}{n}2^{-l\sigma[P]}} = \frac{N - 2}{N - 2 + n2^{l\sigma[P]+1}}\qquad(22)$$

Requiring a certain minimal reproduction-before-deletion probability and solving for the population size N, we derive the following bound:

$$N > \frac{2n2^{l\sigma[P]}}{1 - P(\texttt{rep.before del.})} + 2\qquad(23)$$

This bounds the population size by $O(n2^{l\sigma})$. Since specificity σ can be set proportional to $\sigma = 1/l$, the bound actually diminishes usually. However, in problems in which the problem complexity $k_m > 1$, we have to ensure that classifiers that represent order k_m schemata have reproductive opportunities.

Given a current population-wide specificity $\sigma[P]$, the expected specificity of a representative of an order k_m schema can be estimated by

$$E(\sigma(\texttt{repres.of schema of order } k_m)) = \frac{k_m + (l - k_m)\sigma[P]}{l}, \qquad (24)$$

assuming that the specificity in the other $l - k_m$ attributes equals the specificity in the population. Substituting $\sigma(cl)$ from Equation 23 with this expected specificity of a representative of a schema of order k, the population size N can be bounded by

$$N > 2 + \frac{n2^{l \cdot \frac{k+(l-k)\sigma[P]}{l}+1}}{1 - P(\texttt{rep.before del.})}$$

$$N > 2 + \frac{n2^{k+(l-k)\sigma[P]+1}}{1 - P(\texttt{rep.before del.})} \qquad (25)$$

This bound ensures that the classifiers necessary in a problem of order of difficulty k_m get reproductive opportunities. Once the bound is satisfied, existing representatives of an order k_m schema are ensured to reproduce before being deleted and XCS is enabled to evolve a more accurate population.

Note that this population size bound is actually exponential in schema order k_m and in string length times specificity $l\sigma[P]$. This would mean that XCS scales exponentially in the problem length, which is certainly highly undesirable. However, since specificity in $[P]$ decreases with larger population sizes as shown in Equation 15, we derive a general *reproductive opportunity bound (ROP-bound)* below that shows that population size needs to grow in $O(l^{k_m})$.

Figure 4 shows several settings for the reproductive opportunity bound. We show the resulting dependency of population size N on specificity $\sigma[P]$ requiring a confidence value of .99. In comparison to the covering bound, shown in Figure 2, it can be seen that the reproductive opportunity bound is always stronger than the covering bound making the latter nearly obsolete. However, the covering bound is still useful to set the don't care probability $P_\#$ appropriately. Mutation rate, and thus the specificity the population converges to, however, is bound by the reproductive opportunity bound.

General Reproductive Opportunity Bound While the above bound ensures the reproduction of *existing* classifiers that represent a particular schema, it does not assure the actual presence or generation of such a classifier. Thus, we need to *combine schema bound* and *reproductive opportunity bound*. Figure 5 shows a *control map* of certain reproductive opportunity bound values and schema bound values requiring a probability of success of 50% (plotting equations 10 and 25). The corresponding intersections denote the best value of specificity and population size to ensure supply and growth with high probability. Initial specificity may be set slightly larger than the value of the intersection to boost initial performance as long as the covering bound is not violated.

We can also quantify the interaction. Substituting the O-notated specificity bound in Equation 15 of the schema bound in the O-notated dependence of the

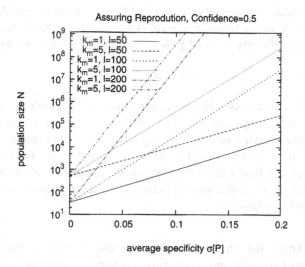

Fig. 4. To ensure successful identification and reproduction of better classifiers, population size needs to be set high enough with respect to a given specificity.

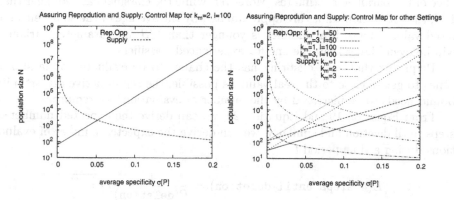

Fig. 5. The shown control maps clarify the competition between reproduction and supply. The shown boundaries assure a 50% probability of success. High specificity ensures supply but may hinder reproduction. Vice versa, low specificity ensures reproduction but lowers the probability of supply.

representative bound on string length l ($N : O(2^{l\sigma[P]})$) and ignoring additional constants, we can derive the following enhanced population size bound:

$$N > 2^{l(\frac{n}{N})^{1/k_m}}$$

$$\log_2 N > l\left(\frac{n}{N}\right)^{1/k_m}$$

$$N^{1/k_m}\log_2 N > ln^{1/k_m}$$

$$N(\log_2 N)^{k_m} > nl^{k_m} \qquad (26)$$

This general *reproductive opportunity bound* (ROP-bound) essentially shows that population size N needs to grow approximately exponentially in the minimal order of problem difficulty k_m and polynomially in the string length.

$$N : O(l^{k_m}) \tag{27}$$

Note that in the usual case, k_m is rather small and can often be set to one. Essentially, when k_m is greater than one, other classification systems in machine learning have also shown a similar scale up behavior. For example, the inductive generation of a decision tree in $C4.5$ (27) would not be able to decide on which attribute to expand first (since any expansion leads to the same probability distribution and thus no information gain) and consequently would generate an inappropriately large tree.

Sufficiently Accurate Values Although we usually assume that classifier parameter estimates are sufficiently accurate, we need to note that this assumption does not hold necessarily. All classifier parameters are only an approximation of the average value. The higher parameter β is set, the higher the expected variance of the parameter estimates. Moreover, while the classifiers are younger than $1/\beta$, the estimation values are approximated by the average of the so far encountered values. Thus, if a classifier is younger than $1/\beta$, its parameter variances will be even higher than the one for experienced classifiers.

Requiring that each offspring has the chance to be evaluated at least $1/\beta$ times to get as close to the real value as possible, the reproductive opportunity bound needs to be increased by the number of evaluations we require.

From Equation 19 and Equation 20, we can derive the expected number of steps until deletion and similarly, we can derive the expected number of evaluations during a time period t.

$$E(\text{\# steps until deletion}) = \frac{1}{P(\text{deletion})}/2 = \frac{N}{2} \tag{28}$$

$$E(\text{\# of evaluations in } t \text{ steps}) = P(\text{in } [A]) \cdot t = \frac{1}{n}0.5^{l\sigma(cl)}t \tag{29}$$

The requirement for success can now be determined by requiring that the number of evaluations before deletion must be larger than some threshold Θ where Θ could for example be set to $1/\beta$.

$$E(\text{\# of evaluations in (\# steps until deletion)}) > \Theta$$
$$\frac{1}{n}0.5^{l\sigma(cl)}\frac{N}{2} > \Theta$$
$$N > \Theta n 2^{l\sigma(cl)+1} \tag{30}$$

Setting Θ to one, Equation 30 is basically equal to Equation 25 since one evaluation is equal to at least one reproductive opportunity disregarding the confidence value in this case. It can be seen that the sufficient evaluation bound only increases the reproductive opportunity bound by a constant. Thus, scale-up behavior is not affected.

As a last point in this section, we want to point out that fitness is actually not computed by averaging, but the Widrow-Hoff delta rule is used from the beginning. Moreover, fitness is usually set to 10% of the parental fitness value (to prevent disruption). Thus, fitness is derived from two approximations and it starts off with a disadvantage so that the early evolution of fitness strongly depends on fitness scaling and on accurate approximations of the prediction and prediction error estimates. To ensure a fast detection and reproduction of superior classifiers it is consequently necessary to choose initial classifier values as accurate as possible. Alternatively, the expected variance in fitness values could be considered to prevent potential disruption. For example, 28) suggests the usage of variance sensitive bidding. Accordingly using the estimated expectable variance, the fitness of young classifiers could be modified for selection to prevent disruption but also to enable the earlier detection of better classifiers.

Bound Verification The derived bounds are experimentally confirmed elsewhere (23; 10; 9). In essence, it was shown that that the derived bounds hold using a Boolean function problem of order of difficulty k_m. The hidden parity function (a set of k of the l attributes are evaluated by a parity (or XOR) operator), originally investigated in XCS in 29), is very suitable to manipulate k_m since at least all k relevant parity bits need to be specified to increase classification accuracy. Thus, the minimal order of problem difficulty equals the size of the parity ($k_m = k$)..

The results confirmed the computational dependency on minimal order of difficulty k_m as well as the problem length l and the chosen specificity reflected in the mutation probability μ. In particular, population size needs to grow polynomially in the string length l as well as exponential in k_m in order to assure quick and reliable learning.

4.4   Estimating Learning Time

Given that schema, covering, and reproductive opportunity bounds are satisfied, we addressed three facets of the four aspects introduced: Problem initialization is appropriate if the covering bound is satisfied. Schema supply is assured if the schema bound is respected. Growth can be assured if the reproductive opportunity bound is considered. Thus, it is assured that better classifier structures can grow in the population.

However, it was not addressed, yet, how long it may take to evolve a complete problem solution by the means of this growing process. In essence, the reproductive opportunity bound ensures that better classifiers *can* grow. This section investigates *how long* it takes to grow a complete solution from this growing process.[3]

---

[3] Related publications of parts of this and the following section can be found in 11).

Assuming that the other problem bounds are satisfied by the choice of mutation and population size, we can estimate how long it takes to discover successively better classifiers until the maximally general, accurate classifiers are found. To do this, we assume a domino convergence model (30) estimating the time until each relevant attribute can be expected to be specialized to the correct value. Considering mutation only, we estimate the time until reproduction and the time until generation of the next best classifier in each problem niche. Using this approach we can show that learning time scales polynomially in problem length and problem complexity.

Time Bound Derivation To derive our learning time bound, we estimate the time until reproduction of the current best classifier as well as the time until creation of the next best classifier via mutation given a reproductive event of the current best classifier. The model assumes to start with an initially completely general population (that is, $P_\# = 1.0$). Initial specializations are randomly introduced via mutation. Problem-specific initialization techniques or higher initial specificity in the population may speed-up learning time (as long as the covering bound is not violated).

Further assumptions are that the current best classifier is not lost and selected as the offspring when it is part of an action set (assured by the ROP-bound in conjunction with tournament selection). The time model assumes domino convergence (30) in which each attribute is successively specified. This means that only once the first attribute is correctly specified in a classifier, the second attribute influences fitness and so forth.

Using the above assumptions, we can estimate the probability that mutation correctly specifies the next attribute

$$P(\texttt{perfect mutation}) = \mu(1 - \mu)^{l-1} \tag{31}$$

where l specifies the number of attributes in a problem instance. This probability can be relaxed in that we only require that the k already correctly set features are not unset (changed to don't care), the next feature is set, and we do not care about the others:

$$P(\texttt{good mutation}) = \mu(1 - \mu)^k \tag{32}$$

Whereas Equation 31 specifies the lower bound on the probability that the next best classifier is generated, Equation 32 specifies an optimistic bound.

As seen before, the probability of reproduction can be estimated by the probability of occurrence in an action set. The probability of taking part of an action set again, is determined by the current specificity of a classifier. Given a classifier which specifies k attributes, the probability of reproduction is

$$P(\texttt{reproduction}) = \frac{1}{n}\frac{1}{2}^k , \tag{33}$$

where n denotes the number of actions in a problem. The best classifier has a minimal specificity of k/l. With respect to the current specificity in the population $\sigma[P]$, the specificity of the best classifier may be expected to be $k+\sigma[P](l-k)$

assuming a uniform specificity distribution in the other $l - k$ attributes. Taking this expected specificity into account, the probability of reproduction is

$$P(\text{rep.in } [P]) = \frac{1}{n}\frac{1}{2}^{k+\sigma[P](l-k)} \tag{34}$$

Since the probability of a successful mutation assumes a reproductive event, the probability of generating a better offspring than the current best is determined by:

$$P(\text{generation of next best cl.}) = P(\text{rep.in } [P]) \, P(\text{good mutation}) =$$
$$\frac{1}{n}\frac{1}{2}^{k+\sigma[P](l-k)} \mu(1-\mu)^{l-1} \tag{35}$$

Since we assume uniform sampling from all possible problem instances, the probability of generating a next best classifier conforms to a geometric distribution (memoryless property, each trial has an independent and equally probable distribution), the expected time until the generation of the next best classifier is

$$E(\text{time until gen.of next best cl.}) =$$
$$1/P(\text{time until gen.of next best cl.}) =$$
$$\frac{1}{\frac{1}{n}\frac{1}{2}^{k+\sigma[P](l-k)}\mu(1-\mu)^{l-1}} = \frac{n2^{k+\sigma[P](l-k)}}{\mu(1-\mu)^{l-1}} \leq \frac{n2^{k+\sigma[P]l}}{\mu(1-\mu)^{l-1}} \tag{36}$$

Given now a problem in which k_d features need to be specified and given further the domino convergence property in the problem, the expected time until the generation of the next best classifier can be summed to derive the time until the generation of the global best classifier:

$$E(\text{time until generation of maximally accurate cl}) =$$
$$\sum_{k=0}^{k_d-1} \frac{n2^{k+\sigma[P]l}}{\mu(1-\mu)^{l-1}} = \frac{n2^{\sigma[P]l}}{\mu(1-\mu)^{l-1}} \sum_{k=0}^{k_d-1} 2^k < \frac{n2^{k_d+\sigma[P]l}}{\mu(1-\mu)^{l-1}} \tag{37}$$

This time bound shows that XCS needs an exponential number of evaluations in the order of problem difficulty k_d. As argued above, the specificity and consequently also mutation needs to be decreased indirect proportional to the string length l. In particular, since specificity $\sigma[P]$ grows as $O((\frac{n}{N})^{\frac{1}{k_m}})$ (Equation 15) and population size grows as $O(l^{k_m})$ (Equation 27), specificity essentially grows as

$$O(\frac{n}{l}) \tag{38}$$

Using the O-notation and substituting in Equation 37, we derive the following adjusted time bound making use of the inequality $(1 + \frac{n}{l})^l < e^n$:

$$O\left(\frac{l2^{k_d+n}}{(1-\frac{n}{l})^{l-1}}\right) = O\left(\frac{l2^{k_d+n}}{e^{-n}}\right) = O\left(l2^{k_d+n}\right) \tag{39}$$

Thus, learning time in XCS is bound mainly by the order of problem difficulty k_d and the number of problem classes n. It is linear in the problem length l. This derivation essentially also validates Wilson's hypothesis that XCS learning time grows polynomially in problem complexity as well as problem length (15). The next section experimentally validates the derived learning bound.

Experimental Validation In order to validate the derived bound, performance was evaluated on an artificial problem in which domino convergence is forced to take place. Similar results are expected in typical Boolean function problems in which similar fitness guidance is available, such as in the layered multiplexer problem (3; 9). In other problems, additional learning influences may need to be considered such as the influence of crossover or the different fitness guidance in the problem (9).

The results in 11) as well as the further experimental evaluations in 22) confirmed the learning time bound. Essentially, the dependency on problem difficulty k_d and on mutation rate μ was confirmed. Moreover, it was shown that the population size needs to be chosen sufficiently high to satisfy the above problem bounds. Moreover, it was shown that mutating the action part as well as allowing free mutation was able to further speed-up learning. A higher GA threshold slightly delayed learning speed as expected.

4.5   Assuring Solution Sustenance: The Niche Support Bound

The above bounds assure that problem subsolutions represented by individual classifiers evolve. The time bound additionally estimates how long it takes the evolutionary process to evolve a complete problem solution. Since the time bound and all other bounds consider individual classifiers integrated in the whole population, the population as a whole is required to evolve a complete problem solution supplying, evaluating, and growing currently best subsolutions in parallel. What remains to be assured is that the final problem solution, represented by a set of maximally accurate and maximally general classifiers, can be sustained in the population. This is expressed in the sixth point of the facetwise theory approach to LCS success: Niching techniques need to assure the sustenance of a complete problem solution.

Thus, we now derive a population size bound that assures the niche support of all necessary problem subsolutions with high probability. To derive the niche support bound, we develop a simple Markov chain model of classifier support in XCS. Essentially, we model the change in niche size of particular problem subsolutions (that is, niches) using a Markov chain.

To derive the bound, we focus on the support of one niche only disregarding potential interactions with other niches. Again we assume that problem instances are encountered according to a uniform distribution over the whole problem space. Additionally, we assume random deletion from a niche. Given the Markov chain over the niche size, we then determine the steady state distribution that estimates the expected niche distribution.

Using the steady state distribution, we derive the probability that a niche is lost. This probability can be bound by minimizing this loss probability. The result is a final population size bound. The bound assures the maintenance of a low-error solution with high probability. The experimental evaluations show that the assumptions hold in non-overlapping problems. In problems that require overlapping solution representations, the population size may need to be increased further.

Markov Chain Model As already introduced for the schema bound in Section 4.2, we define a problem niche by a schema of order k. A representative of a problem niche is defined as a classifier that specifies at least all k bits correctly. The Markov chain model constructs a Markov chain over the number of classifier representatives in a particular problem niche.

Suppose we have a particular problem niche represented by k classifiers; let p be the probability that an input belonging to the niche is encountered, and let N be the population size. Since classifiers are deleted with probability proportional to their action set size estimate, a classifier will be deleted from the niche with probability k/N. Assuming that the GA is always applied (this can be assured by setting $\theta_{GA} = 0$) and disregarding any disruptive effects due to mutation or crossover, the probability that a new classifier is added to the niche is exactly equal to the probability p that an input belonging to the niche is encountered.

However, overgeneral classifiers might inhabit the niche as well so that also an overgeneral classifier might be chosen for reproduction decreasing the reproduction probability p of a niche representative. However, as shown elsewhere (22), due to the action set size relative tournament selection, the probability of selecting a niche representative for reproduction is larger than some constant dependent on the relative tournament size τ. Given that τ is chosen sufficiently large and given further that the population mainly converged to the niche representatives, the probability approaches one.

In the Markov chain model, we assume that at each time step both the GA reproduction and the deletion are applied. Accordingly, we derive three transition probabilities for a specific niche. Given the niche is currently represented by j classifiers, at each time step, (i) with probability r_j the size of the niche is increased (because a classifier has been reproduced from the niche, while another classifier has been deleted from another niche); (ii) with probability d_j the size of the niche is decreased (because genetic reproduction took place in another niche, while a classifier was deleted from this niche); (iii) with probability s_j the niche size remains constant (either because no classifier has been added nor deleted from the niche or because one classifier has been added to the niche while another one has been deleted from the same niche).

The Markov chain associated to the model is depicted in Figure 6. States in the model indicate the niche size determined by the number of representatives in a niche. Arcs labeled with r_j represent the event that the application of the GA and deletion results in an increase of the niche size. Arcs labeled with s_j represent the event that the application of the genetic algorithm and deletion

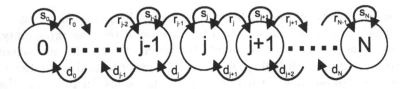

Fig. 6. Markov chain model for the support of one niche in XCS: j is the number of classifiers in the niche; N is the population size; r_j is the probability that a classifier is added to a niche containing j representatives; s_j is the probability that the niche containing j representatives is not modified through reproduction; d_j is the probability that a classifier is deleted from a niche containing j representatives.

results in no overall effect on the niche size. Arcs labeled with d_j represent the event that the application of the genetic algorithm and deletion results in a decrease of the niche size.

More formally, since the current problem instance is part of a particular niche with probability p, a niche representative will be generated via GA reproduction with approximately probability p. Assuming random deletion, a representative of a niche is deleted with probability j/N since there are by definition j representatives in the current population of size N. Accordingly, we compute the probabilities r_j, s_j, and d_j as follows:

$$r_j = p\left(1 - \frac{j}{N}\right) \tag{40}$$

$$s_j = (1-p)\left(1 - \frac{j}{N}\right) + p\frac{j}{N} \tag{41}$$

$$d_j = (1-p)\frac{j}{N} \tag{42}$$

For $j = 0$ we have $r_0 = p$, $s_0 = 1 - p$, and $d_0 = 0$. When $j = 0$, the niche is not represented in the population, therefore: (i) when an input belonging to the niche is presented to the system (with probability p), one classifier is generated through *covering*, therefore $r_0 = p$; (ii) since the niche has no classifiers, deletion cannot take place, therefore $d_0 = 0$; finally, (iii) the probability that the niche remains unrepresented is $1 - r_0 - s_0$, that is $s_0 = 1 - p$. Similarly, when $j = N$ all the classifiers in the population belong to the niche, accordingly: (i) no classifier can be added to the niche, therefore $r_N = 0$; (ii) with probability p an input belonging to the niche is encountered so that a classifier from the niche is reproduced while another one from the niche is deleted, leaving the niche size constant, therefore $s_N = p$; (iii) when an input that does not belong to the niche is presented to the system (with probability $1 - p$), a classifier is deleted from the niche to allow the insertion of the new classifier to the other niche, therefore $d_N = 1 - p$. Thus, for $j = N$ we have $r_N = 0$, $s_N = p$, and $d_N = 1 - p$.

Note that our approach somewhat brushes over the problem of overgeneral classifiers in that overgeneral classifiers are not considered as representatives of any niche. In addition, covering may not be sufficient in the event of an empty niche since overgeneral classifiers might still be present so that $r_0 = p$ is an approximation. However, as pointed out in 31), as long as a sufficiently large population size is chosen, chopping off or approximating the quasi absorbing state r_0 approximates the distribution accurately enough. This is also confirmed by our experimental investigations. However, overgeneral classifiers and more importantly overlapping classifiers can influence the distribution as evaluated below. Given the above assumptions, we are now able to derive a probability distribution over niche support.

Steady State Derivation To estimate the distribution over the number of representatives of a problem niche, we derive the distribution when the Markov chain is in steady state. Essentially, we derive probabilities u_j that the niche has j representatives. To derive the steady state distribution, we first write the fixed point equation for our Markov chain:

$$u_j = r_{j-1}u_{j-1} + s_j u_j + d_{j+1}u_{j+1},$$

which equates the probability that the niche has j representatives with the probability that the niche will have j representatives in the next time step. In the next time step, three events will contribute to the probability of having j representatives: (i) reaching state j from state $j-1$, with probability $r_{j-1}u_{j-1}$, by means of a reproductive event; (ii) remaining in state j with probability $s_j u_j$; (iii) reaching state j from state $j+1$, with probability $d_{j+1}u_{j+1}$, by means of a deletion event. The same equation can be rewritten by acknowledging the fact that in steady state the incoming proportion needs to be equal to the outgoing proportion in each state in the Markov chain:

$$(r_j + d_j)u_j = r_{j-1}u_{j-1} + d_{j+1}u_{j+1} \tag{43}$$

Replacing d_{j+1}, d_j, r_j, and r_{j+1} with the actual values from the previous section (equations 40, 41, and 42) we get the following:

$$\left[p\left(1 - \frac{j}{N}\right) + \frac{j}{N}(1-p)\right]u_j =$$
$$(1-p)\left(\frac{j+1}{N}\right)u_{j+1} + p\left(1 - \frac{j-1}{N}\right)u_{j-1} \tag{44}$$

Equation 44 is a second order difference equation whose parameters are dependent on j, i.e., on the current state. We use Equation 44 and the condition:

$$\sum_{j=0}^{j=N} u_j = 1 \tag{45}$$

to derive the steady state distribution. Multiplying Equation 44 by $N/((1-p)u_{j-1})$, we derive the following:

$$\left[\frac{p}{1-p}(N-j) + j\right]\frac{u_j}{u_{j-1}} = (j+1)\frac{u_{j+1}}{u_{j-1}} + \frac{p}{1-p}(N-j+1) \tag{46}$$

To derive the steady state distribution of probabilities u_j we use Equation 46 to derive an equation for the ratio $\frac{u_j}{u_0}$. Next, we use the equation for $\frac{u_j}{u_0}$ and the condition in Equation 45 to derive the steady state distribution.

As the very first step, we write the following fixed point equation for the transition between state 0 and state 1

$$u_0 = s_0 u_0 + d_1 u_1 \tag{47}$$

substituting the values of r_0 and d_1 we obtain:

$$u_0 = (1-p)u_0 + (1-p)\frac{1}{N}u_1$$

$$pu_0 = (1-p)\frac{1}{N}u_1 \tag{48}$$

from which we derive:

$$\frac{u_1}{u_0} = \frac{p}{1-p}N \tag{49}$$

To derive the equation for u_2/u_0 we start from Equation 46 and set $j = 1$:

$$\left[\frac{p}{1-p}(N-1)+1\right]\frac{u_1}{u_0} = 2\frac{u_2}{u_0} + \frac{p}{1-p}N \tag{50}$$

so that,

$$\frac{u_2}{u_0} = \frac{1}{2}\left[\left(\frac{p}{1-p}(N-1)+1\right)\frac{u_1}{u_0}\frac{p}{1-p}N\right] \tag{51}$$

We replace u_1/u_0 with Equation 49:

$$\frac{u_2}{u_0} = \frac{1}{2}\left[\left(\frac{p}{1-p}(N-1)+1\right)\frac{u_1}{u_0}\frac{p}{1-p}N\right]$$

$$= \frac{1}{2}\left[\left(\frac{p}{1-p}(N-1)+1\right)\frac{p}{1-p}N\frac{p}{1-p}N\right]$$

$$= \frac{1}{2}\left[\frac{p}{1-p}(N-1)\frac{p}{1-p}N\right]$$

$$= \frac{N(N-1)}{2}\left(\frac{p}{1-p}\right)^2$$

$$= \binom{N}{2}\left(\frac{p}{1-p}\right)^2 \tag{52}$$

This leads us to the hypothesis that

$$\frac{u_j}{u_0} = \binom{N}{j}\left(\frac{p}{1-p}\right)^j, \tag{53}$$

which we prove by induction. Using Equation 53, we can first derive that

$$u_{j+1} = \frac{N-j}{j+1} \frac{p}{1-p} u_j \tag{54}$$

$$u_j = \frac{N-j+1}{j} \frac{p}{1-p} u_{j-1} \tag{55}$$

$$u_{j-1} = \frac{1-p}{p} \frac{j}{N-j+1} u_j. \tag{56}$$

With Equation 46 substituting Equation 56 as the inductive step, we now derive

$$u_{j+1} = \frac{\left(\left(\frac{p}{1-p}(N-j)+j\right) \frac{u_j}{u_{j-1}} - \frac{p}{1-p}(N-j+1)\right) u_{j-1}}{j+1}$$

$$= \frac{\left(\frac{p}{1-p}\right)^2 \frac{(N-j)(N-j+1)}{j} u_{j-1}}{j+1}$$

$$= \frac{N-j}{j+1} \frac{p}{1-p} u_j, \tag{57}$$

which proves the hypothesis.

We can now derive the steady state distribution from Equation 45 dividing both sides by u_0

$$\sum_{j=0}^{N} \frac{u_j}{u_0} = \frac{1}{u_0}, \tag{58}$$

substituting Equation 53 we derive

$$\sum_{j=0}^{N} \frac{u_j}{u_0} = \sum_{j=0}^{N} \binom{N}{j} \left(\frac{p}{1-p}\right)^j$$

$$= \left[\sum_{j=0}^{N} \binom{N}{j} p^j (1-p)^{N-j}\right] \frac{1}{(1-p)^N}, \tag{59}$$

where the term $\sum_{j=0}^{N} \binom{N}{j} p^j (1-p)^{N-j}$ equals to $[p + (1-p)]^N$", that is 1, so that:

$$\sum_{j=0}^{N} \frac{u_j}{u_0} = \frac{1}{(1-p)^N}, \tag{60}$$

accordingly,

$$u_0 = (1-p)^N \tag{61}$$

Finally, combining Equation 53 and Equation 61, we derive the steady state distribution over u_j as follows:

$$
\begin{aligned}
u_j &= \binom{N}{j}\left(\frac{p}{1-p}\right)^j u_0 \\
&= \binom{N}{j}\left(\frac{p}{1-p}\right)^j (1-p)^N \\
&= \binom{N}{j} p^j (1-p)^{N-j}
\end{aligned}
\tag{62}
$$

Note that the same derivation is possible noting that the proposed Markov chain results in an *Engset* distribution (32).

Essentially, we see that the constant probability of reproduction p in combination with a linear increasing probability of deletion j/N, results in a binomial distribution over niche support sizes in steady state. In the next sections we validate Equation 62 experimentally, and evaluate the assumptions made such as the influence of mutation, overgeneral classifiers, the r_0 approximation, and overlapping niches.

Evaluation of Niche Support Distribution Evaluations were undertaken in three Boolean function problems including (1) the layered count ones problem, (2) the multiplexer problem, and (3) the carry problem. While the layered count ones problem requires a non-overlapping solution representation, the multiplexer problem allows overlapping subsolutions. The carry problem requires an overlapping solution representation. The experimental evaluations in (12; 22) confirmed the Markov chain model and the resulting niche support bound. Hereby, the chosen selection mechanism slightly influences the result in that tournament selection sometimes facilitated the maintenance of overlapping niches (in the multiplexer problem). In the carry problem where the final subsolutions are overlapping it was possible to confirm the support of one macro-niche that included a class of overlapping niches. Due to the overlap, the support for one niche was smaller that if it was not overlapping. Thus, as mentioned, dependent on the degree of overlap, population size needs to be increased further. Increasing the GA threshold decreases the niche size distribution since the threshold prevents over-reproduction of a frequently occurring niche,

In sum, it was shown that solution spaces interfere with each other during selection in problems that require an overlapping solution representation. The overlap causes a decrease in niche sizes. However, the influence was not as significant as originally feared. Further extensions to balance the niches are imaginable such as taking into consideration the degree of overlap among competing (fitness sharing) classifiers. Nonetheless, the model is able to predict the general behavior of XCS's final solution. Additionally, the model can be used to estimate the probability of a niche loss. The next paragraphs derive this probability and extend the model to a general population size bound that ensures the maintenance of a low-error solution with high probability.

Population Size Bound The reported results show that our model for niche support in XCS can predict the distribution of niche size for problems involving non-overlapping niches. Effectively, our model provides an asymptotic prediction of niche support distribution. It applies once the problem has been learned and there is no further influence from genetic operators. Besides such predictive capabilities, we can use our model to derive a population size bound that ensures that a complete model is maintained with high probability. In particular, from the probability u_0 we can derive a bound to guarantee that a niche is not lost with high probability.

In essence, the model can approximate the probability that a niche is lost. Using Equation 62, we can derive a bound for the population size N that ensures with high probability that XCS does not loose any of the problem niches, that is, any subsolutions. From the derivation of the probability of being in state u_0 (which means, that the respective niche was lost), which is $u_0 = (1-p)^N$, we see that the probability of loosing a niche decreases exponentially with the population size. Given a problem with 2^k problem niches, that is, the perfect solution $[O]$ (33) is represented by 2^k schemata of order k, the probability of loosing a niche equates $u_0 = \left(1 - \frac{1}{2^k}\right)^N$.

Requiring a certainty θ that no niche will be lost (that is, $\theta = 1 - u_0$), we can derive a concrete population size bound

$$N > \frac{\log(1-\theta)}{\log(1-p)} > \frac{\log(1-\theta)}{\log(1-\frac{1}{2^k})}, \tag{63}$$

effectively showing that population size N grows logarithmically in the confidence value and polynomially in the solution complexity 2^k. Figure 7 shows the population size bound that assures niche support. Since the population size scales as the inverse of the probability of niche occurrence, the $\log - \log$-sale shows a straight line.

Thus, the bound confirms that once a problem solution was found, XCS is able to maintain the problem solution with high probability requiring a population size that grows polynomially in solution complexity and logarithmically in the confidence value. This bound confirms that XCS does not need more than a polynomial population size with respect to the solution complexity consequently pointing to the PAC learning capability of XCS confirming Wilson's original hypothesis (15).

4.6 Towards PAC Learnability

The derivations of the problem bounds in the previous sections enables us to connect learning in the XCS classifier system to fundamental elements of computational learning theory (COLT). COLT is interested in showing how much computational power an algorithm needs to learn a particular problem. To derive an overall computational estimate of XCS's learning capabilities, we focus

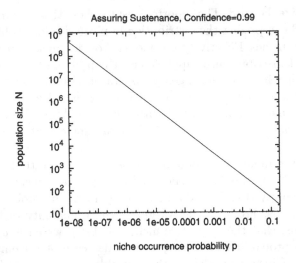

Fig. 7. To sustain a complete solution with high probability, population size needs to grow as the inverse of the niche occurrence probability p. In a uniformly sampled, binary problem, p corresponds to $1/2^k$ where k specifies the minimal number of attributes necessary to do maximally accurate predictions in all solution subspaces.

on the problem of learning k-DNF functions. In particular, we show that k-DNF problems that satisfy few additional properties are PAC-learnable (13; 14) by XCS. In essence, we also confirm Wilson's previous conjecture that XCS scales polynomially in time and space complexity (15).

XCS is certainly not the most effective k-DNF learning algorithm. 16) shows that an algorithm especially targeted to solve noise-free uniformly sampled k-DNF problems is able to reach a much more effective performance. However, this thesis does not only show that XCS is able to PAC-learn k-DNF problems. Rather, it shows that it is able to learn a large variety of problems including nominal and real-valued problems, noisy problems, as well as general RL problems. Restricting the problem to k-DNF problems, we can show that XCS is a PAC-learning algorithm confirming the effectiveness as well as the generality of XCS's learning mechanism.

To approach the PAC-learning bound, we reflect on the previous bounds evaluating their impact on computational complexity. The successive chapters provide a variety of evidence for XCS's successful and broad applicability as well as its effective learning and scale-up properties.

Problem Bounds Revisited In Section 3, we analyzed how the evolutionary pressures in XCS bias learning towards the evolution of a complete, maximally accurate, and maximally general problem solution ensuring *fitness guidance* and *appropriate generalization*. This chapter investigated the requirement on population size and learning time in order to supply better classifiers, make time to detect and grow those better classifiers until the final population is reached, and finally, to sustain the final problem solution with high probability. Satisfying these bounds, we can ensure with few additional constraints that XCS learns the underlying problem successfully.

We now revisit the bounds considering their resulting computational requirement.

Covering Bound. The covering bound ensures that the GA is taking place establishing a covering probability (Equation 6). To ensure a high probability of covering, the specificity can be chosen very low by setting the initial specificity (controlled by $P_\#$) as well as mutation rate sufficiently low. Given a fixed specificity that behaves in $O(\frac{n}{l})$ as necessary to supply better classifiers, as derived above (Equation 38), the population size can be bounded as follows using the approximation $x < -\log(1-x)$:

$$\frac{-\log(1-P(\text{cov.}))}{-\log\left(1-\left(\frac{2-\sigma[P]}{2}\right)^l\right)} < \frac{-\log(1-P(\text{cov.}))}{\left(1-\frac{n}{2l}\right)^l} < -\log(1-P(\text{cov.}))e^{n/2} < N \quad (64)$$

Thus, to satisfy the covering bound, the population size needs to grow logarithmically in the probability of error and exponentially in the number of problem classes n. With respect to PAC-learnability, the bound shows that to ensure that the GA is successfully applied in XCS with probability $1 - \delta_P$ (where $\delta_P = (1 - P(\text{cov.}))$) the population size scales logarithmically in the error probability δ_P as well as exponentially in the number of problem classes.

Schema Bound. The *schema bound* ensures that better classifiers are available in the population. Given a problem with a minimal order of difficulty k_m and requiring a high probability that representatives of this order are supplied (Equation 8), we were able to bound the population size N in Equation 10 showing that population size N needs to grow logarithmically in the probability of error δ_P (that is, $\delta_P = 1 - P(\text{rep.exists})$) and exponentially in the order of the minimal order complexity k_m given a certain specificity and thus polynomial in concept space complexity.

Reproductive Opportunity Bound. In addition to the existence of a representative, we showed that it is necessary to ensure reproduction and thus growth of such representatives. This is ensured by the general reproductive opportunity bound which was shown to require a population size growth of $O(l^{k_m})$ (Equation 27) with respect to the minimal order complexity k_m of a problem. The reproductive opportunity bound was generated with respect to one niche. However, since XCS is evolving the niches in parallel and the probability of niche occurrence as well as the probability of deletion underly a geometric distribution (memoryless property and approximately equal probabilities), we can assure with high confidence $1 - \delta_P$, that all relevant niches receive reproductive opportunities. Thus, we can assure with high probability that lower-order representatives grow leading to higher accuracy within a complexity that is polynomial in the concept space complexity.

Time Bound. The time bound estimates the number of problem instances necessary to learn a complete problem solution using XCS. Given a problem that requires classifiers of maximal schema order k_d (a k_d-conjunction in a k-*DNF*) and given further the domino convergence property in the problem, the expected time until generation of an optimal classifier was approximated in Equation 37 yielding a population size requirement of $O(l2^{k_d+n})$ (Equation 39). The estimation approximates the expected time til creation of the best classifier of a problem niche of order k_d. As in the reproductive opportunity bound, we can argue that since XCS evolves all problem niches in parallel and since the generation of the next best classifier underlies a geometric distribution (memoryless property and equal probability), given a certain confidence value δ, the time until a particular classifier of order k is generated with high confidence δ grows within the same limits. Similarly, assuming a probability p of problem occurrence, we can bound the time requiring a maximal error in the final solution ϵ with low probability δ by $O(l\frac{1}{\epsilon}2^n)$.

Niche Support Bound. To ensure the support of the final solution, we finally established the niche support bound. Given a problem whose solution is expressed as a disjunction of distinct *subsolutions*, XCS tends to allocate distinct rules to each *subsolution*. To ensure a complete problem solution, it needs to be assured that all subsolutions are represented with high probability. Deriving a Markov model over the number of representatives for a particular niche, we were able to derive the steady state niche distribution given a niche occurrence probability p (Equation 62).

Requiring that all niches with at least niche occurrence probability p are expected to be present in the population with high probability, we were able to derive a bound on the population size N requiring that the probability of no representative in a niche with more than p occurrence probability is sufficiently low. With respect to PAC-learnability this bound requires that with high probability $1 - \delta$ we assure that our solution has an error probability of less than ϵ (where ϵ is directly related to p). Using this notation, we can derive the following equation from Equation 61 substituting ϵ for p and δ for u_0 using again the approximation $x < -\log(1 - x)$:

$$\frac{\log \delta}{\log(1 - \epsilon)} < -\frac{1}{\epsilon} \log \frac{1}{\delta} < N \tag{65}$$

This bound essentially bounds the population size showing that it needs to grow logarithmically in $\frac{1}{\delta}$ and linear in $\frac{1}{\epsilon}$. Approximating ϵ by $(\frac{1}{2})^{k_d}$ assuming a uniform problem instance distribution, we see that to prevent niche loss, the population size needs to grow linearly in the concept space complexity 2^{k_d}.

PAC-Learning with XCS With the bounds above, we are now able to bound computational effort and number of problem instances necessary to evolve with high probability $(1 - \delta)$ a low error ϵ solution of an underlying k-*DNF* problem.

Additionally, the k-DNF problem needs to be maximally of minimal order of difficulty k_m as discussed in sections 4.2 and 4.3.

Thus, Boolean function problems in k-DNF form with l attributes and a maximal order of problem difficulty k_m are PAC-learnable by XCS using the ternary representation of XCS conditions. That is, XCS evolves with high probability $(1 - \delta)$ a low error ϵ solution of the underlying k-DNF problem in time polynomial in $1/\delta$, $1/\epsilon$, l, and concept space complexity.

The bounds derived in Section 4.6 show that the computational complexity of XCS, which is bounded by the population size N, is linear in $1/\delta$, $1/\epsilon$ and l^{k_m}. Additionally, the time bound shows that the number of problem instances necessary to evolve a low-error solution with high probability grows linearly in $1/\delta$, $1/\epsilon$, l and the solution complexity. Consequently, we showed that Boolean functions that can be represented in k-DNF and have a maximal order of problem difficulty k_m, are PAC-learnable by XCS using the ternary representation of XCS conditions as long as the assumptions in the bound derivations hold.

The following further assumptions about the interaction of the bounds have been made. First, crossover is not modeled in our derivation. While crossover can be disruptive as already proposed in 4), crossover may also play an important innovative role in recombining currently found subsolutions effectively as proposed in 19) and experimentally confirmed for XCS in 9). The next chapter provides a more detailed investigation on the impact of crossover.

Second, the specificity derivation from the mutation rate assumes no actual fitness influence. Subtle interactions of various niches might increase specificity further. Thus, problems in which the specificity assumption does not hold might violate the derived reproductive opportunity bound.

Third, if the probability of reproduction p is approximated by $(\frac{1}{2})^{k_d}$, niche support assumes a non-overlapping representation of the final solution. Thus, overlapping solution representations require an additional increase in population size as evaluated in Section 4.5.

5   Summary and Conclusions

This chapter showed *when* XCS is able to learn a problem. Along our facetwise theory approach to LCSs, we derived population size specificity, and time bounds that assure that a complete, maximally accurate, and maximally general problem solution can evolve and can be sustained.

In particular, we derived a *covering bound* that bounds population size and specificity to ensure proper XCS initialization making way for classifier evaluation and GA application. Next, we derived a *schema bound* that bounds population size and specificity to ensure *supply* of better classifiers. Better classifiers were defined as classifiers that have higher accuracy on average. They can be characterized as *representatives* of minimal order schemata or BBs—those BBs, that increase classification accuracy in the problem at hand. Next, we derived a *reproductive opportunity bound* that bounds specificity and population size to assure *identification* and *growth* of better classifiers. The subsequently derived *time bound* estimates the learning time needed to evolve a complete problem solution given the other bounds are satisfied. Finally, we derived a *niche bound* that assures the *sustenance* of a low-error solution with high probability.

Along the way, we defined two major problem complexities: (1) the minimal order complexity k_m, which specifies the minimal number of features that need to be specified to decrease class entropy (that is, increase classification accuracy), and (2) the general problem difficulty k_d, which specifies the maximal number of attributes necessary to specify a class distribution accurately. While the former is relevant for supply and growth, the latter is relevant for the sustenance of a complete problem solution.

Putting the bounds together, we showed that XCS can *PAC-learn* a restricted class of *k-DNF* problems. However, the reader should keep in mind that XCS is an online generalizing, evolutionary-based RL system and is certainly not designed to learn *k-DNF* problems particularly well. In fact, XCS can learn a much larger range of problems including DNF problems but also multistep RL problems as well as datamining problems as validated in subsequent chapters.

Before the validation, though, we need to investigate the last three points of our facetwise LCS theory approach for single-step (classification) problems. Essentially, it needs to be investigated if search via mutation and recombination is effective in XCS and how XCS distinguishes between local and global solution structure. The next chapter consequently considers problems which are hard for XCS's search mechanism since whole BB structures need to be processed to evolve a complete problem solution. We consequently improve XCS's crossover operator using statistical methods to detect and propagate dependency structures (BBs) effectively.

Acknowledgment

We are grateful to Xavier Llorà, Kei Onishi, Kumara Sastry, and the whole IlliGAL lab for their help and the useful discussions. We are also in debt to Stewart W. Wilson who initiated the analysis of XCS with the covering and schema considerations. The work was sponsored by the Air Force Office of Scientific Research, Air Force Materiel Command, USAF, under grant F49620-03-1-0129. The US Government is authorized to reproduce and distribute reprints for Government purposes notwithstanding any copyright notation thereon. Additional funding from the German research foundation (DFG) under grant DFG HO1301/4-3 is acknowledged. Additional support from the Computational Science and Engineering graduate option program (CSE) at the University of Illinois at Urbana-Champaign is acknowledged. The views and conclusions contained herein are those of the authors and should not be interpreted as necessarily representing the official policies or endorsements, either expressed or implied, of the Air Force Office of Scientific Research or the U.S. Government.

References

[1] Holland, J.H.: Adaptation. In Rosen, R., Snell, F., eds.: Progress in theoretical biology. Volume 4. Academic Press, New York (1976) 263–293

[2] Holland, J.H., Reitman, J.S.: Cognitive systems based on adaptive algorithms. In Waterman, D.A., Hayes-Roth, F., eds.: Pattern directed inference systems. Academic Press, New York (1978) 313–329

[3] Wilson, S.W.: Classifier fitness based on accuracy. Evolutionary Computation **3** (1995) 149–175

[4] Holland, J.H.: Adaptation in Natural and Artificial Systems. University of Michigan Press, Ann Arbor, MI (1975) second edition, 1992.

[5] Goldberg, D.E.: Genetic Algorithms in Search, Optimization and Machine Learning. Addison-Wesley, Reading, MA (1989)

[6] Bernadó, E., Llorà, X., Garrell, J.M.: XCS and GALE: A comparative study of two learning classifier systems and six other learning algorithms on classification tasks. In Lanzi, P.L., Stolzmann, W., Wilson, S.W., eds.: Advances in Learning Classifier Systems (LNAI 2321). Springer-Verlag, Berlin Heidelberg (2002) 115–132

[7] Dixon, P.W., Corne, D.W., Oates, M.J.: A preliminary investigation of modified XCS as a generic data mining tool. In Lanzi, P.L., Stolzmann, W., Wilson, S.W., eds.: Advances in learning classifier systems: Fourth international workshop, IWLCS 2001 (LNAI 2321). Springer-Verlag, Berlin Heidelberg (2002) 133–150

[8] Bernadó-Mansilla, E., Garrell-Guiu, J.M.: Accuracy-based learning classifier systems: Models, analysis, and applications to classification tasks. Evolutionary Computation **11** (2003) 209–238

[9] Butz, M.V., Goldberg, D.E., Tharakunnel, K.: Analysis and improvement of fitness exploitation in XCS: Bounding models, tournament selection, and bilateral accuracy. Evolutionary Computation **11** (2003) 239–277

[10] Butz, M.V., Kovacs, T., Lanzi, P.L., Wilson, S.W.: Toward a theory of generalization and learning in XCS. IEEE Transactions on Evolutionary Computation **8** (2004) 28–46

[11] Butz, M.V., Goldberg, D.E., Lanzi, P.L.: Bounding learning time in XCS. Proceedings of the Sixth Genetic and Evolutionary Computation Conference (GECCO-2004): Part II (2004) 739–750

[12] Butz, M.V., Goldberg, D.E., Lanzi, P.L., Sastry, K.: Bounding the population size to ensure niche support in XCS. IlliGAL report 2004033, Illinois Genetic Algorithms Laboratory, University of Illinois at Urbana-Champaign (2004)

[13] Valiant, L.: A theory of the learnable. Communications of the ACM **27** (1984) 1134–1142

[14] Mitchell, T.M.: Machine Learning. McGraw-Hill, Boston, MA (1997)

[15] Wilson, S.W.: Generalization in the XCS classifier system. Genetic Programming 1998: Proceedings of the Third Annual Conference (1998) 665–674

[16] Servedio, R.A.: Efficient Algorithms in Computational Learning Theory. PhD thesis, Harvard University, Cambridge, MA (2001)

[17] Butz, M.V., Wilson, S.W.: An algorithmic description of XCS. Soft Computing 6 (2002) 144–153

[18] Lanzi, P.L.: Learning classifier systems from a reinforcement learning perspective. Soft Computing: A Fusion of Foundations, Methodologies and Applications 6 (2002) 162–170

[19] Goldberg, D.E.: The Design of Innovation: Lessons from and for Competent Genetic Algorithms. Kluwer Academic Publishers, Boston, MA (2002)

[20] Kovacs, T.: Strength or Accuracy: Credit Assignment in Learning Classifier Systems. Springer-Verlag, Berlin Heidelberg (2003)

[21] Butz, M.V., Sastry, K., Goldberg, D.E.: Tournament selection in XCS. Proceedings of the Fifth Genetic and Evolutionary Computation Conference (GECCO-2003) (2003) 1857–1869

[22] Butz, M.V.: Rule-based evolutionary online learning systems: Learning bounds, classification, and prediction. PhD thesis, University of Illinois at Urbana-Champaign, Urbana, IL (2004)

[23] Butz, M.V., Kovacs, T., Lanzi, P.L., Wilson, S.W.: How XCS evolves accurate classifiers. Proceedings of the Third Genetic and Evolutionary Computation Conference (GECCO-2001) (2001) 927–934

[24] Holland, J.H.: Processing and processors for schemata. In Jacks, E.L., ed.: Associative Information Techniques, New York, American Elsevier (1971) 127–146

[25] Butz, M.V., Pelikan, M.: Analyzing the evolutionary pressures in XCS. Proceedings of the Third Genetic and Evolutionary Computation Conference (GECCO-2001) (2001) 935–942

[26] Butz, M.V., Goldberg, D.E.: Bounding the population size in XCS to ensure reproductive opportunities. Proceedings of the Fifth Genetic and Evolutionary Computation Conference (GECCO-2003) (2003) 1844–1856

[27] Quinlan, J.R.: C4.5: Programs for Machine Learning. Morgan Kaufmann, San Francisco, CA (1993)

[28] Goldberg, D.E.: Probability matching, the magnitude of reinforcement, and classifier system bidding. Machine Learning 5 (1990) 407–425

[29] Kovacs, T., Kerber, M.: What makes a problem hard for XCS? In Lanzi, P.L., Stolzmann, W., Wilson, S.W., eds.: Advances in learning classifier systems: Third international workshop, IWLCS 2000 (LNAI 1996). Springer-Verlag, Berlin Heidelberg (2001) 80–99

[30] Thierens, D., Goldberg, D.E., Pereira, A.G.: Domino convergence, drift, and the temporal-salience structure of problems. In: Proceedings of the 1998 IEEE World Congress on Computational Intelligence, New York, NY, IEEE Press (1998) 535–540

[31] Horn, J.: Finite Markov chain analysis of genetic algorithms with niching. Proceedings of the Fifth International Conference on Genetic Algorithms (1993) 110–117

[32] Kleinrock, L.: Queueing Systems: Theory. John Wiley & Sons, New York (1975)

[33] Kovacs, T.: XCS classifier system reliably evolves accurate, complete, and minimal representations for boolean functions. In Roy, Chawdhry, Pant, eds.: Soft computing in engineering design and manufacturing. Springer-Verlag, London (1997) 59–68

An Analysis of Continuous-Valued Representations for Learning Classifier Systems

Christopher Stone and Larry Bull

Faculty of Computing, Engineering and Mathematical Sciences
University of the West of England
Bristol, BS16 1QY, United Kingdom
christopher.stone@uwe.ac.uk
larry.bull@uwe.ac.uk

1 Introduction

Learning Classifier Systems [11] typically use a ternary representation to encode the environmental condition that a classifier matches. However, many real-world problems are not conveniently expressed in terms of a ternary representation and several alternate representations have been suggested to allow Learning Classifier Systems to handle these problems more readily [1, 3, 6, 15].

XCS [22] is a Learning Classifier System in which a classifier's fitness for the Genetic Algorithm (GA) [10] is based on the accuracy of its predictions rather than its ability to receive reward. The XCS algorithm is described in detail in [9].

This chapter investigates a representation for continuous-valued inputs [23] introduced by Wilson for XCS and also a representation for integer data [25], which has also been used for function approximation with XCS [24]. Both of these representations replace the standard ternary representation. The only other changes made to XCS to accommodate the new representations are in the cover, mutation and GA subsumption operators. Two-point crossover is retained.

Most aspects of continuous-valued representation are unaffected by the type of GA used in the Learning Classifier System. However, the operation of crossover with continuous-valued representations is affected by the type of GA used. Modern Learning Classifier Systems use both panmictic[1] [21] and niche[2] [22] GAs and an interval representation has been applied to the ZCS [21] architecture on a real robot [12]. ZCS uses a panmictic GA, so we therefore study the effects of crossover and continuous-valued representations with both types of GA in the present work.

Although we examine a one of m binary encoding for real numbers, the issues are also relevant to a floating point encoding of real numbers.

The remainder of this chapter is organized as follows. Section 2 introduces interval predicates, terminology and the real multiplexer problem. Sections 3 and 4

---

[1] In a panmictic GA, the GA operates over the entire population on each invocation.
[2] A niche GA is one operating over the Learning Classifier System's match or action set.

C. Stone and L. Bull: *An Analysis of Continuous-Valued Representations for Learning Classifier Systems*, StudFuzz **183**, 127–175 (2005)
www.springerlink.com

study the two representations for continuous-valued data introduced by Wilson, Centre-Spread Representation (CSR) and Ordered Bound Representation (OBR), by examining their properties and operators. In Section 5 we introduce a new representation, Unordered Bound Representation (UBR) and analyse it in the same way. Section 6 studies crossover with interval representations in more detail. Section 7 looks further at the real multiplexer problem and Section 8 extends this discussion to hyper-rectangles, the decision surfaces constructed by interval predicates. In Section 9, we introduce a new test problem, the checkerboard problem and use both this and the real multiplexer problem in Section 10 to compare representations and operators. Section 11 provides conclusions to the work.

2   Interval Predicates

2.1   Motivation

XCS has been shown to generate complete and maximally general maps [13] for ternary representations. There is evidence [23, 25] to suggest that XCS is able to do this for continuous and integer-valued domains.

Thus, XCS approximates the function mapping $X \times A \Rightarrow P$ where X represents the environment, A is the set of possible actions and P is the payoff received for executing a particular action in an environmental state. In this chapter, we consider n-dimensional continuous-valued environments, $X \in \Re^n$ and Boolean actions, $A \in \{0, 1\}$.

Learning Classifier Systems in general and XCS in particular, typically use a ternary representation to encode the environmental condition that a classifier matches. Bits in the condition string of a classifier are allocated to represent the state of a single environmental variable, x_i. Exact matching in this way is generally not suitable for a continuous-valued environment, where real-valued data over a range must be represented. One possibility for continuous-valued environments is to encode the environment in the form of inequalities, $x_i < \theta_i$. The decision surface represented by a classifier is then a hyperplane in the n-dimensional solution space.

The representations considered here replace the $\{0, 1, \#\}$ classifier predicate with one representing a half-open interval $[p_i, q_i)$. This interval matches the environment if $p_i \le x_i < q_i$. The classifier condition is a vector of length n, each element of which encodes such an interval. A classifier with such a representation describes a hyper-rectangle in solution space. Representations describing hyper-rectangles are also used in other branches of machine learning. See, for example, [19]. A Learning Classifier System with an interval representation was also presented in [4].

2.2   Terminology

To avoid confusion and aid precision, we adopt the following notation throughout this chapter.

1. Intervals in phenotype space are tagged with the subscript p, e.g., $[0,1)_p$
2. Intervals in genotype space are tagged with the subscript g, e.g., $[0,n]_g$
3. Tuples are distinguished from intervals by the absence of a subscript.

The solution space is $[p_{min}, q_{max})_p^n$, where p_{min} and q_{max} are the minimum and supremum of the interval. For clarity of presentation and without loss of generality, we assume that p_{min} and q_{max} are the same for all dimensions i of the solution space.

2.3   The Real Multiplexer

The Boolean multiplexer is a standard benchmark problem for Learning Classifier System evaluation. Wilson [23] introduced the real multiplexer as a test problem for Learning Classifier Systems with continuous-valued inputs. Each 'bit' of the Boolean multiplexer is presented as a value x_i in the $[0,1)_p$ interval, with $x_i < \theta_i$ meaning binary 0 and $x_i \geq \theta_i$ meaning binary 1. The value θ_i is a control parameter that may be varied to provide problems of varying difficulty. By default $\theta_i = 0.5 \ \forall \ i \leq n$ and this is the threshold used in this chapter.

Experiments on XCS were performed using the 6-bit real multiplexer. XCS was presented with randomly generated (6 element) vectors of real numbers in the interval $[0,1)$. For each of these random vectors, XCS suggested a binary action representing the output value of the multiplexer. For this, it was rewarded with a payoff of 1000 for the correct action and 0 otherwise.

XCS settings used for all real multiplexer experiments in this chapter were $N = 800$, $\beta = 0.2$, $\alpha = 0.1$, $\varepsilon_0 = 10$, $\nu = 5$, $\theta_{GA} = 12$, $\chi = 0.8$, $\mu = 0.04$, $\theta_{del} = 20$, $\delta = 0.1$, $\theta_{sub} = 20$, $p_I = 10$, $\varepsilon_I = 0$, $f_I = 0.01$, $\theta_{mna} = 2$, $m = 0.1$, $s_0 = 1.0$. These match the settings published in [23]. GA subsumption, but no action set subsumption, was used. All experimental results presented are the average of 10 runs using alternate explore and exploit trials. A 16-bit binary encoding was used for real numbers.

Wilson showed that XCS was able to solve almost optimally the 6-bit real multiplexer using Centre-Spread Representation. A duplicate of these results is shown in Fig. 1. This shows the *system performance*, the fraction of correct actions, the *system error*, the absolute difference of the payoff and the predicted payoff, the *macroclassifier fraction*, the number of unique classifiers in the population divided by the population size limit and *mean interval width*, the mean width of all intervals in the population. These metrics are averaged over the previous 50 exploit trials. Fig. 1 also shows other information that will be referred to later in the chapter.

More recently, Bull et al. [7] have shown that XCS can solve almost optimally the 11-bit real multiplexer.

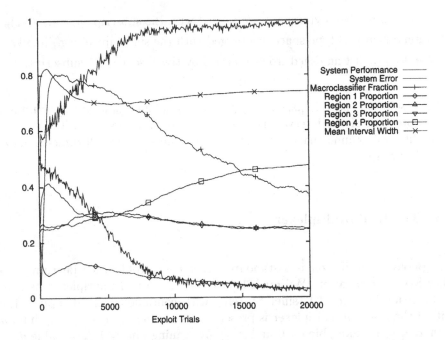

Fig. 1. 6-bit real multiplexer with Centre-Spread Representation, standard cover with $s_0 = 1$, 2-point crossover and standard mutation

3   Centre-Spread Representation

3.1   Background

To extend XCS into continuous-valued environments, Wilson [23] introduced the Centre-Spread Representation. Centre-Spread Representation provides a form of receptive field for Learning Classifier Systems. An interval predicate, $[p_i, q_i)_p$, is represented as a tuple (c_i, s_i) where $c_i, s_i \in \Re$. c_i encodes the centre of the interval and s_i encodes the spread (or width) of the interval. The interval is decoded as follows:

$$p_i = \min(p_{\min}, c_i - s_i)$$
$$q_i = \max(q_{\max}, c_i + s_i)$$

Use of Centre-Spread Representation thus involves a genotype (c_i, s_i) to phenotype $[p_i, q_i)_p$ mapping, or gene expression.

Wilson does not provide details of the encoding used for real numbers in the Centre-Spread Representation. However, given that the solution space is bounded, we assume a one of m binary encoding where the real values for both the centre and spread are encoded into binary integers of length k using the equation

$$\left\lfloor \frac{(2^k - 1)(p_i - p_{\min})}{q_{\max} - p_{\min}} \right\rfloor$$

With this scheme, the real values for centres and spreads are discretized into one of m possible values upon encoding. There are 2^{2k} possible centre-spread combinations and each possible centre-spread is represented exactly once. Because of the discretization of the phenotype, the half-open solution space $[p_{min}, q_{max})_p$ in phenotype space may be regarded as the closed solution space $[0, 2^k - 1]_g$ in genotype space.

The ternary representation used in most Learning Classifier Systems has an explicit 'don't care' value in the form of the '#' allele. Centre-Spread Representation does not have any explicit 'don't care' scheme. Instead, the maximally general interval $[p_{min}, q_{max})_p$ provides an implicit 'don't care' mechanism by matching all possible environmental inputs. An implication of this is that the proportion of maximally general intervals introduced into the population is not directly controllable by a system parameter, as is normally the case with a ternary representation.

3.2 Properties

As the solution space is half-open, the centre-spread genotype must be limited upon expression in order to restrict the range of the phenotype to the interval $[p_{min}, q_{max})_p$. We refer to this process as *truncation*. Truncation means that the genotype to phenotype $(g \rightarrow p)$ mapping is non-linear and many to one. In short, it is possible for a phenotype to be represented by more than one genotype. As an example, consider the solution space interval $[0, 1)_p$. The phenotype $[0.5, 1)_p$ may be represented as centre-spread tuples (0.75, 0.25), (0.8, 0.3), (0.9, 0.4), (1, 0.5) or any number of other tuples.

In practice, the number of possible centre-spread tuples representing an interval is finite, due to the discretization necessary when representing real numbers. The number of possible genotypes for a particular phenotype is therefore determined by both the phenotype itself and the details of the encoding of real numbers employed.

If there were no truncation on $g \rightarrow p$ mapping (and thus a one to one $g \rightarrow p$ mapping), there would be 2^{2k} possible phenotypes. However, given the need for truncation, certain phenotypes are expressed from multiple genotypes. There are therefore less than 2^{2k} unique phenotypes with the Centre-Spread Representation. Certain phenotypes are 'missing' and we refer here to these missing phenotype to genotype $(p \rightarrow g)$ mappings as *holes*.

The above two properties mean that using Centre-Spread Representation:

1. Expression of random genotypes results in increased frequency of expression of certain phenotypes over other possible phenotypes.
2. The phenotype space contains holes where certain phenotypes are missing, as they are not expressible.

To examine these phenomena in more detail we enumerated the $g \rightarrow p$ mapping for one of m binary encodings of length $2 \leq k \leq 12$. Without loss of generality, $p_{min} = 0$ and $q_{max} = 2^k - 1$.

Table 1. Phenotype frequency matrix for Centre-Spread Representation with $k = 3$

		q_i							
		0	1	2	3	4	5	6	7
	0	1	1	2	2	3	3	4	20
	1		1	0	1	0	1	0	4
	2			1	0	1	0	1	3
p_i	3				1	0	1	0	3
	4					1	0	1	2
	5						1	0	2
	6							1	1
	7								1

As a readily understandable example of one of these enumerations, Table 1 shows the frequency of each possible phenotype for $k = 3$, a 3-bit one of m binary encoding of real values[3].

The phenotypic frequency in Table 1 shows several interesting properties:

1. The frequency of all possible phenotype intervals is not uniform (as already discussed).
2. Certain phenotype intervals have zero frequency (as already discussed).
3. The frequency of an exact number $[p_i, p_i]_p$ is always 1.
4. The frequency of intervals of the form $[p_{min}, q_i)_p$ increases as q_i increases and the frequency of intervals of the form $[p_i, q_{max})_p$ increases as p_i decreases.
5. The frequency of the $[p_{min}, q_{max})_p$ interval is much greater than any other.

All encodings $2 \leq k \leq 12$ show the same patterns and properties. Only the specific frequencies vary. Space does not permit the publication of the details of each of these, but as a further example and aid to visualization, Fig. 2 shows the frequency matrix for $k = 5$ plotted as a surface that may be viewed as the landscape of the $g \rightarrow p$ mapping for that particular encoding.

From these results, we can state certain properties of the Centre-Spread Representation.

Property 1: Many to one genotype to phenotype mapping. This property is the key property from which all others derive and has already been discussed in detail.

Property 2: Incomplete phenotype to genotype mapping. A corollary of the discretization of the centre and spread and the many to one $g \rightarrow p$ mapping (Property 1) is that the $p \rightarrow g$ mapping is undefined for certain phenotypes. Holes arise because of the discretization of the encoding:

1. The centre must be located at a point that can be represented using the discrete encoding (i.e., it must be integer-valued).

---

[3] That is, 3 bits each for centre and spread.

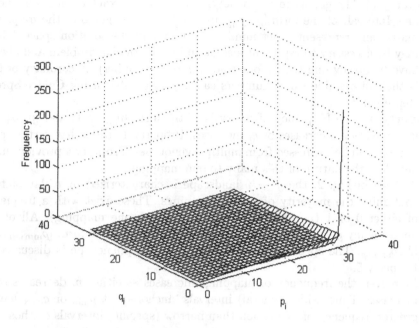

Fig. 2. Phenotype frequency landscape for Centre-Spread Representation with $k = 5$

2. The spread must be able to be represented using the discrete encoding (it must also be integer-valued).

For example, consider interval $[1,2]_p$ in Table 1. This interval cannot be represented using Centre-Spread Representation and an encoding of length $k = 3$. Neither of the two requirements can be met for the interval above. The centre would need to be located at point 1.5 with a spread of 0.5.

In general, any interval where $q_i - p_i$ is odd is unable to be represented using Centre-Spread Representation. Because of this, where holes exist in the $g \rightarrow p$ mapping, they are uniformly distributed around the solution space, such that they are neighbours with a $g \rightarrow p$ mapping that has non-zero frequency (i.e., is a one-to-one or many-to-one mapping). No two holes are ever situated next to each other.

The presence of holes is undesirable in the $g \rightarrow p$ mapping, since it means that certain phenotypes cannot be expressed. This, in turn, means that the accuracy of an expressed phenotype is lower than it would otherwise be since the effective discretization of the phenotype is coarser than desired. However, the fact that holes are always located next to a phenotype that can be expressed places a lower limit on the effective discretization of the Centre-Spread Representation.

Holes are an artefact from using one of m binary encoding. With a floating point encoding, the holes would be small enough to cause no practical problems.

Property 3: The genotype to phenotype mapping of exact numbers is one to one. Intervals of the form $[p_i, p_i]_p$ are the leading diagonal of the frequency matrix and represent exact numbers or points in the solution space. These may be necessary to represent a solution in a particular problem and always have frequency one (i.e., a one-to-one $g \rightarrow p$ mapping). A corollary of this is that all possible exact numbers can be represented using Centre-Spread Representation.

Property 4: The frequency of intervals of the form $[p_{min}, q_i)_p$ and $[p_i, q_{max})_p$ increases as p_i decreases or q_i increases. Property 1 states that certain phenotypes can be expressed from multiple genotypes. Property 4 provides more detail on the nature of this many to one mapping.

As seen in Fig. 2, the $g \rightarrow p$ landscape is characterized by a flat plateau containing the majority of $g \rightarrow p$ mappings. These exist with a frequency of either 0 or 1 (i.e., not expressible, or one-to-one mapping). All of the one-to-many $g \rightarrow p$ mappings occur in phenotypes of the form $[p_{min}, q_i)_p$ or $[p_i, q_{max})_p$. The special case of the $[p_{min}, q_{max})_p$ phenotype is discussed in Property 5.

Moreover, the frequency of mappings increases as either p_i decreases or q_i increases. Thus, wide (general) intervals 'anchored' at p_{min} or q_{max} have a greater frequency of expression than narrow (specific) intervals or those not anchored at p_{min} or q_{max}.

Property 5: The frequency of the $[p_{min}, q_{max})_p$ interval is much greater than that of any other interval. Property 4 states that the expression frequency of intervals of the form $[p_{min}, q_i)_p$ and $[p_i, q_{max})_p$ increases as p_i decreases or q_i increases. A special case of this is the interval $[p_{min}, q_{max})_p$. The expression frequency of this interval is much larger than any other and it completely dominates the $g \rightarrow p$ landscape, as shown in Fig. 2.

The $[p_{min}, q_{max})_p$ interval has special significance for interval-based representations, as it describes the maximally general interval predicate. A consequence of Property 5 is that randomly generated genotypes will be expressed as the maximally general interval predicate with a far higher frequency than would otherwise be expected.

To shed more light on these five properties, it is desirable to pursue a quantitative approach. In particular, we wish to understand the nature of the increased expression frequency of $[p_{min}, q_i)_p$, $[p_i, q_{max})_p$ and $[p_{min}, q_{max})_p$ intervals.

From examination of the results of enumerating the expression frequencies for real encodings of length $2 \leq k \leq 12$, we can derive equations for the total frequency of expression of all possible intervals of the form $[p_{min}, q_i)_p$ and $[p_i, q_{max})_p$:

$$f_{p_{min}, q_i} = 2^{2(k-1)} \ \forall \ p_{min} \leq q_i < q_{max}$$

$$f_{p_i, q_{max}} = 2^{2(k-1)} \ \forall \ p_{min} < p_i \leq q_{max}$$

and frequency of expression of the $[p_{min}, q_{max})_p$ interval:

$$f_{p_{min}, q_{max}} = 2^{2(k-1)} + 2^{k-1}$$

The number of possible $g \to p$ mappings is 2^{2k}, so the probabilities $P_{p_{\min}, q_i}$ and $P_{p_i, q_{\max}}$ of expression of intervals of the form $[p_{\min}, q_i)_p$ and $[p_i, q_{\max})_p$ respectively, is given by

$$P_{p_{\min}, q_i} = \frac{2^{2(k-1)}}{2^{2k}}$$

$$= \frac{2^{k-1}}{2^{k+1}}$$

$$= 0.25$$

and similarly, $P_{p_i, q_{\max}} = 0.25$

The probability $P_{p_{\min}, q_{\max}}$ of expression of the maximally general interval $[p_{\min}, q_{\max})_p$ is given by

$$P_{p_{\min}, q_{\max}} = \frac{2^{2(k-1)} + 2^{k-1}}{2^{2k}}$$

$$= \frac{2^{k-1} + 1}{2^{k+1}}$$

$$\lim_{k \to \infty} = \frac{2^{k-1}}{2^{k+1}} \qquad (1)$$

$$= 0.25$$

Although Equation 1 describes limiting behaviour for infinite length encoding of real numbers, actual values of $P_{p_{\min}, q_{\max}}$ are close to 0.25 for values of k likely to be used in practice (8, 16 or 32 bit encodings). For example, for $k = 8$, $P_{p_{\min}, q_{\max}} = 0.25195$ and for $k = 16$, $P_{p_{\min}, q_{\max}} = 0.25001$.

$P_{p_{\min}, q_{\max}}$ is the probability of the interval $[p_{\min}, q_{\max})_p$ being expressed on $g \to p$ mapping of a random genotype. Centre-Spread Representation thus includes a form of implicit 'don't care' mechanism, similar to that of the '#' allele in ternary representations. However, unlike ternary representations this $P_{\#}$ value is fixed at 0.25 and so is not adjustable to suit different problems.

The probability of expression of an interval of the form

$$[p_i, q_i)_p \; \forall \; p_i = p_{\min} \lor q_i = q_{\max}$$

is

$$P_{p_{\min}, q_i} + P_{p_i, q_{\max}} + P_{p_{\min}, q_{\max}} = 0.75$$

This value is essentially independent of encoding length.

So these 'special' phenotypes constitute 75% of all $g \to p$ mappings, yet only comprise $2^{k+1} - 1$ of the 2^{2k} possible $g \to p$ mappings. Their frequency therefore far exceeds what might be reasonably expected for a $g \to p$ mapping. In contrast, all remaining $g \to p$ mappings of the form $[p_i, q_i)_p \; \forall \; p_i > p_{\min} \land q_i < q_{\max}$ (the plateau in the $g \to p$ landscape) constitute only the remaining 25% of all mappings.

Table 2. Phenotype regions and their structural forms

	q_i	
p_i	Region 2 $[p_{min}, q_i)_p$	Region 4 $[p_{min}, q_{max})_p$
	Region 1 $[p_i, q_i)_p$	Region 3 $[p_i, q_{max})_p$

We can partition the phenotype space into four regions corresponding to the four different structural forms of interval predicate resulting from the properties of the encoding. Table 2 shows the four structural forms of interval predicate, together with the region number we shall, for convenience, assign to them. This diagram mimics the shape of the phenotype frequency matrix and shows the allocation of $g \rightarrow p$ mappings by region. In the diagram and the rest of this chapter, unless otherwise noted,

$$p_{min} < p_i \leq q_{max} \ \wedge \ p_{min} \leq q_i < q_{max}$$

3.3   Operators

Since the real multiplexer is essentially a binary problem in disguise, solutions to the real multiplexer are expressed by an alphabet of three possible interval predicates directly corresponding to the Boolean multiplexer's $\{0,1,\#\}$ alphabet. For the real multiplexer, the solution interval predicates are $\{[0, \theta_i)_p, [\theta_i, 1)_p, [0, 1)_p\}$. However, these are exactly the forms of interval predicate found in regions 2, 3 and 4 of Table 2 that exhibit many to one $g \rightarrow p$ mappings and account for 75% of all $g \rightarrow p$ mappings! We may therefore expect that the choice of Centre-Spread Representation has a bearing on XCS' ability to solve the real multiplexer problem.

There are four places where the influence of the representation is felt in a Learning Classifier System: initialization, covering, crossover and mutation. We note that, for continuous-valued representations, GA subsumption is performed at the level of the phenotype and is independent of the representation in use.

Initialization Where a population is generated at random by genotype, a non-uniform $g \rightarrow p$ mapping will affect the proportion of phenotypes expressed by the population. For Centre-Spread Representation, generation of random genotypes at initialization time will provide a population containing on average a proportion of

$$1 - \frac{1}{4^n} \tag{2}$$

of classifiers with one or more intervals of the correct structural form to solve the real multiplexer problem, i.e., those in regions 2, 3 and 4. These can then be

recombined by the GA to provide complete solutions. For the 6-bit real multiplexer considered here, this equates to 0.9998. In contrast, a one to one $g \rightarrow p$
mapping would provide a proportion of

$$\frac{(n-2)^2 + 1}{n^2}$$

For $n = 6$ this is 0.472. We note here that Fig. 1 shows results obtained
without an initial population.

Covering In XCS' cover operator, the centre of the interval is fixed by the environmental state. For the real multiplexer problem, the environmental state is
externally generated from the uniform probability distribution $U[0, 1)$. The cover
operator for Centre-Spread Representation generates the spread from the uniform probability distribution $U[0, s_0)$. In Wilson's real multiplexer experiments
$s_0 = 1$, so any spread $0 \leq s_i < 1$ is equally possible. Therefore, both centre
and spread are drawn from $U[0, 1)$, so all possible centre-spreads are equally
probable and the probabilities $P_{p_{\min}, q_i}$, $P_{p_i, q_{\max}}$ and $P_{p_{\min}, q_{\max}}$ also apply during
covering. The cover operator, like initialization, thus generates classifiers with a
0.75 probability of being in region 2, 3 or 4 and that have a probability given by
Equation 2 of one or more intervals of the correct structural form to solve the
problem.

Crossover We have determined the probability distribution for new classifiers
generated by the initialization and cover operators. As it is applied with high
probability, crossover has the opportunity to affect this distribution by its production of offspring. We examine the impact of crossover on a single interval
predicate, represented as a centre-spread tuple. We are only interested in the
action of crossover *when it occurs* for a specific interval predicate. For crossover
to alter that interval, a crossover point must occur within the interval. All the
crossover operators considered in this chapter that allow a crossover point within
an interval restrict the crossover point to occur between the two alleles representing the interval. If the crossover point happens to occur between intervals,
the interval itself survives unscathed, though it is likely to be paired with other
intervals during crossover in a multi-dimensional solution space.

As mathematical analysis of crossover is difficult, we enumerate centre-spread
combinations for two parents. We enumerate only those parental intervals that
have at least one point in common with each other, since XCS uses a niche GA. A
factor in the enumeration is the length of the real encoding used. For consistency
with the results presented for mutation, we used an encoding of length $k = 8$ and
crossed over a single centre and spread with a fixed crossover point between the
alleles. For each combination, we noted the region(s) that the parents occupied
and the region(s) occupied by their children. Enumeration of all possible centre-
spread combinations implies an equal probability of parental intervals across
the four regions. As already discussed, this is the case for intervals generated
by initialization and covering. From this we may readily calculate the expected

Table 3. Phenotype proportions for Centre-Spread Representation with $k = 8$ and crossover within an interval

Region	1	2	3	4
Form	$[p_i, q_i)_p$	$[p_{min}, q_i)_p$	$[p_i, q_{max})_p$	$[p_{min}, q_{max})_p$
Parent proportion	0.25	0.25	0.25	0.25
Niche GA	0.225	0.25	0.25	0.275
Panmictic GA	0.248	0.25	0.25	0.252

proportions of offspring across regions. This is shown for both panmictic and niche GAs in Table 3.

For both niche and panmictic GAs, crossover with Centre-Spread Representation tends to preserve the distribution of intervals across regions, with a small bias from region 1 to region 4. Note that this effect only occurs for an interval that has a crossover point within the interval predicate. For crossover points between interval predicates the interval is unchanged, and so the distribution of intervals, and hence regions, cannot alter. The probability of an interval being disrupted in this way depends on n, the number of interval predicates in the classifier's condition and the type of crossover operator used.

Mutation The mutation operator for XCS with Centre-Spread Representation mutates a classifier by adding or subtracting with equal probability an amount m_i drawn from $U[0, m)$. A setting of $m = 0.1$ was used for Wilson's real multiplexer experiments.

We examine the behaviour of mutation over the four regions by studying the probability of an offspring occupying a region, given each possible parental region. To do this, we enumerate all possible mutations for an interval using a setting of $m = 0.1$ for every centre-spread combination over a range of possible values of k, the real encoding length. In the actual Learning Classifier System, the alleles corresponding to the interval's centre and spread are independently mutated with probability μ, so we examine these separately. Mutation of the centre allele shifts the centre by the amount of the mutation, viz:

$$p_i = c_i - s_i + m_i$$
$$q_i = c_i + s_i + m_i$$

Mutation of the spread allele alters the width of the interval:

$$p_i = c_i - s_i - m_i$$
$$q_i = c_i + s_i + m_i$$

For brevity, and since both centre or spread alleles have equal probability of mutation, we average the individual results from these enumerations to provide a picture of the effect of a single mutation on the interval predicate[4]. Mutations

---

[4] In fact, results for the centre and spread alleles differed only slightly and the combined results shown here are also indicative of the individual results.

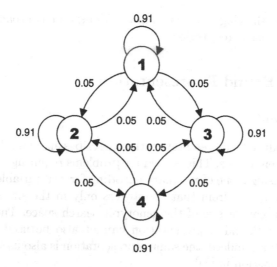

Fig. 3. Region transition diagram for Centre-Spread Representation with $k = 8$ and standard mutation. Regions are represented by states and transition probabilities by arrows

for values of $4 \leq k \leq 8$ were examined, and a common pattern was seen across all enumerations. As an illustration of the results found, Fig. 3 shows the transition diagram of a single mutation of the interval predicate for $k = 8$. In the diagram, the states are the four possible regions that the parent could occupy and the numbers next to the arrows indicate the probability of a transition to a particular region for the offspring. As these probabilities are rounded to two decimal places, we do not distinguish here between very low probabilities (<0.01) and zero probability.

We can see from the diagram that the vast majority of all possible mutations do not produce any migration of region between the parent and offspring (where a region maps to itself). A parent tends to generate offspring that occupy the same region as itself, so parents with the correct structural form of the real multiplexer solution, $[p_{\min}, q_i)_p$, $[p_i, q_{\max})_p$ and $[p_{\min}, q_{\max})_p$, will overwhelmingly produce offspring with these characteristics. This mutation operator therefore has the effect of refining the details of an interval, but is unlikely to cause a change of region from parent to offspring. It is possible for mutation to produce offspring that occupy a different region to that of the parent, but this occurs with a low probability. Migration between regions is possible via the two pathways $1 \leftrightarrow 2 \leftrightarrow 4$ and $1 \leftrightarrow 3 \leftrightarrow 4$ through which offspring are likely to progress through multiple mutation steps. These represent a partial ordering of regions. It is possible to travel in both directions along these pathways with roughly equal probability, showing that this mutation operator is broadly neutral with respect to transitions across regions.

4   Ordered Bound Representation

4.1   Background

Wilson [25] introduced a representation that describes an interval predicate by its lower and upper endpoints. This is used for problems requiring integer variables. However, when using a one of m binary encoding for real variables, the encoding for real numbers differs from that of integers only in the size of the alphabet used and the consequent size of the genotypic search space. There is no reason why the Ordered Bound Representation cannot also be used for continuous-valued problems and, indeed, the same representation is also used for real-valued function approximation in [24].

Ordered Bound Representation stores an interval predicate $[p_i, q_i)_p$ as a tuple (l_i, u_i) where for real-valued problems $l_i, u_i \in \Re$. l_i is the lower bound of the interval and u_i is the upper bound. We again assume that the alleles are encoded using a one of m binary encoding of length k.

One issue with Ordered Bound Representation is the ordering imposed on the alleles representing an interval predicate by the restriction that $l_i \leq u_i$. Wilson does not describe how this restriction is addressed, but essentially, any operator that could cause a situation where $l_i > u_i$ must check each interval predicate in a classifier's condition and swap the lower and upper alleles if the ordering $l_i \leq u_i$ is violated as a result of the operation. This affects the crossover and mutation operators with a Learning Classifier System using a panmictic GA (such as ZCS), but only mutation is affected when using a niche GA, as is the case with XCS.

Apart from discretization in the real encoding, there is a direct mapping between the elements of the genotype and phenotype, so that $l_i \equiv p_i$ and $u_i \equiv q_i$. As a result, no explicit gene expression is necessary and, at the level of the representation, a one to one mapping exists for all possible phenotypes and their corresponding genotypes. No truncation occurs upon the expression since it is not possible to represent values outside the endpoints of the phenotype interval $[p_{min}, q_{max})_p$. The issues arising from the many to one $g \to p$ mapping with Centre-Spread Representation cannot arise with Ordered Bound Representation.

4.2   Properties

In Section 3.2 we stated certain properties that exist due to the many to one $g \to p$ mapping of Centre-Spread Representation. For reference, the equivalent properties of Ordered Bound Representation are:

Property 1: One to one genotype to phenotype mapping.
Property 2: Complete phenotype to genotype mapping.

Property 3: The genotype to phenotype mapping of exact numbers is one to one.

Property 4: The frequency of intervals of the form $[p_{min}, q_i)_p$ and $[p_i, q_{max})_p$ is constant for all p_i and q_i.

Property 5: The frequency of the $[p_{min}, q_{max})_p$ interval is the same as that of any other interval.

These are all due to the one to one $g \to p$ mapping that exists for Ordered Bound Representation. The representation stores all possible interval predicates with equal frequency and shows no bias towards certain types of interval predicate. This suggests that Ordered Bound Representation may be more suited for continuous-valued domains where the structure of the problem is *a priori* unknown.

The maximally general interval $[p_{min}, q_{max})_p$ is represented by a single tuple in Ordered Bound Representation, so given a random genotype, an interval predicate will be maximally general with probability

$$\frac{1}{2^{2k}}$$

As the size k of the real number encoding increases and thus its granularity becomes finer, the chances of a maximally general interval appearing in the initial population becomes exponentially lower. This is in contrast to the Centre-Spread Representation, which results in the maximally general interval being represented with an essentially fixed probability of 0.25. For this reason, Ordered Bound Representation effectively provides no implicit 'don't care' mechanism analogous to the '#' allele in a ternary representation.

4.3   Operators

We examined the effects of the interaction between Centre-Spread Representation and its operators in Section 3.3. Here we investigate Ordered Bound Representation and its associated operators.

Initialization Where a population is generated at random by genotype, a uniform $g \to p$ mapping across intervals means that all possible phenotypes will exist in the population with identical probability. The frequency of expression of intervals of the form $[p_{min}, q_i)_p$, $[p_i, q_{max})_p$ and $[p_{min}, q_{max})_p$ (regions 2, 3, and 4) can be calculated as

$$f_{p_{min},q_i} = 2^k - 1 \ \forall \ p_{min} \leq q_i < q_{max}$$
$$f_{p_i,q_{max}} = 2^k - 1 \ \forall \ p_{min} < p_i \leq q_{max}$$
$$f_{p_{min},q_{max}} = 1$$

The number of possible $g \to p$ mappings where $l_i \leq u_i$ is $2^{k-1}(2^k + 1)$, so the frequency of intervals of the form $[p_i, q_i)_p \ \forall \ p_i > p_{min} \wedge q_i < q_{max}$ (region 1) is given by

$$f_{p_i,q_i} = 2^{k-1}(2^k + 1) - 2(2^k - 1) - 1$$
$$= (2^k - 1)(2^{k-1} - 1)$$

The probability of expression of an interval of this form is

$$P_{p_i,q_i} = \frac{(2^k - 1)(2^{k-1} - 1)}{2^{k-1}(2^k + 1)}$$
$$= \frac{2^{2k-1} - 3(2^{k-1}) + 1}{2^{2k-1} + 2^{k-1}}$$
$$\lim_{k \to \infty} = 1$$

Initialization therefore generates classifiers essentially exclusively in region 1. No classifiers are to be expected in a small, finite, population with the correct structural form of the solution as occurs in Centre-Spread Representation.

Covering The cover operator generates a classifier containing intervals with the (l_i, u_i) tuples given by

$$l_i = x_i - U[0, s_0)$$
$$u_i = x_i + U[0, s_0)$$

To match the experiments performed with Centre-Spread Representation [5] $s_0 = 1$. Note that, unlike the case with Centre-Spread Representation, the resulting interval is not generally centred on the environmental variable, x_i. This is the method adopted by Wilson, which we also use here. However, it would be trivial to alter the algorithm to emulate Centre-Spread Representation strategy by using the same random spread for both endpoints and this is examined in Section 10.2. In either case, truncation is necessary when mapping from the generated interval to the genotype. This truncation causes similar effects to those seen for Centre-Spread Representation. For example, Table 4 shows the frequency matrix for all possible intervals generated by the cover operator using a real encoding of length of $k = 3$.

This shows increased frequency of regions 2, 3 and 4 similar to that of Centre-Spread Representation. Furthermore, region 1 also shows increased mapping frequency as the interval width $q_i - p_i$ increases. Studies of such matrices for encoding lengths of $2 \le k \le 8$ all showed the same effects. From these we can derive the frequencies of expression of the region 2, 3 and 4 intervals:

$$f_{p_{\min},q_i} = \frac{2^{3k} - 2^k}{3}$$
$$f_{p_i,q_{\max}} = \frac{2^{3k} - 2^k}{3}$$
$$f_{p_{\min},q_{\max}} = \frac{(2^k + 1)^3 - 2^k - 1}{6}$$

[5] [25] refers to the cover spread as r_0. For consistency we use s_0 for all representations.

Table 4. Phenotype frequency matrix for Ordered Bound Representation with $k = 3$ and standard cover with $s_0 = 1$

		q_i							
		0	1	2	3	4	5	6	7
	0	8	15	21	26	30	33	35	120
	1		1	2	3	4	5	6	35
	2			1	2	3	4	5	33
p_i	3				1	2	3	4	30
	4					1	2	3	26
	5						1	2	21
	6							1	15
	7								8

and the probabilities $P_{p_{\min},q_i}$, $P_{p_i,q_{\max}}$ and $P_{p_{\min},q_{\max}}$ for the cover operator with Ordered Bound Representation:

$$P_{p_{\min},q_i} = \frac{2^{3k} - 2^k}{3(2^{3k})}$$

$$= \frac{1}{3} - \frac{1}{3(2^{2k})}$$

$$\lim_{n \to \infty} = \frac{1}{3}$$

and similarly, $P_{p_i,q_{\max}} = \frac{1}{3}$

$$P_{p_{\min},q_{\max}} = \frac{(2^k + 1)^3 - 2^k - 1}{6(2^{3k})}$$

$$\lim_{n \to \infty} = \frac{1}{6}$$

thus $P_{p_i,q_i} = \frac{1}{6}$

So, even though Ordered Bound Representation has no intrinsic bias, the truncation necessary when using the cover operator introduces bias. This is shown in Fig. 4 for $k = 5$.

Note that the amount of bias is determined by the setting of the s_0 parameter, as instances of covering where truncation is not necessary cannot introduce bias. The setting of $s_0 = 1$ provides a good bias for the real multiplexer problem since it generates classifiers with one or more intervals of the correct structural form (i.e., those in regions 2, 3 and 4) with a probability of

$$1 - \frac{1}{6^n}$$

For the 6-bit real multiplexer, this probability is 0.99998.

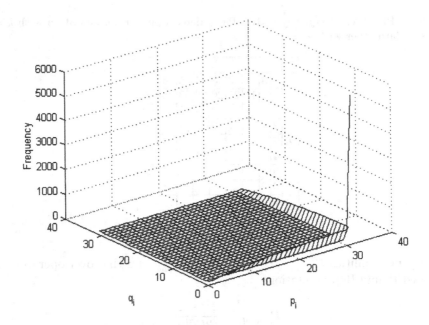

Fig. 4. Phenotype frequency landscape for Ordered Bound Representation with $k = 5$ and standard cover with $s_0 = 1$

Table 5. Phenotype proportions for Ordered Bound Representation with $k = 8$ and crossover within an interval

Region	1	2	3	4
Form	$[p_i, q_i)_p$	$[p_{min}, q_i)_p$	$[p_i, q_{max})_p$	$[p_{min}, q_{max})_p$
Parent proportion	0.25	0.25	0.25	0.25
Niche GA	0.984	0.008	0.008	< 0.001
Panmictic GA	0.984	0.008	0.008	< 0.001

Crossover Analysis of crossover with Ordered Bound Representation for an encoding of length $k = 8$ as described in Section 3.3 yields the results shown in Table 5.

Crossover under Ordered Bound Representation tends to produce offspring in region 1 at the expense of those in regions 2, 3 and 4. This affects the probability distribution of the population across regions and tends to remove offspring in regions 2, 3 and especially 4 (the maximally general interval) from the population. As such, crossover appears to provide the same bias as that of initialization for Ordered Bound Representation (Section 4.3). For simplicity, the analysis assumes a parent population with a uniform distribution across regions, which is not generally the case. But, although exact details will vary, the general trends seen here should still apply.

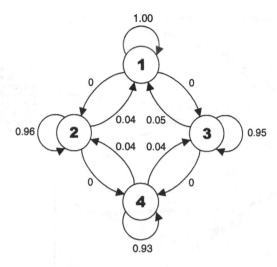

Fig. 5. Region transition diagram for Ordered Bound Representation with $k = 8$ and standard mutation. Regions are represented by states and transition probabilities by arrows

Mutation Mutation for Ordered Bound Representation was studied in the same way as for Centre-Spread Representation (Section 3.3). The transition diagram for $k = 8$ is shown in Fig. 5.

This displays similar characteristics to those of Centre-Spread Representation. Most mutations cause no change of region from parent to offspring, but simply refine the details of the interval within the region. When a transition does occur, the transition probabilities for Ordered Bound Representation essentially only allow transitions away from anchored (region 2, 3 and 4) intervals.

5   Unordered Bound Representation

5.1   Background

Ordered Bound Representation provides a one to one $g \rightarrow p$ mapping, but the $l_i \leq u_i$ ordering restriction is unnecessary. If this restriction is lifted, the phenotype can still be directly encoded using the endpoints of the interval, but without an ordering requirement. Thus, an interval $[p_i, q_i)_p$ may be encoded as the tuples (p_i, q_i) or $(q_i, p_i) \ \forall \ p_i \neq q_i$. There are thus exactly two equivalent genotypes for each phenotype except where $p_i = q_i$ when there is exactly one genotype for each phenotype. In other words, the $g \rightarrow p$ mapping is normally two to one, except for exact numbers, which have a one to one mapping. We do not expect the resulting small bias in favour of intervals over exact numbers to substantially affect the performance of a Learning Classifier System using Unordered Bound

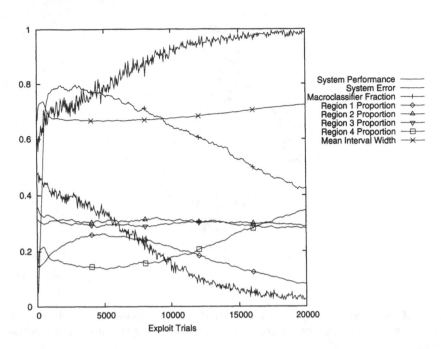

Fig. 6. 6-bit real multiplexer with Unordered Bound Representation, standard cover with $s_0 = 1$, 2-point crossover and standard mutation

Representation. Indeed, the desire for an interval-based phenotype suggests that the solution to a problem using an interval-based representation is best expressed in the form of a vector of intervals, rather than simple inequalities, so any resulting slight performance differences compared to Ordered Bound Representation should, if anything, be advantageous. In any event, the bias induced by the Unordered Bound Representation's $g \rightarrow p$ mapping is negligible compared to the major disparities in phenotype expression frequency seen using Centre-Spread Representation.

The advantage of Unordered Bound Representation over Ordered Bound Representation is that it avoids the additional operator complexity associated with swapping the endpoints of an interval if the $l_i \leq u_i$ ordering restriction is violated. Although this may seem trivial, the presence of the ordering restriction constitutes a form of epistasis between the l_i and u_i alleles, as their values are mutually dependent. A resulting swap may generate great change in a particular locus when viewed before and after the operation that caused the swap to occur. This cannot occur using Unordered Bound Representation, since no ordering of endpoints exists for the interval predicate at the level of the genotype.

Fig. 6 shows the results of using Unordered Bound Representation on the 6-bit real multiplexer problem.

5.2   Properties

Properties of Unordered Bound Representation are:

Property 1: Two to one genotype to phenotype mapping for intervals (but not exact numbers.)

Property 2: Complete phenotype to genotype mapping.

Property 3: The genotype to phenotype mapping of exact numbers is one to one.

Property 4: The frequency of intervals of the form $[p_{min}, q_i)_p$ and $[p_i, q_{max})_p$ is constant for all p_i and q_i.

Property 5: The frequency of the $[p_{min}, q_{max})_p$ interval is the same as that of any other interval.

The maximally general interval $[p_{min}, q_{max})_p$ may be represented by two possible tuples in Unordered Bound Representation, so given a random genotype, an interval will be maximally general with probability

$$\frac{1}{2^{2k-1}}$$

Thus, Unordered Bound Representation, like Ordered Bound Representation, provides no implicit 'don't care' mechanism.

5.3   Operators

In this section we investigate Unordered Bound Representation and the operators adapted for it.

Initialization The frequency of expression of intervals of the form $[p_{min}, q_i)_p$, $[p_i, q_{max})_p$ and $[p_{min}, q_{max})_p$ (regions 2, 3, and 4) can be calculated as

$$f_{p_{min}, q_i} = 2^{k+1} - 3 \; \forall \; p_{min} \leq q_i < q_{max}$$
$$f_{p_i, q_{max}} = 2^{k+1} - 3 \; \forall \; p_{min} < p_i \leq q_{max}$$
$$f_{p_{min}, q_{max}} = 2$$

The number of possible $g \to p$ mappings is 2^{2k}, so the frequency of intervals of the form $[p_i, q_i)_p \; \forall \; p_i > p_{min} \wedge q_i < q_{max}$ (region 1) is given by

$$f_{p_i, q_i} = 2^{2k} - 2(2^{k+1} - 3) - 2$$
$$= (2^k - 2)^2$$

The probability of expression of an interval of this form is

$$P_{p_i, q_i} = \frac{(2^k - 2)^2}{2^{2k}}$$
$$= \frac{2^{2k} - 2(2^{k+1}) + 4}{2^{2k}}$$
$$\lim_{n \to \infty} = 1$$

The implication of this is that, like Ordered Bound Representation, initialization generates classifiers essentially exclusively in region 1.

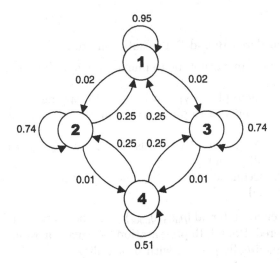

Fig. 7. Region transition diagram for Unordered Bound Representation with $k = 8$ and standard mutation. Regions are represented by states and transition probabilities by arrows

Covering The cover operator for Unordered Bound Representation is the same as that for Ordered Bound Representation, with the addition that, to avoid unnecessary bias, it encodes the endpoints of the generated interval in a random order. This does not affect its operation, so the results presented for Ordered Bound Representation in Section 4.3 also apply here.

Crossover Analysis of crossover with Unordered Bound Representation for an encoding of length $k = 8$ as described in Section 3.3 yields the same results as those shown in Table 5. The comments made in Section 4.3 for crossover with Ordered Bound Representation therefore also apply to Unordered Bound Representation.

Mutation Mutation for Unordered Bound Representation was studied in the same way as for Centre-Spread Representation (Section 3.3). The region transition diagram for $k = 8$ is shown in Fig. 7.

This shows that the mutation operator for Unordered Bound Representation acts to provide a strong pressure away from region 4. For region 4 intervals, there is only a 0.51 probability of staying in region 4 upon mutation, with a transition to regions 2 or 3 equally likely. Similarly for a region 2 or 3 interval, a transition to region 1 is possible with a probability of 0.25.

6   Crossover with Interval Representations

6.1   Mean Interval Width

Analysing the intervals resulting from crossover by region is useful, but it does not tell the whole story of the effects of crossover. It is possible for region 1, 2 and 3 intervals to be generated by crossover with varying widths and to understand the effects of crossover further, we must study the widths of offspring intervals.

Using Centre-Spread Representation, an interval is represented as a centre and spread tuple (c_i, s_i) where c_i encodes the centre (or position) of the interval and s_i encodes its spread (or width). During crossover within an interval predicate, centre and spread alleles are exchanged:

$$[c_i^1, s_i^1]_g \times [c_i^2, s_i^2]_g \to [c_i^1, s_i^2]_g, [c_i^2, s_i^1]_g$$

This process can be viewed one of two ways – either that the interval positions are invariant from parent to offspring with the interval widths changing, or that the interval widths are invariant from parent to offspring with the position of the intervals changing. In practice, both of these transformations occur simultaneously. From the perspective of interval width, crossover has the effect of specializing or generalizing intervals by altering their widths (but not positions), whereas from the perspective of interval position, crossover alters the position, but maintains the widths of intervals. The epistasis between centre and spread alleles means that the degree of difference in position and width between parents and offspring depends on the amount of variance between the centres and spreads of the respective parents. Where there is a large difference between parental centres and/or spreads, offspring will have little in common with their parents, so large jumps in interval position or width are possible under crossover. A metric that can be readily observed experimentally is mean interval width. This may be measured for the entire population, for the match set or, as we use it here, for the two parental intervals undergoing crossover and for the resulting two offspring. For Centre-Spread Representation, all pairs of offspring produced with Centre-Spread Representation preserve the mean interval width of the parents.

With Ordered Bound Representation, an interval is represented by the tuple (l_i, u_i) where l_i is the lower bound and u_i is the upper bound of the interval. Crossover within interval predicates swaps the two alleles representing an interval:

$$(l_i^1, u_i^1) \times (l_i^2, u_i^2) \to (l_i^1, u_i^2) \times (l_i^2, u_i^1)$$

The mean width of the parental intervals is

$$\frac{(u_i^1 - l_i^1) + (u_i^2 - l_i^2)}{2}$$

The mean width of the offspring intervals for Ordered Bound Representation is

Table 6. Parental and offspring genotypes for Unordered Bound Representation. All four parental genotypes express to the same pair of phenotypes. Offspring genotypes express to two distinct phenotype pairs, A and B

	Parent 1	Parent 2	Offspring 1	Offspring 2	
1	$[l_i^1, u_i^1]_g$	$[l_i^2, u_i^2]_g$	$[l_i^1, u_i^2]_g$	$[l_i^2, u_i^1]_g$	A
2	$[l_i^1, u_i^1]_g$	$[u_i^2, l_i^2]_g$	$[l_i^1, l_i^2]_g$	$[u_i^2, u_i^1]_g$	B
3	$[u_i^1, l_i^1]_g$	$[l_i^2, u_i^2]_g$	$[u_i^1, u_i^2]_g$	$[l_i^2, l_i^1]_g$	B
4	$[u_i^1, l_i^1]_g$	$[u_i^2, l_i^2]_g$	$[u_i^1, l_i^2]_g$	$[u_i^2, l_i^1]_g$	A

$$\frac{(u_i^2 - l_i^1) + (u_i^1 - l_i^2)}{2}$$

$$= \frac{(u_i^1 - l_i^1) + (u_i^2 - l_i^2)}{2}$$

So, the mean width of the two offspring generated by Ordered Bound Representation crossover within interval predicates is the same as that of the two parental intervals.

For Unordered Bound Representation, the situation is more complex. Because, in general, there are two possible genotypes, $[l_i, u_i]_g$ and $[u_i, l_i]_g$ for a particular phenotype, intervals produced by crossover with Unordered Bound Representation depend on the ordering of the parental genotypes. Table 6 shows the possible parental genotype orderings for a crossing of parental phenotypes $[l_i^1, u_i^1]_p \times [l_i^2, u_i^2]_p$ and the resulting offspring genotypes. The lack of an ordering restriction on genotypes with Unordered Bound Representation means that offspring genotypes 1 and 4 in Table 6 are equivalent, as are offspring genotypes 2 and 3, so there are exactly two forms of offspring phenotype, one where parental lower and upper bounds are paired and the other where a parental lower (upper) bound is paired with another parental lower (upper) bound. Note that crossover with Ordered Bound Representation can only produce genotypes of type A.

The only ordering restriction imposed at the phenotype level is

$$l_i^1 \leq u_i^1 \ \wedge \ l_i^2 \leq u_i^2$$

We cannot assume anything about the relative values of the bounds of one interval compared to the interval with which it is paired. To proceed, we must consider the possible basic configurations of pairs of intervals.

6.2 Crossover Configurations

Fig. 8 shows the three possible configurations of parental interval and the offspring configurations that result after crossover if the relative orderings of parental interval bounds are taken into account. Other symmetries of the basic configurations are possible. Parental configurations 1 and 2 occur with both niche and

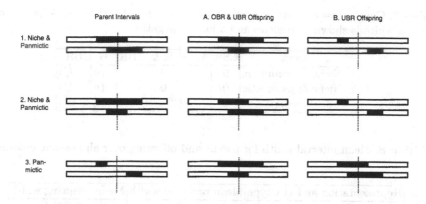

Fig. 8. Crossover configurations. Parental configurations are rows in the diagram and offspring configurations are columns. Each entry shows two intervals (shaded) in a single dimension of the solution space. The dotted line represents the environmental variable x_i

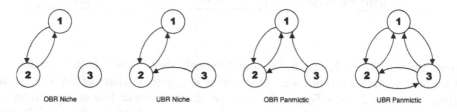

Fig. 9. Transitions possible from parental to offspring configurations. Numbered states correspond to the configurations shown in Fig. 8

panmictic GAs, as the parental intervals share at least one point in common, namely the environmental variable x_i. Configuration 3 can only occur with a panmictic GA as the parental intervals do not overlap. Offspring configurations in column A represent offspring intervals with phenotypes of the form $[l_i^1, u_i^2)_p$ or $[l_i^2, u_i^1)_p$, as in Table 6. As already explained, these can occur with both Ordered Bound Representation and Unordered Bound Representation. The additional offspring configurations resulting from the use of unordered tuples with Unordered Bound Representation appear in offspring column B. These configurations cannot occur with Ordered Bound Representation as they have the form $[l_i^1, l_i^2)_p$ or $[u_i^1, u_i^2)_p$.

Fig. 9 shows the possible transitions from the three parental configurations to offspring configurations, where offspring are also classified according to their configuration, 1-3. This shows that only Ordered Bound Representation with a niche GA and Unordered Bound Representation with a panmictic GA are symmetrical and allow all possible transitions between parental and offspring configurations.

It is straightforward to compute the gain or loss of mean interval width from parents to offspring for each of the configurations that appear in Fig. 8. These

Table 7. Gain or loss of mean interval width from parent to offspring. A gain in mean interval width is shown as positive and a loss as negative

Parent	CSR	A. OBR & UBR	B. UBR
1. Niche & panmictic	0	0	$-(u_i^1 - l_i^2)$
2. Niche & panmictic	0	0	$-(u_i^2 - l_i^2)$
3. Panmictic	0	$+(l_i^2 - u_i^1)$	$+(l_i^2 - u_i^1)$

Table 8. Mean interval width for parent and offspring over all possible crossings

Representation and GA type	Mean parental width	Mean offspring width
CSR niche	0.699	0.699
CSR panmictic	0.665	0.665
OBR niche	0.400	0.400
OBR panmictic	0.333	0.401
UBR niche	0.401	0.301
UBR panmictic	0.335	0.335

are shown in Table 7, which, for completeness, also shows this information for Centre-Spread Representation. This makes it apparent that certain configurations of parental and offspring intervals produce a gain or loss of mean interval width during crossover. We can see that crossover with Ordered Bound Representation and a niche GA is neutral with respect to mean interval width, but with a panmictic GA it possible to produce offspring intervals that are on average, wider than the parents. Crossover with Unordered Bound Representation and a niche GA can produce intervals that are narrower, on average, than the parents, whilst with a panmictic GA it is also possible to produce intervals that are wider, on average, than the parents. These results can also be seen intuitively in Fig. 8 and Fig. 9.

To investigate these effects further, we enumerated all possible combinations of interval for two parents. For niche GA results, we enumerated only those parental intervals that have at least one point in common with each other. For consistency with the results presented in earlier sections, we used an encoding of length $k = 8$. For each combination, we noted the mean interval width of the parents and offspring and averaged these over all combinations of parental interval to provide an indication of the overall effects of crossover with different representations and types of GA. Results are shown in Table 8.

Table 8 shows that over all possible intervals crossover with Centre-Spread Representation is neutral with respect to mean interval width for both niche and panmictic GAs. We can also see from these results that the mean interval width for Centre-Spread Representation is quite wide (around 0.7), in contrast to Ordered Bound Representation and Unordered Bound Representation which width of around 0.33 for a niche GA and 0.4 for a panmictic GA. As expected from Fig. 9 and Table 7, crossover with Ordered Bound Representation and a niche GA causes no change to the mean interval width of parents and offspring, while crossover with Ordered Bound Representation and a panmictic GA tends to generate wider intervals. Conversely, crossover with Unordered Bound Representation and a niche GA tends to generate narrower intervals, whereas with Unordered Bound Representation and a panmictic GA, no change occurs to mean interval width. Because Unordered Bound Representation with a panmictic GA can generate offspring that are both wider and narrower, on average, than those of the parents, it is more difficult to predict the overall direction of the effect. Although this is neutral over all possible crossings, this may not be true for intervals drawn from arbitrary populations.

6.3 Genetic Search

There are differences in the characteristics of intervals produced by crossover with the three representations considered for real numbers in Learning Classifier Systems. Centre-Spread Representation preserves interval position or width from parents to offspring, which, in general, causes the endpoints of parental intervals to be lost under crossover. Ordered Bound Representation and Unordered Bound Representation preserve the endpoints of parental intervals, but shuffle them into new combinations to produce offspring intervals. Unordered Bound Representation offers additional possibilities for endpoint shuffling over Ordered Bound Representation, due to the lack of ordering of bounds in this representation.

The GA in a Learning Classifier System with an interval representation is searching for useful hyper-rectangular decision surfaces. Crossover with Ordered Bound Representation and Unordered Bound Representation appear to facilitate this search by preserving elements of the boundaries of hyper-rectangles represented by high fitness classifiers. In contrast, crossover with Centre-Spread Representation constructs offspring by generalizing, specializing or shifting the position of high fitness hyper-rectangles. However, there would seem to be a degree of randomness involved in this process, due to epistatic interactions between centre and spread alleles that do not exist between alleles in Ordered Bound Representation and Unordered Bound Representation representations.

7 The Real Multiplexer Revisited

7.1 Solving the Real Multiplexer

We have seen how Centre-Spread Representation with the operators and parameter settings being used make it especially suited to solving the real multiplexer problem. Wilson [23] states, "notice how the system has 'sculpted' the predicates and is in effect finding the thresholds. Most predicates either show ranges between 0.0 and 0.5, 0.5 and 1.0 or are 'don't cares'". We suggest that it is the combination of Centre-Spread Representation, the operators and their parameter settings that provides pressure for these effects. Initialization and/or covering generate classifiers containing intervals of the correct structural form, $[p_{min}, q_i)_p$, $[p_i, q_{max})_p$ and $[p_{min}, q_{max})_p$ (regions 2, 3 and 4). Crossover and mutation then refine these by discovering the correct thresholds, θ_i to solve the problem. Having one of the endpoints of an interval correct upon initial generation of a classifier allows a much simpler genetic search compared to having to discover both ends of an interval concurrently. In this way, the representation and/or operators relieve the other mechanisms of XCS from much of the burden of solving the real multiplexer problem because the solution to the problem happens to match the nature of the classifiers being generated. This is especially true for the real multiplexer, because its solution is composed of entirely of region 2, 3 and 4 intervals with only a single threshold, θ_i. For an arbitrary problem, the ability of the representation to generate such intervals may be an advantage or hinderance, depending on the nature of the problem (i.e., whether the solution consists of region 2 and 3 intervals) and the degree of general (region 4) intervals in the solution.

7.2 Sampling Bias

This hypothesis suggests that the time taken to solve the real multiplexer problem should be independent of the threshold, θ_i. We repeated the real multiplexer experiment in [23] where $\theta_i = 0.75$ and, like Wilson, were unable to solve the problem in 20,000 exploit trials. However, we found that if XCS is allowed to run for 50,000 trials, it does solve the problem. In fact, XCS takes approximately 2.5 times longer to solve the real multiplexer with $\theta_i = 0.75$, than when $\theta_i = 0.5$, even if $\theta_i = 0.75$ does not alternate across values of i. Further experimentation revealed that this is because $[0, 0.75]_p$ intervals are sampled with three times the frequency than that of $[0.75, 1)_p$ intervals. If both intervals are sampled with equal frequency, then XCS solves the problem in the same number of trials as for when $\theta_i = 0.5$ (not shown) and this is the explanation that Wilson suggests. Importantly, the difference in performance solely arises due to sampling bias and not from any representation or operator bias present.

7.3 Relationship to Integer Results

Even with a neutral representation, such as Ordered Bound Representation or Unordered Bound Representation, the present cover operator still generates classifiers containing intervals with the correct structural form with an increased frequency. The Random-Data2 and Random-Data9 test problems [25] exhibit the same characteristics as described for the real multiplexer; that is, solutions to the problem are of the form $[p_{min}, q_i)_p$, $[p_i, q_{max})_p$ and $[p_{min}, q_{max})_p$ (regions 2, 3, and 4). It would appear from Wisconsin Breast Cancer results[6] that this problem also has these characteristics. Wilson asks why XCS solves the Random-Data9 problem within a factor of 10 of the simpler Random-Data2 problem when the input space is exponentially larger (10^9 versus 10^2). We hypothesize that the

---

[6] Figure 5 in [25].

covering bias described plays a part in this anomaly by generating classifiers containing intervals of the correct structural form. It is not unreasonable then to assume that the additional effort for crossover and mutation to refine these is better than exponential. Further work is necessary to validate this hypothesis with the above test problems.

7.4 Interval Predicates and the Real Multiplexer

We suggest that the real multiplexer problem is a poor choice of test problem for Learning Classifier Systems operating with continuous-valued data and interval predicates, since its solutions all have one endpoint in common with the maximally general interval in the solution space. Because of this, it is not representative of the broader class of problem where solutions are not, in general, closely aligned to the representation of the 'don't care' state.

The benefits of interval predicates are that they are able to represent hyper-rectangular decision surfaces in solution space. These benefits only accrue if (i) the problem solution requires a hyper-rectangle, rather than a hyperplane decision surface or (ii) the form of the problem solution is not known *a priori*. The real multiplexer problem can be solved using a hyperplane decision surface since all of the hyper-rectangles needed for the solution are anchored at a boundary of the solution space. It does not strictly require the presence of interval predicates to represent the solution and consequently cannot adequately test the general operation and performance of representations that use interval predicates.

We argue that the real benefit of the use of interval predicates is their ability to represent arbitrary intervals in solution space. This provides a richer expressive power that cannot be achieved using hyperplane Decision Surfaces and potentially allows a broader class of problem to be solved. In many real-world problems, the form of the solution is unknown *a priori* and test problems for Learning Classifier Systems using interval predicates must be flexible enough to explore all aspects of their operation and performance. This is not the case with the real multiplexer problem.

8   Hyper-Rectangles

8.1   Full Environmental Map

XCS attempts to build a full environmental map of the problem in order to cover the solution space with classifiers. The map takes the form of the population of classifiers, with individual classifiers representing portions of the map.

Classifiers using an interval-based representation construct hyper-rectangular decision surfaces in solution space. For all of the problems discussed in Section 7, the decision surface can be represented by a hyperplane and so one of the faces of the hyper-rectangle is always at the boundary of the solution space. This face simply serves to specify the direction of the inequality otherwise represented by the hyperplane.

Fig. 10. Rectangle centred in a 2-dimensional solution space. The decision surface is shaded

Fig. 11. The four types of decision surface representing the solution interval (shaded) possible in a 1-dimensional solution space

A hyper-rectangle can approximate more complex decision surfaces than a hyperplane. In this case the decision surface is closed and will have faces that are not at solution space boundaries. This would seem to be a disadvantage for representation and operator combinations that provide bias towards the solution space boundaries. However, since XCS builds a complete map of the solution space, for each classifier representing a closed decision surface, there are multiple classifiers representing the solution space outside the closed decision surface. For example, consider a rectangle centred in a 2-dimensional solution space (Fig. 10).

There are at least four other rectangles outside this rectangle mapping the solution space. Each of these touches the bounds of the solution space and presumably gains benefit from any bias of the representation and operators towards solution space boundaries. In general, the balance of this benefit will depend on the shape of the decision surface.

8.2   1-Dimensional Solution Space

A single closed decision surface can be created in a 1-dimensional solution space by dividing the solution space into non-overlapping hyper-rectangles in four ways (Fig. 11).

In case 1 three hyper-rectangles (i.e., interval predicates) must be constructed to cover the solution space. Two of these have their faces at the solution space boundary (the unshaded rectangles in the diagram). This is the general problem $\theta_l \leq x_i < \theta_u$, where θ_l and θ_u are the lower and upper bounds of the solution interval.

For cases 2 & 3 the solution space is covered by two hyper-rectangles, both of which have one of their faces at the solution space boundary. These are the cases for the real multiplexer and the other experiments discussed in Section 7.

Case 4 shows a hyper-rectangle covering all of one dimension of the solution space, with both faces at the boundary of the solution space. This represents the maximally general 'don't care' interval.

Notice that the four cases shown correspond to the four structural forms (regions) of interval predicate previously described.

Thus, for a single closed decision surface representing the solution interval, there are always more hyper-rectangles requiring faces at the solution space boundary than those that do not. This is because XCS 'fills in' the missing parts of the solution space when building its complete environmental map. Strictly, this closed decision surface is all that is required for a classifier in a traditional (non-accuracy based) classifier system. However, XCS also generates the other hyper-rectangles to complete the map.

Even if a dimension of the solution space is divided into multiple Decision Surfaces representing the solution interval, then excluding the maximally general interval (case 4 above), there will always be exactly two hyper-rectangles with faces at the solution space boundary. The number of hyper-rectangles without faces at solution space boundaries exceeds those with faces at solution space boundaries only when a dimension of the solution space is divided into a total of five or more hyper-rectangles. Thus, it would seem reasonable to assume that a bias towards hyper-rectangles with faces at solution space boundaries is an advantage when a 1-dimensional solution space is divided into less than five hyper-rectangles.

8.3 Multidimensional Solution Space

Of course, in a multidimensional solution space, the influence of the hyper-rectangle complexity of all dimensions must be taken into account. Assuming that each dimension, n, of the solution space is divided into the same number of hypercubes, n_d, the proportion of hypercubes with one or more faces at the solution space boundary is given by

$$\frac{n_d^n - (n_d - 2)^n}{n_d^n} \; \forall \, n_d \geq 2$$

This is plotted in Fig. 12, which shows that the proportion of hypercubes with one or more faces at the solution space boundary depends primarily on the dimensionality of the problem and only secondarily on the number of hypercubes into which each dimension is divided. For almost all problems, this proportion is greater than 0.5, while for problems with several dimensions (i.e., most real-world problems) the number of hypercubes with no face at the solution space boundary becomes insignificantly small. As a result, these hypercubes are likely to have little influence on the performance of XCS when constructing its environmental map. Therefore, even problems where the solution is of the form $[p_i, q_i)_p \; \forall \, p_i > p_{\min} \wedge q_i < q_{\max}$ (region 1) should benefit from the representation and operator bias studied here, as the solution to the problem is dominated by the search for intervals in regions 2, 3 and 4 – precisely those for which bias exists.

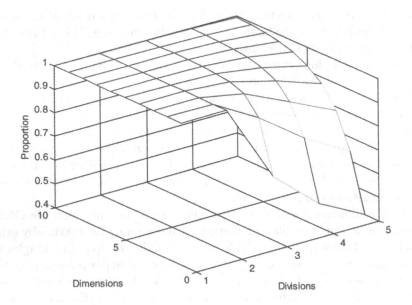

Fig. 12. Proportion of hypercubes with one or more faces at the solution space boundary for problems of dimension n, with each dimension divided into n_d hypercubes

Fig. 13. 2-dimensional checkerboard with $n_d = 5$

9   The Checkerboard Problem

9.1   Description

To circumvent the limitations of the real multiplexer problem, we use a new abstract single-step test problem, the checkerboard problem. This problem divides up the n-dimensional solution space into equal sized hypercubes. Each hypercube is assigned a 'colour' black or white, with the colours alternating in all dimensions. For $n = 2$ the solution space takes on the appearance of a chess or checkers board. The problem difficulty is controlled by both the dimensionality of the solution space, n and the number of divisions of each dimension of the solution space, n_d. To allow the colours to alternate in all dimensions, n_d must be an odd number. Fig. 13 shows a 2-dimensional checkerboard with $n_d = 5$.

On each trial, the Learning Classifier System is presented with a vector of n random real numbers in the interval $[0, 1)_p$, representing a point in the solution space. The Learning Classifier System then attempts to assign an action, 0 or 1 depending on whether the point is contained in a black (0) or white (1) hypercube. The classifiers generated by the Learning Classifier System thus correspond directly to hypercubes in the solution space.

The solution to the checkerboard problem, as presented, requires no maximally general intervals due to the presence of alternating hypercubes. Although we do not use it here, a controlled number of maximally general intervals may be added to the checkerboard problem by making black entire hyper-rows and hyper-columns of the checkerboard. The number of hyper-rows and hyper-columns generalized in this way is controlled by a parameter, n_g with the maximally general intervals being allocated uniform randomly among dimensions and divisions of the problem.

The checkerboard problem is analogous to the test suite for ternary representations detailed in [14].

9.2   Checkerboard with Initial Population

Fig. 14 and Fig. 15 show the performance of Centre-Spread Representation and Unordered Bound Representation on the checkerboard problem with $n = 3$ and $n_d = 3$. The solution to this problem consists of 27 hypercubes, so XCS needs 54 classifiers to construct a full map. In these experiments, an initial population of 2000 classifiers was used. Other settings are as for the real multiplexer experiments. We did not test the performance of Ordered Bound Representation due to its similarity to Unordered Bound Representation.

Here, the initial proportions of intervals in each region of the population match well the theoretically predicted values for initialization. In these experiments, proportions of intervals in each region are measured with reference to the macroclassifier population.

By observing the proportions of intervals in the population occupying the four regions, it is possible to gain some insight into the dynamics occurring as XCS solves the problem. Notice in Fig. 14 (Centre-Spread Representation) how the proportion of each region diverges from the initial value of 0.25. Compare this to Fig. 15 (Unordered Bound Representation), where the proportions converge from values of 0 (regions 2 and 3) and 1 (region 1).

The expected proportions of each region may be calculated for the checkerboard problem with $n = 3$ and $n_d = 3$ by counting the number of hypercubes at the corners, edges, faces and centre of the solution space. Each of these types of hypercube is represented by a specific combination of interval regions. For example, a hypercube at a corner of the solution space is represented by three region 2 or 3 intervals, while a hypercube at an edge is represented by two region 2 or 3 intervals and one region 1 interval. From this, we find that the expected proportion of each of region 1, 2 and 3 classifiers is $\frac{1}{3}$. It is clear from Fig. 15 that although a solution to the problem appears to have been found, the proportion of region 1 classifiers is too high, whereas the proportion of region 2 and 3 classifiers is too low. This is because, apart from a low probability of mutation, the only pressure towards generalization is that provided by the environment

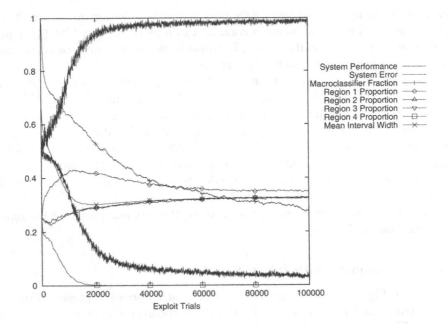

Fig. 14. Checkerboard problem with Centre-Spread Representation, an initial population, standard cover with $s_0 = 1$, 2-point crossover 'within' and standard mutation

via cover spread. With an environment that presents uniform random values, as classifiers become more general (i.e., have wider intervals), the probability of encountering an environmental input that is outside of an existing interval's range becomes lower and asymptotically approaches zero. Generalization pressure is thus variable and diminishes as XCS gets closer to solving the problem. As such, the mean interval width of intervals in the population at the end of the runs (0.329) is very close to, but not exactly, the expected mean interval width of 0.333.

Although the dynamics of execution may differ, there is no great difference in System Performance and System Error between representations. We found that the presence of an initial population tended to mask the differences between representations. For this reason, we now focus on comparing representations using experiments without an initial population.

9.3   Checkerboard with no Initial Population

Fig. 16 and Fig. 17 show the same experiments with no initial population. Again, these results show initial proportions of intervals in each region close to the predicted values for the cover operator, given the small sample size due to the empty initial population.

It is immediately apparent that XCS with Centre-Spread Representation makes no inroads towards solving the problem, whereas XCS with Unordered

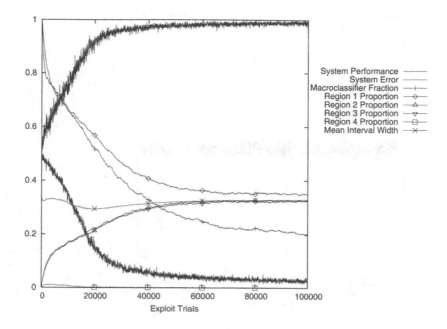

Fig. 15. Checkerboard problem with Unordered Bound Representation, an initial population, standard cover with $s_0 = 1$, 2-point crossover 'within' and standard mutation

Bound Representation comes much closer. This is due to the abnormally high number of region 4 intervals and low number of region 1 intervals in the population with Centre-Spread Representation. For Centre-Spread Representation, covering is used only during the first 50 trials, during which time the number of region 4 (maximally general) intervals rises in the population. For the remaining trials the region 4 intervals in the population cover all environmental inputs and covering is unnecessary. These intervals take over the population and stall the search. In contrast, the search using Unordered Bound Representation makes progress from the start, correctly promoting region 1 intervals at the expense of those in region 4. In addition, the proportion of region 2 and 3 intervals after 100,000 exploit trials (0.356) is similar to the expected proportion of 0.33, suggesting that many of the cubes at the boundaries of the solution space have been identified. As suggested in Section 8.3, it appears that the bias of the Unordered Bound Representation operators and parameter settings better match the type of intervals needed to solve the problem than those of Centre-Spread Representation. In fact, covering generates region 2 and 3 intervals with a probability of $\frac{1}{3}$, which is exactly the right proportion required for the solution to the problem.

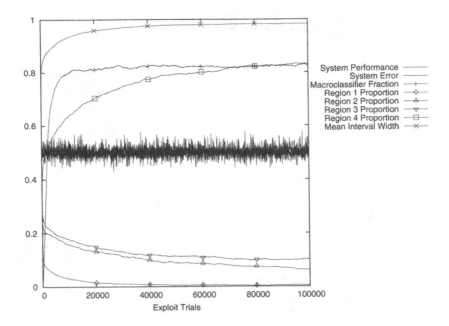

Fig. 16. Checkerboard problem with Centre-Spread Representation, standard cover with $s_0 = 1$, 2-point crossover 'within' and standard mutation

10   Comparing Representations and Operators

10.1   Background

We have seen that the choice of representation can make a large performance difference even when using the same system parameters. This difference can only arise from the action of the operators working on the representation. In order to isolate the reasons for any performance differences, we must systematically constrain operators to behave identically for both representations.

Without an initial population, the operators responsible for any performance differences are covering, crossover and mutation. GA subsumption is performed at the level of the phenotype, so no performance differences can arise from this operator.

For Unordered Bound Representation, truncation during covering and mutation occurs on the alleles representing the lower and upper bound of the interval. This means that an interval in Unordered Bound Representation is limited to $[p_{min}, q_{max})_p$. When Centre-Spread Representation is used, it is the centre and spread alleles that are truncated during covering and mutation. Therefore, for Centre-Spread Representation, intervals in the underlying population are in the range $[2p_{min} - q_{max}, 2q_{max} - p_{min})_p$ and further truncation must be applied upon expression to limit these intervals to $[p_{min}, q_{max})_p$. The cover, crossover and mutation operators all work at a genotypic level, so in the case of Centre-Spread

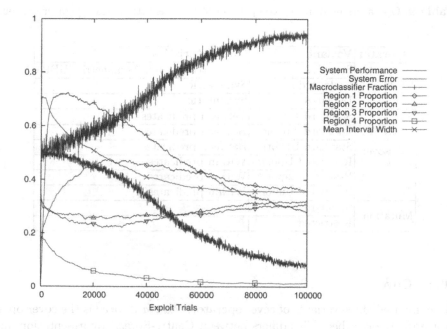

Fig. 17. Checkerboard problem with Unordered Bound Representation, standard cover with $s_0 = 1$, 2-point crossover 'within' and standard mutation

Representation, it is possible for intervals to be maintained in the population that are outside the range of the phenotype, but which are available for crossover and mutation to manipulate, and from which 'useful' intervals within range may subsequently emerge. This feature is not available with Unordered Bound Representation, where all intervals in the population are restricted to the range of the phenotype.

To design operators with identical characteristics for both representations, we need to limit the range of intervals in the population to that of the solution space. We thus refer to these as *restricted* operators. To test the restricted operators and verify that XCS behaves identically when using them, we ran experiments using restricted cover, no crossover and restricted mutation with both representations on the 6-bit real multiplexer and $n = 3$, $n_d = 3$ checkerboard problem. These showed the same results and population dynamics for both representations (not shown).

To compare the effects of different operator choices, we performed extensive experimentation using the Centre-Spread Representation and Unordered Bound Representation operators and variants. These are listed in Table 9 and described in more detail in the following sections. Space precludes detailed examination of every combination of representation, operator and problem, so we focus instead only on general trends and results of particular interest.

Table 9. Operators used during comparison of representations and operator variants

Operator	Variant	Characteristics
Cover	Standard	Symmetric (CSR), Asymmetric (UBR)
	Restricted	Symmetric
	Unbiased	Symmetric
Crossover	Standard 1-point	Between predicates
	Standard 2-point	Between predicates
	Standard Uniform	Between predicates
	Restricted 1-point	Within predicates
	Restricted 2-point	Within predicates
	Restricted Uniform	Within predicates
Mutation	Standard	
	Restricted	

10.2   Cover

We compared three variants of cover operator. *Standard cover* is the cover operator already described. This differs between Centre-Spread Representation and Unordered Bound Representation in two ways:

1. The Centre-Spread Representation cover operator is symmetric, since by definition, the spread must be equal on both sides of the centre. The Unordered Bound Representation cover operator, as presented, is asymmetric.
2. The Centre-Spread Representation cover operator generates intervals in the range $[2p_{min} - q_{max}, 2q_{max} - p_{min})_p$ while the Unordered Bound Representation cover operator generates intervals in the range $[p_{min}, q_{max})_p$.

Restricted cover is symmetric and generates intervals in the range $[p_{min}, q_{max})_p$ for both representations. The properties of restricted cover are the same as those described in Section 3.3. In the case of Unordered Bound Representation, the only change to the covering algorithm is to apply the same random spread to both sides of the environmental variable being covered. The algorithm for the Centre-Spread Representation restricted cover operator is more complex:

$$s_i = U[0, s_0)$$
$$l_i = EncodeandTruncate(x_i - s_i)$$
$$u_i = EncodeandTruncate(x_i + s_i)$$
$$s_i = \frac{(u_i - l_i)}{2} + (u_i - l_i) \bmod 2$$
$$c_i = l_i + s_i$$

The algorithm generates an interval as a lower and upper bound so that truncation occurs as for Unordered Bound Representation. It then converts the encoded interval back to an encoded centre and spread as needed for Centre-Spread Representation. The spread is incremented by one if it is an odd number to ensure that region 4 intervals are generated in the correct proportion. This is necessary because using a one of m binary encoding, the range of the maximally general interval $[p_{min}, q_{max})_p$ is always odd. It cannot be represented in Centre-Spread Representation without truncation, as only even ranges can be represented.

Unbiased cover is simply a variant of standard cover with a symmetric spread that is limited to $U[0, \min(x_i - p_{min}, q_{max} - x_i))_p$. This avoids the need for truncation, as the spread is limited to the bounds of the solution space.

We found performance differences between standard Unordered Bound Representation (asymmetric) cover and restricted (symmetric) cover with certain combinations of problem, operators and parameter settings. It is possible that such variation in performance arises simply because of the differing nature of the bias of the two types of cover, as seen in Section 3.3, Section 5.3 and below. Alternatively, it could be related to a bias caused by the fact that asymmetric cover chooses a different random spread for each side of the environmental state, whereas symmetric cover produces an interval that is (excepting truncation) centred on the environmental input. Further work is necessary to understand these performance differences in more detail.

We also examined the effect of variations of the cover spread parameter, s_0. We used a value of $s_0 = 0.5$ for these experiments. This value allows all possible intervals to be generated with the exception of the maximally general interval, but results in a minimal amount of truncation. Table 10 and Table 11 show the phenotype frequency matrices for symmetric and asymmetric cover with $s_0 = 0.5$. It can be seen that although there are no region 4 intervals generated, for both operators there is still an increased probability of region 2 and 3 intervals. Moreover, the most frequent many to one $g \rightarrow p$ mappings in region 2 and 3 occur around the median values of p_i and q_i with the frequencies ramping up to these values from the solution bounds. This means that covering is more likely to generate region 2 and 3 intervals with ranges around the median than those with very large or small ranges. In addition, the asymmetric cover operator shows a similar effect for region 1 intervals, which does not occur with the symmetric cover operator.

For both representations, the smaller cover spread was an advantage for the checkerboard problem, but produced poorer performance on the real multiplexer problem. This difference arises because of the need for maximally general intervals in the real multiplexer problem that is not present in the checkerboard problem. If the cover operator is able to generate intervals in region 4, this aids XCS in solving the real multiplexer problem. In contrast it is a handicap for the checkerboard problem, where no region 4 intervals are necessary to solve the problem.

The performance difference obtained by simply altering the cover spread parameter can be quite spectacular. Fig. 18 shows the performance of XCS on the checkerboard problem with Centre-Spread Representation and $s_0 = 0.5$. A comparison of these results with those of Fig. 16 reveals a major difference in performance, yet the only parameter change was to alter the cover spread

Table 10. Phenotype frequency matrix for symmetric cover (Centre-Spread Representation and Unordered Bound Representation) with $k = 3$ and $s_0 = 0.5$

		q_i							
		0	1	2	3	4	5	6	7
	0	1	1	2	2	2	1	1	0
	1		1	0	1	0	1	0	1
	2			1	0	1	0	1	1
p_i	3				1	0	1	0	2
	4					1	0	1	2
	5						1	0	2
	6							1	1
	7								1

Table 11. Phenotype frequency matrix for asymmetric cover (Unordered Bound Representation) with $k = 3$ and $s_0 = 0.5$

		q_i							
		0	1	2	3	4	5	6	7
	0	4	7	9	10	6	3	1	0
	1		1	2	3	4	3	2	1
	2			1	2	3	4	3	3
p_i	3				1	2	3	4	6
	4					1	2	3	10
	5						1	2	9
	6							1	7
	7								4

from 1 to 0.5. The two sets of results show very different dynamics with respect to the evolution of the proportions of intervals of each region in the population. Although the performance differences between different values of cover spread are not always so great, the value of the cover spread does have a significant effect on system dynamics and, ultimately, on performance. For reference and comparison with Fig. 17, Fig. 19 shows the results for Unordered Bound Representation with $s_0 = 0.5$.

Unbiased cover showed similar effects with both representations. Whilst its performance on the checkerboard problem (Fig. 20) was better than covering with $s_0 = 1$, it proved totally unsuitable for the real multiplexer problem (Fig. 21). In Fig. 20 it is possible to see how the proportion of region 1 intervals starts at 100% and then decreases as region 2 and 3 intervals are discovered. This happens only very slowly for the real multiplexer. Here, the proportion of region 2, 3 and 4 intervals needed to solve the problem is very low, even after 20,000 runs, when the problem would have been solved with a biased cover operator (Fig. 1).

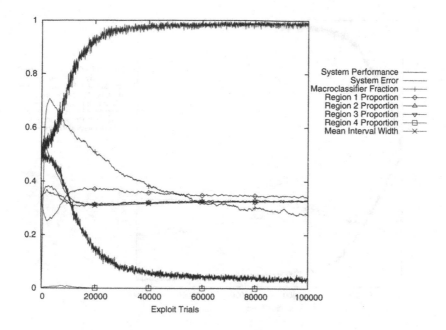

Fig. 18. Checkerboard problem with Centre-Spread Representation, standard cover with $s_0 = 0.5$, 2-point crossover 'within' and standard mutation

10.3   Crossover

As crossover is so tightly coupled with the representation, it is difficult to provide a restricted crossover operator that behaves identically for both Centre-Spread Representation and Unordered Bound Representation. However, if crossover operates only between predicates, it manipulates entire intervals and the underlying representation should be irrelevant. In this case, no performance difference is to be expected between representations. We refer to this as crossover *between* predicates. The standard crossover operators for Centre-Spread Representation and Unordered Bound Representation work *within* predicates, where the crossover point may be between any two alleles. As well as minimizing any differences due to representation, crossover between predicates allows us to see the benefits or otherwise compared to crossover within predicates.

We experimented with 1-point, 2-point and uniform crossover operators, both within and between predicates. These experiments were performed with restricted cover and restricted mutation to minimize differences between representations due to cover and mutation. Experiments were performed with both $s_0 = 1$ and $s_0 = 0.5$.

In general, we found little to choose between Centre-Spread Representation and Unordered Bound Representation, except on the checkerboard problem with $s_0 = 1$, where Unordered Bound Representation crossover within predicates produced consistently better results than Centre-Spread Representation crossover

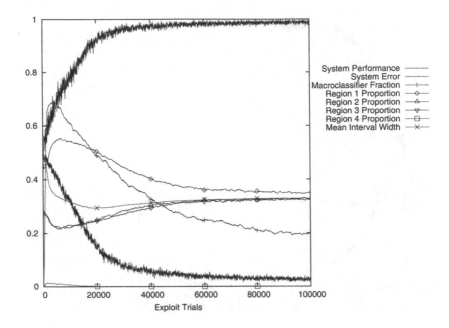

Fig. 19. Checkerboard problem with Unordered Bound Representation, standard cover with $s_0 = 0.5$, 2-point crossover 'within' and standard mutation

within predicates. We attribute this to the nature of the intervals produced by covering and the bias of crossover within intervals. With $s_0 = 1$, a relatively high proportion of region 4 intervals are introduced into the population. Crossover within predicates for Centre-Spread Representation does not materially affect this proportion (Sections 3.3 and 6.2) and the proportion of region 4 intervals remains high. In contrast, the bias of crossover within predicates for Unordered Bound Representation (Sections 5.3 and 6.2) reduces the proportion of region 4 intervals in the population to a small amount so that classifiers with maximally general intervals are unable to dominate action sets. Inspection of the results shows that the proportion of region 4 intervals is higher for Centre-Spread Representation (Fig. 22) than that of Unordered Bound Representation (Fig. 23) and that this is at the expense of the proportion of region 1, 2 and 3, which are needed to solve the problem.

We found that crossover between predicates tended to produce better results than crossover within predicates for the real multiplexer problem, but that the converse was true for the checkerboard problem (not shown). These results occurred for both Centre-Spread Representation and Unordered Bound Representation with settings of $s_0 = 0.5$ and $s_0 = 1$. In all cases examined, performance correlated with the ability of the operators to generate or remove from the population region 1 and 4 intervals as needed by the problem. Proportions of region 2 and 3 intervals appear to be less critical to performance. As discussed in Section 6.2, it is possible that the recombination of centres and spreads is

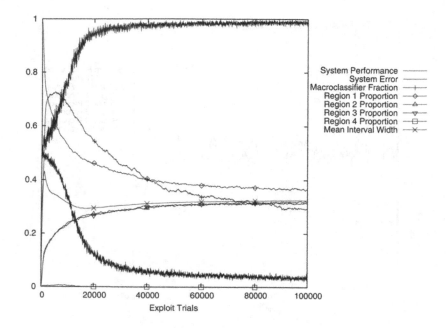

Fig. 20. Checkerboard problem with Centre-Spread Representation, unbiased cover, 2-point crossover 'within' and standard mutation

disruptive, as centre and spread alleles are mutually dependent, and further work is necessary to understand these results more fully.

10.4   Mutation

We used two types of mutation for the experiments. The first was the standard mutation operator already described. Although this is essentially the same algorithm for both representations, the alleles undergoing mutation differ for the two representations and the details of truncation differ between representations:

1. Mutation for Centre-Spread Representation creates a shift of the centre or a change in the size of the spread. Mutation for Unordered Bound Representation changes the value of the lower or upper bound.
2. The Centre-Spread Representation mutation operator generates intervals in the range $[2p_{min} - q_{max}, 2q_{max} - p_{min})_p$ while the Unordered Bound Representation mutation operator generates intervals in the range $[p_{min}, q_{max})_p$. This difference is apparent when the neutrality of the Centre-Spread Representation mutation operator with respect to region (Fig. 3) is compared with the bias of the Unordered Bound Representation mutation operator (Fig. 7).

To allow comparison between representations, we also implemented a restricted mutation operator. This mutates the effective centre or spread as per

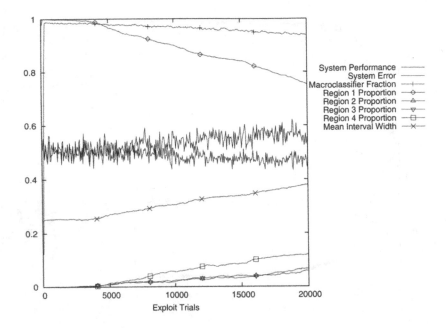

Fig. 21. 6-bit real multiplexer with Centre-Spread Representation, unbiased cover, 2-point crossover 'within' and standard mutation

Centre-Spread Representation, but limits the resulting lower and upper bounds as per Unordered Bound Representation. For Unordered Bound Representation, this means that both alleles in an interval predicate are altered upon each mutation. Restricted mutation was used only to allow meaningful comparisons between variants of cover and crossover and was not intended for comparison with the standard mutation operator.

We compared the standard mutation operators for Centre-Spread Representation and Unordered Bound Representation. In these experiments, mutation operated in conjunction with the restricted cover and restricted crossover operators. However, we found no evidence suggesting that one mutation operator was superior to the other. Indeed, it is shown in [20] that varying the mutation rate has little effect on XCS' performance at solving the 6-bit real multiplexer with either Centre-Spread Representation or Ordered Bound Representation.

11   Conclusions

We showed that the Centre-Spread representation has a many to one $g \rightarrow p$ mapping that affects the proportions of intervals in the population. As a result, operators typically used with this representation provide bias in the intervals they generate. This bias is caused by the need to truncate both the interval itself (during gene expression) and the alleles representing the interval (within operators) to allow only legal ranges to be produced. If the solution space was

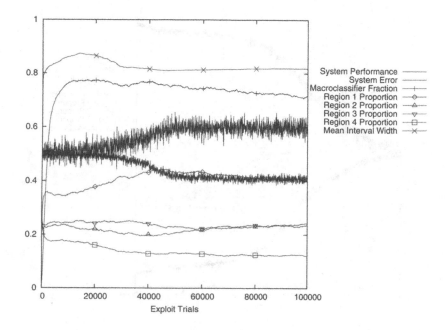

Fig. 22. Checkerboard problem with Centre-Spread Representation, restricted cover with $s_0 = 1$, 2-point crossover 'within' and restricted mutation

unbounded, truncation would be unnecessary and no such bias would exist. We have not yet experimented with unbounded solution spaces.

Ordered Bound Representation has a one to one $g \rightarrow p$ mapping, but the need for truncation in its operators still causes bias. The ordering requirement within tuples with this representation motivated us to introduce a new representation, Unordered Bound Representation, which obviates problems caused by the ordering requirement, yet retains the desirable features of Ordered Bound Representation.

We hypothesized that such representational and operator bias aids the solution of the real multiplexer problem because the intervals favoured by the bias correspond closely to those needed for the solution to the real multiplexer problem. Consequently, we introduced a new test problem for continuous-valued domains, the checkerboard problem, which has a solution that is not closely correlated with the biased intervals and which matches more closely that of real-world problems. The checkerboard problem typically showed performance differences between operators and representations better than the real multiplexer.

Testing with two representations and different variants of the standard cover operator showed that the type and amount of bias introduced by the representation and operators used does affect the performance of XCS. In particular, the spread parameter of the cover operator can make a huge difference in performance, because this parameter acts as a control over the distribution of intervals

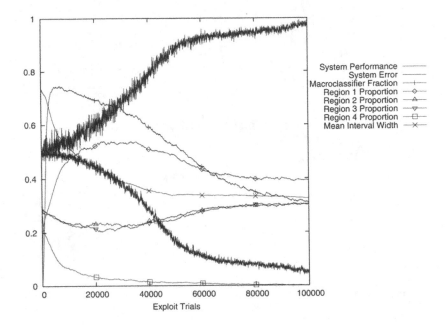

Fig. 23. Checkerboard problem with Unordered Bound Representation, restricted cover with $s_0 = 1$, 2-point crossover 'within' and restricted mutation

introduced into the population. One idea that we have not yet tried is to augment the cover and mutation operators with an explicit mechanism to introduce maximally general intervals into the population in a probabilistic manner similar to that used by a ternary representation. This may allow more control over this aspect of the distribution of intervals in the population.

In general, our experimental results support the hypothesis that representation and operators aid performance by generating intervals that are useful to solve the problem. As a result, representation and operators must be matched to the problem at hand in order to achieve the best results. If this does not occur, the Learning Classifier System may not be able to solve the problem. These results have similarities with those reported in [8] for XCS with a discrete representation with respect to the impact of generalization upon system performance.

We also found that sampling bias affects system performance. It is possible to solve the real multiplexer problem when $\theta_i = 0.75$ in the same number of trials as for $\theta_i = 0.5$ by sampling solution intervals with equal frequency. Bias caused by unbalanced training examples is a well-known problem in machine learning [5].

In general these results and conclusions apply to any Learning Classifier System architecture using the representations studied. In particular, all of the analysis of representation and operator bias is applicable to any architecture. However, the outcome of these biases with architectures that do not build a complete environmental map may not correspond to those seen here for XCS. This is because the arguments presented in Section 8 relating to the proportion of hyper-rectangles at the solution boundary do not apply unless a complete map is built. In architectures where this does not occur, the relative desirability of intervals will differ from that seen for XCS and performance may be affected in a different manner.

Bibliography

[1] M. Ahluwalia and L. Bull. A Genetic Programming-based classifier system. In Banzhaf et al. [2], pages 11–18.

[2] W. Banzhaf, J. Daida, A. E. Eiben, M. H. Garzon, V. Honavar, M. Jakiela, and R. E. Smith, editors. *GECCO-99: Proceedings of the Genetic and Evolutionary Computation Conference*, San Francisco, CA, 1999. Morgan Kaufmann.

[3] A. Bonarini. An introduction to Learning Fuzzy Classifier Systems. In Lanzi et al. [16], pages 83–104.

[4] L. Booker. Representing attribute-based concepts in a classifier system. In G. J. E. Rawlins, editor, *Proceedings of the First Workshop on Foundations of Genetic Algorithms*, pages 115–127, San Mateo, CA, 1991. Morgan Kaufmann.

[5] L. Breiman, J. H. Friedman, R. A. Olshen, and C. J. Stone. *Classification and Regression Trees*. Chapman & Hall, New York, NY, 1984.

[6] L. Bull and T. O'Hara. Accuracy-based neuro and neuro-fuzzy Classifier Systems. In W. B. Langdon, E. Cantú-Paz, K. Mathias, R. Roy, D. Davis, R. Poli, K. Balakrishnan, V. Honavar, G. Rudolph, J. Wegener, L. Bull, M. A. Potter, A. C. Schultz, J. F. Miller, E. Burke, and N. Jonoska, editors, *GECCO-2002: Proceedings of the Genetic and Evolutionary Computation Conference*, pages 905–911, San Francisco, CA, 2002. Morgan Kaufmann.

[7] L. Bull, D. Wyatt, and I. Parmee. Initial modifications to XCS for use in Interactive Evolutionary Design. In Merelo et al. [18], pages 568–577.

[8] M. V. Butz, T. Kovacs, P. L. Lanzi, and S. W. Wilson. Theory of generalization and learning in XCS. Technical Report 2002011, Illinois Genetic Algorithms Laboratory, University of Illinois at Urbana-Champaign, May 2002.

[9] M. V. Butz and S. W. Wilson. An algorithmic description of XCS. In Lanzi et al. [17], pages 253–272.

[10] J. H. Holland. *Adaptation in Natural and Artificial Systems*. The University of Michigan Press. Republished by MIT Press, 1992, Ann Arbor, MI, 1975.

[11] J. H. Holland. Escaping brittleness: The possibilities of general-purpose learning algorithms applied to parallel rule-based systems. In R. S. Michalski, J. G. Carbonell, and T. M. Mitchell, editors, *Machine Learning, an Artificial Intelligence Approach. Volume II*, pages 593–623, Los Altos, CA, 1986. Morgan Kauffmann.

[12] J. Hurst, L. Bull, and C. Melhuish. TCS Learning Classifier System controller on a real robot. In Merelo et al. [18], pages 588–600.

[13] T. Kovacs. Evolving optimal populations with XCS classifier systems. Master's thesis, School of Computer Science, University of Birmingham, Birmingham, UK, 1996. Also Technical Report CSR-96-17 and CSRP-96-17, School of Computer Science, University of Birmingham.

[14] T. Kovacs and M. Kerber. What makes a problem hard for XCS? In Lanzi et al. [17], pages 80–99.

[15] P. L. Lanzi. Extending the representation of classifier conditions, part ii: From messy coding to s-expressions. In Banzhaf et al. [2], pages 345–352.

[16] P. L. Lanzi, W. Stolzmann, and S. W. Wilson, editors. *Learning Classifier Systems. From Foundations to Applications*, volume LNAI-1813 of *Lecture Notes in Artificial Intelligence*, Berlin, 2000. Springer.

[17] P. L. Lanzi, W. Stolzmann, and S. W. Wilson, editors. *Advances in Learning Classifier Systems. Proceedings of the Third International Workshop (IWLCS-2000)*, volume LNAI-1996 of *Lecture Notes in Artificial Intelligence*, Berlin, 2001. Springer.

[18] J. Merelo, P. Adamidis, H.-G. Beyer, J.-L. Fernandez-Villacanas, and H.-P. Schwefel, editors. *Parallel Problem Solving from Nature - PPSN VII*, Berlin, 2002. Springer.

[19] S. Salzberg. A nearest hyperrectangle learning method. *Machine Learning*, 6(3):251–276, 1991.

[20] A. Wada, K. Takadama, K. Shimohara, and O. Katai. Analyzing parameter sensitivity and classifier representations for real-valued XCS. Technical Report NIS-0001, ATR, 2004.

[21] S. W. Wilson. ZCS: A zeroth order classifier system. *Evolutionary Computation*, 2(1):1–18, 1994.

[22] S. W. Wilson. Classifier fitness based on accuracy. *Evolutionary Computation*, 3(2):149–175, 1995.

[23] S. W. Wilson. Get real! XCS with continuous-valued inputs. In Lanzi et al. [16], pages 209–219.

[24] S. W. Wilson. Function approximation with a classifier system. In L. Spector, E. D. Goodman, A. Wu, W. B. Langdon, H.-M. Voigt, M. Gen, S. Sen, M. Dorigo, S. Pezeshk, M. H. Garzon, and E. Burke, editors, *Proceedings of the Genetic and Evolutionary Computation Conference, GECCO-2001*, pages 974–981, San Francisco, CA, 2001. Morgan Kaufmann.

[25] S. W. Wilson. Mining oblique data with XCS. In Lanzi et al. [17], pages 158–174.

Section II

Credit Assignment

Reinforcement Learning: A Brief Overview

Jeremy Wyatt

School of Computer Science, University of Birmingham, Edgbaston, Birmingham, B15 2TT, jlw@cs.bham.ac.uk

1 Introduction

Learning techniques can be usefully grouped by the type of feedback that is available to the learner. A commonly drawn distinction is that between *supervised* and *unsupervised* techniques. In supervised learning a teacher gives the learner the correct answers for each input example. The task of the learner is to infer a function which returns the correct answers for these exemplars while generalising well to new data. In unsupervised learning the learner's task is to capture and summarise regularities present in the input examples. Reinforcement learning (RL) problems fall somewhere between these two by giving not the correct response, but an indication of how good an response is. The learner's task in this framework is to learn to produce repsonses that maximise goodness.

Most commonly in RL we are concerned with an *agent* acting in an *environment*, where the principal form of feedback is a measure of *immediate* performance, and the goal of the agent is to learn to act so as to maximise some *long term* measure of performance based on this. There are further important differences between this and typical problems in supervised or unsupervised learning. Because the agent is selecting actions *while* it is learning the actions it chooses affect the learning experiences it will have in the future. Furthermore, the outcomes of actions are not certain so that the agent cannot select its next experience, only influence it. We concentrate on RL problems with both these characteristics in this chapter. The RL framework as described here is suited in some respects to studying problems properly characterised as involving on-going interaction: such as those in robotics, animal learning, optimal foraging, and optimal learning.

There is a comprehensive body of mathematics for modelling stochastic interactions between agents and their environments. It is this that underpins current work in RL and while it has been set out previously elsewhere (see [40, 15, 24, 6]) this chapter summarises the main results and algorithms. The environment-agent interaction is typically modelled as a Markov decision process (MDP), or a partially observable MDP (POMDP) in which the agent observes and controls the process. I shall describe methods for prediction and control in both known and unknown MDPs. The prediction problem is the problem of inferring the long term behaviour of the process in terms of reward, and the control problem is that in which we must determine which actions maximise the agent's performance. Solution methods can be seen as falling into three categories, policy modification techniques, value function based techniques, and model based techniques. I have not made any assumptions in terms of mathematical knowledge other than a grasp of basic probability theory.

2   Markov Processes

Markov Processes are a form of stochastic process, and a stochastic process is simply a sequence of random events. Stochastic processes can be used to model many phenomena: the motion of particles in a liquid or gas; the fluctuations of the stock market; the motion of a robot; or the sequence of moves in a game of chance like backgammon or cards. We are concerned here with random processes that evolve in discrete time, and which have a countable number of outcomes. As an example let us imagine a frog in a pond full of lily pads. The lily pads are the outcomes (or states) of our process. The frog hops from pad to pad at regular intervals, and which pad it jumps to next is uncertain. We can describe this mathematically.

In a discrete stochastic process like this we take the random variable X_t to denote the outcome at the tth stage or time step. The stochastic process is defined by the set of random variables $\{X_t, t \in T\}$, where $T = \{0, 1, 2 \ldots\}$ is the set of possible times. The domain of X_t is the set of possible outcomes denoted $S = \{s_1, s_2, \ldots s_N\}$. In the general case the outcome at time t is dependent on the prior sequence of outcomes $x_0, x_1, \ldots x_{t-1}$. The likelihood of the outcome at time t being s_j is therefore written:

$$\Pr(X_t = s_j | x_{t-1} \wedge x_{t-2} \wedge \ldots \wedge x_0) \tag{1}$$

A process can be said to be an *independent process* if the outcome at each time t is independent of the outcomes at all prior stages:

$$\Pr(X_t = s_j | x_{t-1} \wedge x_{t-2} \wedge \ldots \wedge x_0) = \Pr(X_t = s_j) \tag{2}$$

A Markov process weakens this independence assumption minimally by requiring that the outcome at time t is independent of all events prior to $t - 1$:

$$\Pr(X_t = s_j | (X_{t-1} = s_i) \wedge x_{t-2} \wedge \ldots \wedge x_0) = \Pr(X_t = s_j | X_{t-1} = s_i) \tag{3}$$

Equation 3 is known as the *Markov property*. The probability $\Pr(X_t = s_j | X_{t-1} = s_i)$ can be regarded as a transition probability from the outcome s_i at $t - 1$ to the outcome s_j at time t, denoted by $s_i \rightsquigarrow s_j$. If the transition probabilities are independent of time then the process is a Markov chain. The possible outcomes are referred to as the *states* of the process. We use the following shorthand to denote the probability of the transition from state $s_i \rightsquigarrow s_j$:

$$p_{ij} = \Pr(X_t = s_j | X_{t-1} = s_i) \tag{4}$$

Given the current state of a Markov chain and its transition probabilities we can predict its behaviour any number of steps into the future. The transition probabilities are represented in the form of a transition matrix, $\mathbf{P}$, the i, j^{th}

element of which is p_{ij}. We also define a probability distribution on the initial states (i.e. when $t = 0$), denoted by the row vector $\mathbf{x}_0 = [\Pr(X_0 = s_1), \Pr(X_0 = s_2), \ldots \Pr(X_0 = s_n)]$, where n is the number of states. I denote the probability distribution on S at time t by $\mathbf{x}_t$. Given $\mathbf{x}_0$ and $\mathbf{P}$, $\mathbf{x}_t$ can be expressed elegantly as the product:

$$\mathbf{x}_t = \mathbf{x}_0 \cdot \mathbf{P}^t \tag{5}$$

The significance of this is that the study of the state of the process n steps into the future is the study of the nth power of the transition matrix. It is worth noting for practical purposes that the notion of the future behaviour of the process being dependent solely on the current state of the process is a representational device. Processes whose future behaviour relies on knowing some or all of the process history can be made to satisfy the Markov property by including sufficient record of that history in the description of the current state. This may be expressed in the following manner. If the description of the state at time t is denoted by the column vector κ_t then we can denote the supplemented description of the current state by the concatenation of two vectors:

$$\kappa'_t = [\kappa_t^T, f(\kappa_{t-1} \ldots \kappa_{t-k})^T]^T \tag{6}$$

where T means transposition and $f(.)$ is a function summarising the process history in the form of a new vector from states as far back in time as necessary, here k steps. In many cases the additional information may not add excessively to the length of the state description. If for example, we wish to predict the trajectory of a ball thrown through the air, then we use first and second order derivatives of position to summarise the history of the process necessary for the prediction of the future. If we use this information to control a process then we say that the controller has state. One of the primary problems with optimization methods relying on the Markov assumption is that we do not always know how much information it is necessary to supplement the description of the current state with. This is referred to as the question of how much state to include in the controller. State that is not directly observable by an agent is referred to as *hidden state*.

It can, however, be seen that this ability in principle to represent *any* stochastic process as a Markov process is a potentially powerful one. The inferential power gained is achieved by the way the Markov property separates the past and the future. The *necessary* history of the process is encapsulated in the description of the current state and this state completely determines future behaviour. We will now outline Markov decision processes.

2.1 Markov Decision Processes

We have said previously that we are interested in problems where the agent can select an action at each step to influence the evolution of the process. To incorporate this the finite state, discrete-time Markov chain model needs to be extended by making the transition matrix at time t depend on an *action* a_t chosen at that time. The set of possible actions may vary from state to state so we write,

- $\mathcal{A}$ for the set of possible actions across all states
- $\mathcal{A}_x \subseteq \mathcal{A}$ for the set of actions allowable in state x.

The transition probabilities that depend on the action chosen are denoted $p_{ij}(a)$, where $a \in \mathcal{A}$. There are now m transition matrices (where the size of the set $\mathcal{A}$ is m), one for each action: $\mathbf{P}_a = [p_{ij}(a)]$. If an action a is not possible in a particular state s_i, then $p_{ij}(a) = 0$. We may regard the transition function $\mathbf{P}$ as a function specified by these m transition matrices, mapping from all possible pairs of states and actions into a probability distribution on the set of states. We denote the transition from s_i to s_j following the selection of action a in state s_i by $s_i \overset{a}{\rightsquigarrow} s_j$. Finally we define a *reinforcement function* R which in the most general case is defined as a mapping from the state, action and next state into a probability density over the set of possible rewards $\mathcal{R} \subseteq \Re$:

$$R : \mathcal{S} \times \mathcal{A} \times \mathcal{S} \times \mathcal{R} \rightarrow [0, 1]$$

At each observation, a *reward* is generated dependent upon the state, the action performed, and the next state. The random variable denoting the reward at time t is $R_t = R(x_t, a_t, x_{t+1})$. The actual reward generated during the transition $x_t \overset{a_t}{\rightsquigarrow} x_{t+1}$ is r_t. For some problems it is important to distinguish whether or not the reward function is known to the agent. Problems of optimal learning are, for example, simpler if the reward function is known to the agent.

Reward functions The simplest possible reward function is when the set of possible rewards is Boolean, $\mathcal{R} = \{0, 1\}$. In this case the reward model is termed a P-model. Any problem with well-defined criteria for success and failure can be represented as a P-model. If, for example, the aim of a process is to track a set-point ω, and a certain magnitude of error ε is acceptable then taking 1 to be success[1] the reward at time t is:

$$r_t = \begin{cases} 1 \text{ if } |\hat{\omega}_t - \omega| \leq \varepsilon \\ 0 \text{ otherwise} \end{cases}$$

where $\hat{\omega}_t$ is the system's approximation to the set point at time t. The minimal extension of this model is to allow any finite number of reward values in the interval $[0, 1]$. Such a model is termed a Q-model. Problems with real-valued rewards can be expressed in this form by means of normalisation and quantisation. The most general case is when the reward can take any real value in the interval $[0, 1]$. Such a reward model is termed an S-model. By normalisation any problem with bounded reward can be expressed as an S-model.

The reward function merely specifies the reward generated at a particular instant. Using the notion of immediate reward we can construct measures of performance over many time steps. Following Barto et al. [4], I refer to a measure of long-term reward as a measure of *return*. $\mathbf{R}_t$ is the random variable denoting

---

[1] Usually in Learning Automata Theory [24] 0 is taken to be success and 1 to be failure. Since the convention in RL is to maximise reward I reverse this for convenience.

return at time t. There are several measures of return investigated in the literature. All can be expressed in terms of a discount vector $G = \{\gamma_0, \gamma_1, \gamma_2, \ldots\}$, where the return at time t is:

$$\mathbf{R}_t = \sum_{k=0}^{\infty} \gamma_k r_{t+k} \tag{7}$$

The vector G may in principle be arbitrary, but in practice one of three *discount* schemes is used. The first is the *finite horizon* model of return, where the horizon is a finite number h of steps into the future:

$$\gamma_k = \begin{cases} 1 \text{ if } k \leq h \\ 0 \text{ if } k > h \end{cases}$$

This model has been studied extensively in the bandit literature.[2] Alternatively we may use the *average-reward* model [29, 31]:

$$\gamma_k = \begin{cases} 1/h, \text{ if } k \leq h \\ 0 \quad \text{ if } k > h \end{cases}$$

By far the most widely studied measure of return however, particularly within work on learning from delayed reinforcement, is an infinite horizon model termed the *geometric discount* model of return:

$$\gamma_k = \gamma^k, \text{ where } 0 \leq \gamma < 1$$

The value of γ chosen determines the relative weighting of short and long term rewards. As $\gamma \to 0$ short-term rewards become more important. When $\gamma = 0$ the only reward that matters is the immediate reward. As well as being attractive for its elegance this method has been shown to make certain problems in learning from reinforcement more tractable, e.g. bandit tasks[12]. The reason for this is that a geometric discount makes the decision problem the agent faces the same at each step, so that optimal policies are not time dependent. Here on I consider only the geometric discount model of return.

The importance of our definition of return is that we now have a model that enables us to take an immediate measure of performance and turn it into a long term measure of performance. It is not a trivial task to design a reward function that will give an appropriate return function for real world tasks. The assumption that this conversion of a short term measure into a long term measure is a useful thing to do lies at the heart of almost all modern work on reinforcement learning. Next we will consider ways an agent can act, and how we can say that one way of acting is better than another according to our measure of long term performance.

---

[2] A k-armed bandit is an MDP with a single state and k actions available. The actions generate stochastic rewards. A bandit problem is one in which the stochastic effects of the actions are unknown, and the learner must maximise its performance while it is learning. This is the simplest problem in the optimal learning literature, also known as the exploration-exploitation trade-off.

3 Policies and Optimal Policies

A policy π is a mapping that specifies the actions the agent takes in each state of the environment, and is thus the sole essential component of an agent. π says what to do in every possible state. Thus it can also be seen as a universal plan, i.e. a plan with no explicitly specified sequence of actions. A *stationary* policy specifies an action to be taken for each state, $\pi : \mathcal{S} \to \mathcal{A}$, this means that the action taken in a state is always the same. A *stochastic* policy is a mapping $\pi : \mathcal{S} \times \mathcal{A} \times \mathcal{S} \to [0, 1]$. π is specified by a stochastic matrix, where the i, j^{th} element of π is,

$$\pi_{ij} = \Pr(a_j | s_i), \text{ and } \sum_j \pi_{ij} = 1, \forall i$$

Under a stochastic policy the action an agent selects in a given state may vary (hence it is not stationary), but is always selected according to the same distribution. A *non-stationary* policy is one in which the policy is indexed by time. In finite horizon models of reward optimal policies are typically non-stationary. All the agents we are interested in modify π directly or indirectly as a function of their experience.

For our purposes it is obviously necessary to be able to order policies according to some index of performance. The reward models I have discussed can be used to derive just such an ordering on the set of possible policies. A policy π_1 is said to be at least as good as a policy π_2 if it has an expected return which is greater than or equal to that of policy π_2 in each of the possible initial states of the process:

$$E[\mathbf{R}(s_i) | \pi_1] \geq E[\mathbf{R}(s_i) | \pi_2], \forall s_i \in \mathcal{S}_0 \Leftrightarrow \pi_1 \succeq \pi_2 \tag{8}$$

where $a \succeq b$ is a preference operator, indicating that a is at least as good as b; and $\mathcal{S}_0 \in \mathcal{S}$ is the set of possible initial states of the process. An *optimal* policy is defined as any policy π^* which is as least as good as any other policy,

$$E[\mathbf{R}(s_i) | \pi^*] \geq E[\mathbf{R}(s_i) | \pi_j], \forall s_i \in \mathcal{S}_0, \pi_j \in \Pi \Leftrightarrow \pi^* \succeq \pi_j, \forall \pi_j \in \Pi \tag{9}$$

where Π is the set of possible policies. It is an important result that for both the finite horizon and geometrically discounted models of return there is at least one stationary policy that is optimal for any completely observable MDP [6]. There may be more than one optimal policy, and so I will denote the set of optimal policies Π^*. Some Partially Observable MDPs (POMDPs) which have no optimal stationary policy have an optimal stochastic policy. For other POMDPs the optimal policy is non-stationary [32]. The aim of any policy-modifying agent is to converge to an optimal policy. Our aim is to design learning agents which converge to an optimal policy quickly and reliably. The first step in designing such agents is to be able to estimate the expected return for a given policy in order that we may compare policies by Equations 8 and 9.

4   Prediction

The prediction problem is concerned with estimating the mean goodness of each state of environment given that we follow a certain policy. Given a policy π, a transition function $\mathbf{P}$, and a reward function R for a Markov decision process we can calculate the expected return. Before discussing this we need some additional notation. The random variable $R^\pi(s_i, n)$ denotes the reward received on the n-th step after starting in s_i and following policy π for n steps. This random variable captures the stochastic effects of the MDP up to n steps into the future. The *value* $V^\pi(s_i)$ of policy π in state s_i is the expected return under that policy. Hence $V^\pi(s_i)$ can be written:

$$V^\pi(s_i) = E[R^\pi(s_i, 1) + \gamma R^\pi(s_i, 2) + \gamma^2 R^\pi(s_i, 3) + \ldots + \gamma^n R^\pi(s_i, n) + \ldots]$$

Where $E[X]$ is the expectation of the random variable X. This can be expressed recursively for all $s_i \in \mathcal{S}$:

$$V^\pi(s_i) = E[R^\pi(s_i, 1)] + \gamma \sum_j p_{ij}(\pi(s_i)) V^\pi(s_j), \forall s_i \in \mathcal{S} \qquad (10)$$

The $V^\pi(s_i)$ for all s_i define the *value function* under the policy π. If we know $E[R^\pi(s_i, 1)]$ as well as $\mathbf{P}$ then the value function can be calculated off-line by solving this set of linear equations. Probably the most widely used technique is some form of dynamic programming. We will return to describe this in detail in Section 5.

Methods employing a known transition function $\mathbf{P}$ to derive the value function are commonly termed *model-based* methods for predicting the value of a policy. To be more accurate we refer to models that explicitly represent the probability distributions over outcomes as *distribution models*. If the agent does not possess a distribution model in advance it may estimate such a model from its own experience as it proceeds through the environment. We can then use the estimated model to estimate the value function. Such estimates are often referred to as *certainty equivalent* estimates, as they assume that the model is essentially correct. A simple approach is to pick the maximum likelihood estimates of the model parameters. The certainty equivalent value function constructed from these will be the maximum likelihood value function. Methods that estimate the value function using either learned or a priori models are also referred to as *indirect methods*.

If the agent has neither $E[R^\pi(s_i, 1)]$ nor $\mathbf{P}$ and we do not want to learn a model then we can use a *direct* or *model-free* method for predicting the value of the policy on-line. Model-free methods build an estimate of the value function directly from their experience, i.e. from the sequence of perceived states and rewards generated. There are two classes of direct methods that have been carefully studied: simple Monte-Carlo methods, and temporal difference methods.

In basic Monte-Carlo approaches we sample a sequence of observations and rewards from the world, and calculate the actual return from each state. Many such samples are taken, and we then calculate the average return from each state over those samples. Such methods are often simply referred to as Monte-Carlo methods. There are two studied Monte-Carlo estimates of expected return: the every visit Monte-Carlo (EVMC) estimate and the first visit Monte-Carlo (FVMC) estimate [30]. The FVMC estimate has been shown to have a connection to the maximum likelihood certainty equivalent estimate of value for an MDP under a given policy. Simple Monte-Carlo methods are important in that they give a performance baseline from which to work, and aspects of them have been important in developing more sophisticated algorithms. In particular temporal difference algorithms can be seen as a combination of ideas from dynamic programming and Monte-Carlo methods. The difficulty with simple Monte-Carlo estimators is that their standard error declines very slowly as the sample size rises.

A well-known and elegant model-free method for estimating expected return for an MDP under a policy is Sutton's *temporal difference* method [36]. This works in roughly the following manner. Given that the transition $x_t \rightsquigarrow x_{t+1}$ occurs, the reward r_t is received during the course of this transition. At time t we have estimates of the value of each state, $\hat{V}_t(x_t)$ and $\hat{V}_t(x_{t+1})$, where we have dropped the explicit reference to the policy π in our notation. It turns out that a better estimate of $V(x_t)$ than $\hat{V}_t(x_t)$ can be provided by:

$$r_t + \gamma \hat{V}_t(x_{t+1})$$

The temporal difference is the difference between these two estimates:

$$r_t + \gamma \hat{V}_t(x_{t+1}) - \hat{V}_t(x_t)$$

The basic temporal difference equation uses this to update the estimate of $V(x_t)$ each time the transition is made:

$$\hat{V}_{t+1}(x_t) = \hat{V}_t(x_t) + \alpha_t[r_t + \gamma \hat{V}_t(x_{t+1}) - \hat{V}_t(x_t)] \tag{11}$$

$0 < \alpha_t \leq 1$, is the learning rate at time t. As $\alpha_t \to 1$ so $\hat{V}_{t+1}(x_t)$ depends more on $r_t + \gamma \hat{V}_t(x_{t+1})$ and less on $\hat{V}_t(x_t)$. α_t acts as a filter damping the variance in $r_t + \gamma \hat{V}_t(x_{t+1})$. As $t \to \infty$ the estimates $\hat{V}_t(x_t)$ are guaranteed to converge to $V_t(x_t)$ if it is the case that $\sum_{t=0}^{\infty} \alpha_t = \infty$; that $\sum_{t=0}^{\infty} \alpha_t^2 < \infty$; and that $\alpha_t > 0$, $\forall t \in T$. Equation 11 forms the basis of the TD(0) algorithm which is specified in Figure 2.

Updating $V(x)$ only on making a transition from x is comparatively inefficient. Because $V(x)$ depends to some extent on $V(y)$ for all y which can be reached eventually from x, $\hat{V}(x)$ may not only be updated when it occurs, but also on the basis of the temporal difference for *any* subsequent transition. Using this insight Sutton generalised TD(0) by defining a class of prediction algorithms called TD(λ) (see Figure 2). In TD(λ) the extent to which a change in $\hat{V}_t(x_{t+1})$ is mirrored in other states is determined by the value of a function $\bar{e}$ called an eligibility trace, defined on the domain $\mathcal{S}$. It is included in the temporal difference update equation as follows:

$$\hat{V}_{t+1}(s) = \hat{V}_t(s) + \alpha_t[r_t + \gamma \hat{V}_t(x_{t+1}) - \hat{V}_t(x_t)]\bar{e}_t(s), \forall s \in \mathcal{S} \tag{12}$$

The value of $\bar{e}_t(s)$ is updated each transition for all $s \in \mathcal{S}$. There are two forms of the update equation. An *accumulating* trace updates the eligibility of a state using:

$$\bar{e}_t(s) = \begin{cases} \gamma\lambda\bar{e}_{t-1}(s) + 1 & \text{if } s = x_t \\ \gamma\lambda\bar{e}_{t-1}(s) & \text{otherwise} \end{cases} \tag{13}$$

A *replacing* trace update is defined by:

$$\bar{e}_t(s) = \begin{cases} 1 & \text{if } s = x_t \\ \gamma\lambda\bar{e}_{t-1}(s) & \text{otherwise} \end{cases} \tag{14}$$

where $0 \le \lambda \le 1$ controls the rate of decay of the trace. Under both mechanisms the eligibility of a state decays away exponentially when the state is unvisited. Under an accumulating trace the eligibility is increased by a constant every time the state is visited, and under a replacing trace the eligibility is reset to a constant on each visit. The effect of each update rule on the eligibility of a state according to the frequency of visits is illustrated qualitatively in Figure 1. An eligibility

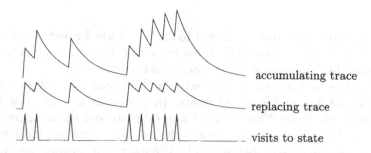

accumulating trace

replacing trace

visits to state

Fig. 1. Behaviour of accumulating and replacing traces.

trace can be thought of as a short-term memory process initiated the first time a state is visited by an agent. The degree of activation depends on the recency of the most recent visit and on the frequency of visits. Thus eligibility traces implement two heuristics, a *recency heuristic* and a *frequency heuristic*. These be can stated informally as saying that reinforcement received now is probably caused to a greater degree by more recently and frequently occurring states than by less recently and frequently occurring states. Accumulating traces implement both these heuristics, while replacing traces implement just the recency heuristic [35]. The rate of decay of the trace is determined by $0 \le \lambda \le 1$. Hence the class of algorithms defined is referred to as TD(λ). If $\lambda = 0$ then Equation 12 simplifies to 11. If $\lambda = 1$ then the estimate of $V(x_t)$ ignores the estimated values $\hat{V}(x_{t+k})$ of any subsequent states and is based entirely on the actual rewards received at each step. If a perceptron is being used for structural credit assignment then using TD(1) makes its updates equivalent to those of the Widrow-Hoff rule. In general high values of λ give fast initial convergence to $V(x)$, and low λ values give low standard error. Better performance than by using fixed λ is therefore obtained

---

Algorithm 1   TD(λ) $0 < \alpha \le 1$

$t := 0$

$\hat{V}_t(x) := 0,\ \bar{e}_0(x) := 0,\ \forall x \in \mathcal{S}$

repeat

    observe the transition $x_t \leadsto x_{t+1}$

    update $\bar{e}(s)$ for all $s \in \mathcal{S}$ according to Eq. 13 or 14

    update $\hat{V}(s)$ for all $s \in \mathcal{S}$ according to

$$\hat{V}_{t+1}(s) := \hat{V}_t(s) + \alpha[r_t + \gamma \hat{V}_t(x_{t+1}) - \hat{V}_t(x_t)]\bar{e}_t(s)$$

    $t := t + 1$

Fig. 2. The TD(λ) algorithm.

---

by declining the value of λ from 1 to 0. Procedures for declining λ systematically have been investigated [39]. These work well for acyclic processes, but do not appear likely to extend to the more general cyclic case.

It has been shown that replacing and accumulating traces converge to different estimates of expected return. In particular accumulating traces are related to the Every Visits Monte Carlo estimate and replacing traces are related to the First Visits Monte Carlo estimate [30]. Some empirical results suggest that replacing traces used with TD(λ) outperform accumulating traces [30]. Although the TD(λ) algorithm applies strictly to estimating the value function for a Markov chain, any Markov decision process under a given policy can be modelled as a Markov chain.

In summary I have reviewed methods for the estimation of the value function for any given policy based on dynamic programming; Monte Carlo methods; temporal difference learning and eligibility traces. It can be seen that in principle such techniques could be used to identify the optimal policy by estimating the value function for all possible policies. Such an approach would be grossly inefficient however, and there are much faster methods which I now discuss.

5   Control

If the prediction problem is concerned with estimating expected future performance the control problem is concerned with finding a policy that optimises our predicted performance. As for the problem of predicting the value function there are both model-based and model-free methods for computing the optimal policy for a Markov decision process. If the agent has a model then we can employ one of a number of dynamic programming techniques to achieve this. I shall initially discuss two of these, *policy iteration* and *value iteration*. These algorithms are suitable for derivation of optimal controllers off-line given accurate process models. They can however, be extended to work on-line employing an adaptive process model.

Policy iteration commences with an arbitrary policy f and calculates the value function V^f by solving the set of linear equations given by Equation 10. It then improves this policy and recalculates the value function, repeating this process until the optimal policy is reached. The key lies in the way f is improved each iteration. Consider a particular state s_i. Suppose the agent takes any action a in state s_i and follows the policy f elsewhere. The value of this modification to policy f is termed the *action-value* of the action a in state s_i with respect to policy f [40], denoted $Q^f(s_i, a)$:

$$Q^f(s_i, a) = E[R^a(s_i, 1)] + \gamma \sum_j p_{ij}(a) V^f(s_j) \tag{15}$$

We choose g as the policy which selects action a in state s_i and follows policy f in all other states, where:

$$a = \arg \max_a \{Q^f(s_i, a)\}$$

Because the set $\mathcal{A}_{s_i}$ includes the best action chosen by policy f we know that g must be at least as good as f. We can carry out this improvement process in all states simultaneously, so that:

$$V^g(x) \geq V^f(x), \forall x \in \mathcal{S}$$

Because a change in the policy at any single state can impact on the value function across the entire state space the entire value-function must then be recalculated for the new policy. This constitutes a single iteration of the policy iteration algorithm (see Figure 3). The algorithm iterates in this manner until no further improvements can be made to the policy in any state. Any such policy must be an optimal policy. Policy iteration can be fast if the action space is quite small, as in such a case it will usually converge in few iterations. Even so recalculating the entire value function from scratch each iteration is expensive. Value iteration avoids this problem by solving the optimality equation for a series of geometrically discounted finite horizon problems defined on the same Markov decision process. Thus it only calculates the value function for the infinite horizon problem once. The value function for the optimal policy π^* is written V^*. Its solution for a problem with horizon $h = 0$ is given by:

$$V_0^*(s_i) = \max_a \{E[R^a(s_i, 1)]\}, \forall s_i \in \mathcal{S} \tag{16}$$

Using Bellman's optimality equation the solution for a finite horizon problem with $h = 1$ is given in terms of V_0^*. In general V_n^* is written in terms of V_{n-1}^*:

$$V_{n+1}^*(s_i) = \max_a \{E[R^a(s_i, 1)] + \gamma \sum_j p_{ij}(a) V_n^*(s_j)\}, \forall s_i \in \mathcal{S} \tag{17}$$

Algorithm 2 (POLICY ITERATION)

$f :=$ arbitrary policy

repeat

    solve $V^f(s_i) = E[R^f(s_i, 0)] + \gamma \sum_j p_{ij}(f(s_i))V^f(s_j)$ $\forall s_i \in \mathcal{S}$

    choose $g(s_i) := \arg\max_a[Q^f(s_i, a)]$ $\forall s_i \in \mathcal{S}$

    where $Q^f(s_i, a) = E[R^a(s_i, 0)] + \gamma \sum_j p_{ij}(a)V^f(s_j)$

    let $f(s_i) := g(s_i)$ $\forall s_i \in \mathcal{S}$

until there is no change in $f(s_i), \forall s_i \in \mathcal{S}$

Fig. 3. The policy iteration algorithm.

As $n \to \infty$ so $V_n^* \to V^*$. The difficulty with value iteration is defining a stopping criterion for n. Recently this problem has been partly solved. A upper bound on the error of the final estimate can be expressed given the maximum error between the last two approximations to the value function [43]. Worst case performance can thus be improved to a measureable point arbitrarily close to the optimum. The value iteration algorithm is given in Figure 4. There is some debate as to whether policy or value iteration converges faster for large problems.

There are a number of variations to value iteration as presented above, which is strictly termed *synchronous value iteration*. Synchronous refers simply to the fact that each iteration all the $V_n^*(x)$ are updated using the estimates $V_{n-1}^*(x)$. Gauss-Seidel iteration changes this by updating the states in a particular order. States updated later in the sequence use the already updated values for other states. Asynchronous value iteration relaxes the rules concerning the order in which the states are updated still further by allowing that an arbitrary subset of states is backed up synchronously each iteration. The set of states may vary in each iteration. In the completely asynchronous case this means that one state is updated each iteration. Both synchronous value-iteration and Gauss-Seidel iteration are guaranteed to converge to the optimal policy; asynchronous value-iteration is only guaranteed to converge if each state has a non-zero probability of being updated each iteration.

The theory of asynchronous dynamic programming was originally developed in order to facilitate multi-processor implementation. More recently this theory has been used to show that dynamic programming algorithms can work on-line, interleaving estimation of the model, policy modification, and control [3]. Asynchronous DP algorithms thus form the basis of model-based reinforcement learning systems. A class of value-iteration algorithms, termed adaptive real-time dynamic programming (ARTDP) algorithms, can be defined (see Figure 5). The

Algorithm 3 (VALUE ITERATION) Choose ϵ so that $\frac{2\epsilon\gamma}{(1-\gamma)}$ is sufficiently small, where $\max_S(|\hat{V}^*(s_i) - V^*(s_i)|) \leq \frac{2\epsilon\gamma}{(1-\gamma)}$.

$V_0 :=$ arbitrary bounded function

$n := 1$

repeat

    let $V_n^*(s_i) := max_a\{E[R^a(s_i, 0)] + \gamma \sum_j p_{ij}(a)V_{n-1}^*(s_j)\}$ $\forall s_i \in \mathcal{S}$

    until $\max_S(|V_n^*(s_i) - V_{n-1}^*(s_i)|) < \epsilon$

    let $\hat{V}^*(s_i) := V_n^*(s_i)$ $\forall s_i \in \mathcal{S}$

Fig. 4. The value iteration algorithm.

set $\mathcal{S}_t$ is the set of states whose values are to be updated at time t. It should be chosen so that the algorithm satisfies the conditions for the convergence of asynchronous value-iteration, and also that there is sufficient computation available to carry out all the updates. In addition because the agent is estimating the model on-line it must choose a suitable sequence of control actions so that it experiences each transition a sufficient number of times for the estimates $\hat{p}_{ij}$ to be good.

In a similar manner to Gauss-Seidel iteration, the choice of $\mathcal{S}_t$ at each stage can speed convergence to the optimal value function. There are a number of possible schemes. One of the simplest is to choose k additional members of $\mathcal{S}_t$ at random[3]. A good choice of $\mathcal{S}_t$ can speed convergence considerably. Moore and Atkeson have published a technique which can be seen as a variant of ARTDP. They call it Prioritised Sweeping [23]. Its central feature is that states leading to states with large changes in the value function are processed first. It is considerably more efficient than the other model-based methods discussed previously. Wiering and Schmidhuber have more recently published a version of the algorithm which is capable of detecting large changes in the value function which occur as a result of a succession of small cumulative changes in the model [42].

There are a number of model-free policy-modification algorithms; the earliest of these [35, 14] used either the TD(λ) algorithm, or temporal difference methods similar to it, to convert a delayed reward signal into a heuristic reward signal (the $\hat{V}^\pi(x_t)$ under the current policy π). The heuristic reward signal is then fed to any algorithm which modifies its policy on the basis of immediate reward, in place of the immediate reward signal r_t. Thus the class of algorithms compatible with this method is large. The policy of the agent is modified using the estimate $\hat{V}^{\pi_t}(x_t)$.

---

[3] See [16] for a description of such an algorithm. They refer to it as a Dyna method, although it is rather different from Sutton's original method of the same name [37].

Algorithm 4 (ADAPTIVE REAL-TIME VALUE ITERATION)

$\hat{Q}_t(x,a) = \bar{r}(x,a) + \gamma \sum_{y \in S} \hat{p}_{xy}(a) \hat{V}_t^*(y)$. explore is a function mapping from estimated Q-values to a probability distribution across actions. $\bar{r}(i,a)$ is an estimate of $E[R^a(i,0)]$.

$t := 0$

$\hat{V}_t^* :=$ arbitrary bounded function

observe x_t

repeat

    choose a_t from $\text{explore}_a(\hat{Q}_t(x_t, a))$

    observe the transition $x_t \overset{a_t}{\leadsto} x_{t+1}$

    update $\hat{p}_{x_t x_{t+1}}(a_t)$ and $\bar{r}(x_t, a_t)$

    for some $S_t \subseteq S$ such that $x_t \in S_t$ update

$$\hat{V}_{t+1}^*(i) := \max_a\{\bar{r}(i,a) + \gamma \sum_{j \in S} \hat{p}_{ij}(a) \hat{V}_t^*(j)\} \qquad \forall i \in S_t$$

$$\hat{V}_{t+1}^*(i) := \hat{V}_t^*(i) \qquad \forall i \notin S_t$$

  $t := t+1$

Fig. 5. The adaptive real-time value iteration algorithm.

In the next step the TD(λ) algorithm estimates V with respect to a different policy π_{t+1}. Thus the value function is changing as the policy changes. Examples of policy-modification algorithms based on the TD(λ) algorithm include [15, 18, 35]. To my knowledge there are currently no proofs for the convergence of any such policy-modification algorithms.

A model-free method which is guaranteed to converge is Q-learning [40]. This is a model-free approximation to adaptive real-time value iteration. The primary structural difference between Q-learning and methods based on the TD(λ) algorithm is that whereas the latter maintain estimates of the values of states under the current policy, Q-learning maintains estimates of action-values. It adjusts these estimates each step using a temporal difference mechanism:

$$\hat{Q}_{t+1}(x_t, a_t) = (1 - \alpha_t)\hat{Q}_t(x_t, a_t) + \alpha_t[r_t + \gamma \max_a\{\hat{Q}_t(x_{t+1}, a)\}] \qquad (18)$$

where $0 < \alpha_t < 1$ is the learning rate at time t. The update equation is fundamentally of the same form as the value iteration update, replacing the estimates $\hat{p}_{ij}$ with α_t. Q-learning is guaranteed to converge asymptotically given that each state-action pair is tried infinitely often, and similar criteria to TD-learning for the reduction of the learning rate [41]. The other nice property of Q-learning is that the estimates of the Q-values are independent of the policy followed by

Algorithm 5 Q(λ)-LEARNING

$\hat{V}^*(x) = \max_a \hat{Q}(x, a)$. $0 \leq \lambda \leq 1$. ε'_t and ε_t are error signals. explore is a function mapping from estimated Q-values to a probability distribution across actions.

$t := 0$

$\hat{Q}(x, a) := 0$ and $\bar{e}_t(x, a) := 0, \forall x, a$

observe x_t

repeat

    choose a_t from $\text{explore}_a(\hat{Q}_t(x_t, a))$

    observe the transition $x_t \overset{a_t}{\leadsto} x_{t+1}$

    $\varepsilon'_t := r_t + \gamma \hat{V}^*_t(x_{t+1}) - \hat{Q}_t(x_t, a_t)$

    $\varepsilon_t := r_t + \gamma \hat{V}^*_t(x_{t+1}) - \hat{V}^*_t(x_t)$

    update $\bar{e}(x, a)$ for all $x \in \mathcal{S}, a \in \mathcal{A}$ according to Eq. 19 or 20

    update $\hat{Q}_{t+1}(x, a)$ for all $x \in \mathcal{S}, a \in \mathcal{A}$ using

        $\hat{Q}_{t+1}(x_t, a_t) := \hat{Q}_t(x_t, a_t) + \alpha_t \varepsilon'_t \bar{e}_t(x_t, a_t)$

        $\hat{Q}_{t+1}(x, a) := \hat{Q}_t(x, a) + \alpha_t \varepsilon_t \bar{e}_t(x, a)$ for all $\hat{Q}(x, a)$ except $\hat{Q}(x_t, a_t)$

    $t := t + 1$

Fig. 6. The $Q(\lambda)$ algorithm.

the agent, the consequence of this being that the agent may deviate from the optimal policy at any stage while still constructing unbiased estimates of the Q-values.

Q-learning may also be extended to take advantage of eligibility traces. Q(λ) [25] contains one-step Q-learning as a special case ($\lambda = 0$) and so I give the full algorithm for this generalised version (Figure 6). Convergence has only been proved for Q(λ) with $\lambda = 0$. The eligibility traces used in Q(λ) are necessarily defined over the domain formed by the Cartesian product $\mathcal{S} \times \mathcal{A}$. The update equations are, however, fundamentally the same:

$$\bar{e}_t(x, a) = \begin{cases} \gamma \lambda \bar{e}_{t-1}(x, a) + 1 & \text{if } x = x_t \text{ and } a = a_t \\ \gamma \lambda \bar{e}_{t-1}(x, a) & \text{otherwise} \end{cases} \tag{19}$$

$$\bar{e}_t(x, a) = \begin{cases} 1 & \text{if } x = x_t \text{ and } a = a_t \\ \gamma \lambda \bar{e}_{t-1}(x, a) & \text{otherwise} \end{cases} \tag{20}$$

The version of Q(λ) learning given in Figure 6 was devised by Peng and Williams and uses the eligibility traces to propagate the temporal difference error for the value function under the optimal policy across the state and action space [25]. This update is incorrect if the agent follows non-greedy actions. This means that the algorithm is not guaranteed to converge. A simple solution is to zero all the eligibilities whenever a non-greedy action is selected. This version of $Q(\lambda)$ learning was originally suggested by Watkins [40].

However this method effectively removes the principle benefit of the idea of combining Q-learning with eligibility traces: that you can learn quickly about the effects of one policy while following another. A different approach to the problem is to use modified Q-learning [27], also known as SARSA. This removes the assumption that the agent follows a greedy policy after executing the current action. It does this by removing the max operator from the temporal difference update rule:

$$\hat{Q}_{t+1}(x_t, a_t) = (1 - \alpha_t)\hat{Q}_t(x_t, a_t) + \alpha_t[r_t + \gamma\hat{Q}_t^\pi(x_{t+1}, a_{t+1})] \qquad (21)$$

Under this rule the agent is now estimating the value of the action given that it follows some policy π after. This policy can obviously be stochastic and non-greedy. SARSA can thus be combined with full eligibility traces in a clean way.

In this section we have concentrated on describing in detail methods that search for a value function in Markovian tasks. These techniques are all subject to Bellman's curse of dimensionality. This arises because as the number of features in a problem increases, the number of combinations of feature values, and hence distinct states rises exponentially. In order to beat this problem reinforcement learning algorithms employ function approximators. Two very simple forms of function approximation are state aggregation and linear function approximators. However, even for these simple function approximators the convergence properties of the algorithms discussed previously can break down quickly. TD learning is known to converge with linear function approximation, but Q-learning is known to diverge with linear function approximators in a number of counter-examples [2] and also often in practice with function approximators that extrapolate. The problem of how to approximate the value function in general is a difficult open problem in reinforcement learning and a large number of papers have been published. There is not space in this chapter to cover the different approaches in any depth. The interested reader is referred to [7] for a comprehensive coverage of these issues.

6   Optimal Learning

So far we have been concerned with algorithms that learn to behave optimally. However, while learning to behave optimally these algorithms may well perform rather badly, particularly in the early stages of learning. Sometimes it matters how well we perform while we are learning. Suppose we are adapting our strategy during a game of robot football, for example, or learning to control a robot while travelling through an office building. In summary how should an agent act *while* it is learning? This question falls into the field studying what is sometimes known as *optimal learning*.

In addition to providing a framework for learning optimal behaviour it conveniently transpires that reinforcement learning also provides an elegant framework for studying optimal learning. Once we have a clear mathematical framework for optimal learning we can ask and sometimes answer questions such as how should I act so as to maximise my performance over a limited lifetime given that I will continue to learn throughout that lifetime? The problem of how to act while learning is a class of optimal control problems with a long history [12, 5], and includes topics such as adaptive dual control from control theory and bandit problems from statistics. In RL it has typically taken the form of two problems: (i) how to act so as to maximise performance during the learning agent's lifetime [20, 15]; and (ii) how to act to identify as good a policy as possible within the learning period [17, 11, 10]. These problems, while related, are not the same [44]. Either is more commonly known as the exploration-exploitation problem, and there is some ambiguity in the RL literature as to which one we are referring when we use this term. The first problem is, however, currently the only one for which we have a clear formulation, and it is therefore the solution of this that I describe here.

Arguably the cleanest framework for understanding optimal learning and the exploration-exploitation trade-off is a Bayesian one. Bayesian approaches model our uncertainty about the form of the transition and reward functions. When we know the transition and reward functions there is after all, no exploration-exploitation trade-off, the optimal way to behave is to act greedily since no information can be gained. If there is uncertainty however, Bayesian approaches can then take that uncertainty into account in calculating value functions which tell us how to act so as to optimise our performance while we are learning.

For MDPs the optimal Bayesian solution to problem (i) is well known, but intractable [20, 5]. Many approximations have been proposed. The domain is a finite state MDP with an unknown transition function P and a known reward function R. This differs slightly from the assumption of some RL researchers that the reward function is essentially part of the environment, and thus unknown (but observable) by the agent. Clearly if we knew P then the problem would reduce to finding the value function using a standard dynamic programming technique and we would have no learning problem at all. Our optimal learning problem is thus concerned with the uncertainty there is about the parameters of P (the transition probabilities) and how this uncertainty changes as the learner gathers information. The essential trade-off occurs because at each step we can choose actions that exploit information we already have about P to gather reward, or actions that explore and gather information about P that we can use later on to gather even higher rewards. In other words our dilemma is "should I sacrifice reward now to gather information that may let me gather greater rewards later on?"

The Bayesian approach is based on there being a space $\mathcal{P}$ of possible transition functions (or models) P for the MDP, and a well-defined prior probability density over that space. The probability density over the space of possible finite state MDPs for a known state space $\mathcal{S}$ is constructed as follows. First let us think about the density over the possible one-step transition functions from a single state action pair. If state $i \in \mathcal{S}$ has N possible succeeding states when action a is taken, then we know already that the transition function from that state action pair is a multinomial distribution over the outcomes:

$$\boldsymbol{p_i}^a = \{p_{i1}^a, p_{i2}^a \cdots p_{iN}^a\} \tag{22}$$

The possible transition functions from i, a are the possible multinomials $\boldsymbol{p_i}^a$. We want a probability density over this space which is closed under sampling from any such multinomial[4]. The Dirichlet density has this property for multinomials:

$$f(\boldsymbol{p_i}^a|\boldsymbol{m_i}^a) = \frac{\Gamma(\sum_{j=1}^{N} m_{ij}^a)}{\prod_{j=1}^{N} \Gamma(m_{ij}^a)} \prod_{j=1}^{N} (p_{ij}^a)^{m_{ij}^a - 1} \tag{23}$$

where $\Gamma(.)$ is the Gamma function. The density is parameterised by the $m_{ij}^a > 0$ for all states j to which the process can transition in one step from state i under action a. Effectively each parameter m_{ij}^a represents the number of observations of that outcome. The experimenter can set a prior $m_{ij}^{a\prime}$ to reflect their beliefs about the likelihood of each outcome before making any actual observations. On making observations the parameter vector is updated as follows: if a single observation of a transition $i \stackrel{a}{\leadsto} j$ is made, then the new density is also Dirichlet with $m_{ij}^{a\prime\prime} = m_{ij}^{a\prime} + 1$. The Bayesian estimate of the likelihood of each transition is simply:

$$\bar{p}_{ij}^a = \frac{m_{ij}^a}{\sum_{k=1}^{N} m_{ik}^a} \tag{24}$$

We can compare this estimate to the maximum likelihood estimate of the transition probabilities which is of the same form, but with the initial $m_{ik}^a = 0$ for all successor states k.

Since the transition probabilities from a single state-action pair are a multinomial distribution, we can see that using a Dirichlet to express the uncertainty we have about the precise transition probabilities is sensible. The density over the space $\mathcal{P}$ of models for the multi-state case follows directly from that for the transitions from a single state action pair. The densities over the one step transition functions for the different state action pairs are mutually independent. The density $f(P|M)$ for a possible transition function $P \in \mathcal{P}$ for the whole MDP (i.e. for all the state action pairs at once) is therefore simply the product of the $f(\boldsymbol{p_i}^a|\boldsymbol{m_{ij}}^a)$ over all i, a. This density is now parameterised by a matrix

---

[4] We say that a family of densities is closed under sampling from a distribution. This means that I have a prior density from a certain family (e.g. Gaussian, Dirichlet) over the (unknown) parameters of the distribution I am sampling from. If I sample from the distribution and incorporate that sample information using Bayes rule then I am guaranteed to end up with a posterior from the same family of densities. Choosing a density that is closed under sampling is appealing because it makes calculating the posterior mathematically straightforward and computationally tractable.

$M = [m_{ij}^a]$, where $M \in \mathcal{M}$. In a Bayesian framework we now choose a prior matrix M', which specifies our prior density over the space of possible models. The additional information from a sequence of observations is captured in a count matrix F. The posterior density given these observations is therefore simply parameterised by $M'' = M' + F$. For convenience the transformation on M due to a single observed transition $i \overset{a}{\leadsto} j$ is denoted $T_{ij}^a(M)$.

Now we have a parametric density over the space of possible MDPs and a clear way of updating its parameters given samples from the true MDP we can turn to the problem of estimating the value function. The value function in an MDP with unknown transition probabilities is clearly a random variable, $\tilde{V}_i$ since it is a function of P which is itself a random variable. Given the usual squared error loss function the Bayesian estimator of expected return under the optimal policy is simply the expectation of $\tilde{V}_i$:

$$V_i(M) = \mathrm{E}[\tilde{V}_i | M] = \int_{\mathcal{P}} V_i(P) f(P | M) dP \qquad (25)$$

where $V_i(P)$ is the value of i given the transition function P. The central result of both Bellman and Martin was that when this integral is evaluated we transform our problem into one of solving an MDP with known transition probabilities, defined on the information space $\mathcal{M} \times \mathcal{S}$:

$$V_i(M) = \max_a \{ \sum_j \bar{p}_{ij}^a(M)(r_{ij}^a + \gamma V_j(T_{ij}^a(M))) \} \qquad (26)$$

where $\bar{p}_{ij}^a(M)$ is the marginal expectation of the Dirichlet as given above, $0 \leq \gamma < 1$ is the discount rate, and r_{ij}^a is the reward associated with the transition $i \overset{a}{\leadsto} j$. The value function $V(M)$ is known as the Bayes value function. The Bayes Q-values are obviously given by the inner part of Equation 26:

$$Q_{ia}(M) = \sum_j \bar{p}_{ij}^a(M)(r_{ij}^a + \gamma V_j(T_{ij}^a(M))) \qquad (27)$$

The Bayes Q-values naturally take into account the uncertainty about the process parameters in P. Thus the Bayesian estimate of value elegantly incorporates the value of future information. The optimal solution to the well-known exploration-exploitation trade-off (problem (i) above) is thus simply to act greedily with respect to the Bayes Q-values. Because the solution involves dynamic programming over a graph of information states the problem of actually obtaining the Bayes Q-values is intractable. A simple approximation to this is the certainty equivalent (CE) estimate constructed by replacing $T_{ij}^a(M)$ with M in (26).

There are a number of possible ways of coming up with an estimate of, or approximation to, the Bayes value function in a reasonable amount of time. One way is to estimate the value of the integral by random sampling [10, 33] directly from the probability density over the space of models. For each sampled model we obtain the Q-value function using dynamic programming and then calculate the average of the sampled Q-value functions for each state of the MDP. Unfortunately this will give us an estimate that has a standard error that declines slowly with the sample size. However, there are now value iteration techniques that work on what are sometimes referred to as structured representations of MDPs. These structured representations encode the states of the MDP as combinations of feature values. It has been shown [9] that the transitions between states can therefore be represented as a dynamic Bayes network (DBN). We can define a density over the space of possible transition models represented by the DBN in exactly the same way as we can for a standard (unfactored) MDP. The space of transition models P is constrained greatly by the structure represented by the DBN, and hence we would expect that the estimates produced by sampling from the resulting density over P would have a lower standard error. It is therefore reasonable to hypothesise that we might improve the performance of Monte Carlo approaches to the optimal learning problem by employing structured models.

Most solutions try to avoid the difficulties of dealing with the information state space in any form and essentially seek to approximate the right answer by solving a different MDP defined on the original state space [23, 15, 22, 42, 45]. The new MDP may differ from the old MDP in a number of ways. Either we can pick a different reward function, which reflects the uncertainty we have about the transition matrix, or we can pick a different transition matrix P. Whichever route we choose we typically select according to the heuristic "be optimistic in the face of uncertainty". If we have a learner which directly estimates the parameters of the MDP then a good solution is to pick a model P which is optimistic with respect to the value function [42, 45]. If we wish to do temporal difference learning then we are probably better off picking a reward function which gives bonuses to states which have been visited infrequently or where there appears to be uncertainty about the value function [22]. It is also worth noting that we do know of quite simple algorithms which are guaranteed to find ϵ-optimal policies for unknown MDPs in strictly polynomial time and space [17]. However the practical performance of these methods is likely to be much worse than that of most heuristic algorithms.

7   Summary

In this chapter we have outlined the simplest mathematical framework within which we can conduct reinforcement learning over many steps, a finite state Markov Decision Process. We have described three approaches to solving the MDP: Monte-Carlo sampling, temporal difference learning and adaptive dynamic programming. Clearly all three are related through Monte-Carlo sampling in the learning case in that we are sampling our experiences directly from a process we believe to be stochastic. Adaptive dynamic programming and temporal difference approaches, however also allow us to take advantage of the conditional probability structure of the Markov chain and so construct estimates of value which have a rapidly decreasing standard error. It is notable that here we have concentrated solely on model-based and value based techniques for solving MDPs. There is a burgeoning subfield working on searching directly in the policy space [28, 38, 34, 13, 26, 1], and this approach may prove to be better for a number of problems. Finally we have also extended the MDP framework to cover the case of MDPs with unknown transition probabilities. This has allowed us to represent the problem of optimal learning, which unlike the problem of learning optimal policies for MDPs is essentially intractable.

Many hard questions remain before we have developed reinforcement learning algorithms for practical tasks. One of the most important questions is whether we can have algorithms that can learn from reinforcement in the face of hidden state. One mathematical framework for such problems is that of partially observable Markov Decision Processes (POMDPs). In these there is an underlying MDP which is not directly observable. Instead each state generates observations probabilistically, and there are typically fewer observations than states; so that many states can generate the same observations. Even solving these for the case where we know the parameters of the POMDP turns out to be intractable [19]. However, for solving known POMDPs we at least have algorithms that are guaranteed to converge in the limit if the value function is representable in finite space. If we want to address the learning case we are trying solve a POMDP where the parameters are unknown. Here the most successful approaches either fall into the model learning or policy modification approaches. Model learning approaches typically try to learn either a hidden Markov model or a kth order Markov model [8, 21]. These rely on statistical tests to try to identify the number of underlying states in the model. However, it now appears that at least for some problems that searching directly in the policy space can produce much better results [13].

Reinforcement learning provides a perspective on learning which combines a number of assumptions. We work with a short term measure of performance and convert this into some long term measure. We assume that the problem is a sequential decision making problem, and try to exploit the conditional independence structure that may exist. Finally we have a class of algorithms that use randomised interaction with the decision process (Monte Carlo sampling) to learn directly or indirectly how to behave. These themes: sequential decision making, randomised interaction, and the exploitation of conditional independence; are ideas that crop up not just in reinforcement learning, but throughout the study of learning mechanisms and artificial intelligence. In reinforcement learning they are studied specifically within the framework imposed by learning from a reward signal. As we have seen this can give us a framework not only for learning optimal behaviour, but also for how to act while learning.

References

1. L. Baird and A. Moore. Gradient descent for general reinforcement learning. In *Advances in Neural Information Processing Systems 11*, 1999.

2. Leemon Baird. Residual algorithms: Reinforcement learning with function approximation. In *Proceedings of 12th International Conference on Machine Learning*, pages 30–37. Morgan Kaufmann, 1995.

3. A.G. Barto, S.J. Bradtke, and S.P. Singh. Learning to act using real-time dynamic programming. Technical Report CMPSCI-TR-93-02, University of Massachusetts, March 1993. Revised version of TR-91-57, Real-time learning and control using asynchronous dynamic programming.

4. A.G. Barto, R.S. Sutton, and C.J.C.H. Watkins. Learning and sequential decision making. COINS Technical Report 89-95, University of Massachussetts, September 1989. Later published in *'Learning and Computational Neuroscience'* edited by M.Gabriel & J.W. Moore.

5. R. E. Bellman. *Adaptive Control Processes: A Guided Tour*. Princeton University Press, 1961.

6. D. Bertsekas. *Dynamic Programming and Stochastic Control*. Academic Press, 1976.

7. Dimitri P. Bertsekas and John N. Tsitsiklis. *Neuro-Dynamic Programming*. Athena Scientific, 1996.

8. Lonnie Chrisman. Reinforcement learning with perceptual aliasing: The perceptual distinctions approach. In *Proceedings of the Tenth National Conference on Artificial Intelligence*, pages 183–188, 1992.

9. Richard Dearden, Craig Boutillier, and Moises Goldszmidt. Stochastic dynamic programming with factored representations. *Artificial Intelligence*, 121(1-2):49–107, 2000.

10. Richard Dearden, Nir Friedman, and David Andre. Model-based Bayesian exploration. In *Proceedings of the Fifteenth Annual Conference on Uncertainty in Artificial Intelligence (UAI-99)*, pages 150–159, San Francisco, CA, 1999. Morgan Kaufmann Publishers.

11. C.N. Fiechter. Expected mistake bound model for on-line reinforcement learning. In Douglas H. Fisher, editor, *Proceedings of the 14th International Conference on Machine Learning*, pages 116–124. Morgan Kaufmann, 1997.

12. J.C. Gittins. *Multi-armed Bandit Allocation Indices*. Interscience Series in Systems and Optimization. John Wiley & Sons, 1989.

13. M.R. Glickman and K. Sycara. Evolutionary search, stochastic policies with memory and reinforcement learning with hidden state. In *Proceedings of the Eighteenth International Conference on Machine Learning*, pages 194–201. Morgan Kaufmann, 2001.

14. J.H. Holland. Escaping brittleness: The possibilities of general purpose learning algorithms applied to parallel rule-based systems. In R.S. Michalski, J.G. Carbonell, and T.M. Mitchell, editors, *Machine Learning II*, pages 593–623. Morgan Kaufmann, 1986.

15. Leslie Pack Kaelbling. *Learning in Embedded Systems*. PhD thesis, Dept of Computer Science, Stanford, 1990.

16. L.P. Kaelbling, M.L. Littman, and A.W. Moore. Reinforcement learning: A survey. *Journal of Artificial Intelligence Research*, 4:237–285, 1995.

17. M. Kearns and S. Singh. Near-optimal reinforcement learning in polynomial time. In Jude Shavlik, editor, *Proceedings of the Fifteenth International Conference on Machine Learning*, pages 260–268. Morgan Kaufmann, 1991.

18. Long-Ji Lin. Self-improving reactive agents based on reinforcement learning, planning and teaching. *Machine Learning*, 8(3/4):293–321, 1992.

19. Michael Littman. *Algorithms for Sequential Decision Making*. PhD thesis, Brown University, March 1996.

20. J.J Martin. *Bayesian Decision Problems and Markov Chains*. Wiley, New York, 1967.
21. Andrew Kachites McCallum. *Reinforcement Learning with Selective Perception and Hidden State*. PhD thesis, University of Rochester, 1996.
22. N. Meuleau and P. Bourgine. Exploration of multi-state environments: Local measures and back-propagation of uncertainty. *Machine Learning*, 35:117–154, 1999.
23. Andrew W. Moore and Christopher G Atkeson. Prioritised sweeping: Reinforcement learning with less data and less time. *Machine Learning*, 13(1):103–130, 1993.
24. K. Narendra and M.A.L. Thathachar. *Learning Automata: An Introduction*. Prentice-Hall, 1989.
25. Jing Peng and Ronald J. Williams. Incremental multi-step Q-learning. In W.W.Cohen and H.Hirsh, editors, *Machine Learning: Proceedings of the 11th International Conference*, pages 226–232, 1994.
26. L. Peshkin, K. Kim, N. Meuleau, and L. Kaelbling. Learning to cooperate via policy search. In *Proceedings of the Sixteenth Conference on Uncertainty in Artificial Intelligence*, pages 307–314. Morgan Kaufmann, 2000.
27. Gavin Rummery. *Problem Solving with Reinforcement Learning*. Ph.D. dissertation, University of Cambridge, 1995.
28. J. Schmidhuber and J. Zhao. Direct policy search and uncertain policy evaluation. Technical Report IDSIA-50-98, IDSIA, August 1998.
29. A Schwartz. A reinforcement learning method for maximising undiscounted rewards. In *Machine Learning: Proceedings of the Tenth International Conference*. Morgan Kaufmann, 1993.
30. Satinder Singh and Richard Sutton. Reinforcement learning with replacing eligibility traces. *Machine Learning*, (22):123–158, 1996.
31. S.P. Singh. Reinforcement learning algorithms for average-payoff Markovian decision processes. In *Proceedings of the Twelfth National Conference on Artificial Intelligence*. AAAI Press/MIT Press, 1994.
32. S.P. Singh, T. Jaakkola, and M. Jordan. Learning without state-estimation in partially observable markovian decision processes. In *Proceedings of the Eleventh International Conference on Machine Learning*. Morgan Kaufmann, 1994.
33. Malcolm Strens. A Bayesian framework for reinforcement learning. In Pat Langley, editor, *Proceedings of the Seventeenth International Conference on Machine Learning*, pages 943–950. Morgan Kaufmann, 2000.
34. Malcolm Strens and Andrew Moore. Direct policy search using paired statistical tests. In Andrea Danyluk and Carla Brodley, editors, *Proceedings of the Eighteenth International Conference on Machine Learning*, pages 545–552. Morgan Kaufmann, 2001.
35. R.S. Sutton. *Temporal Credit Assignmenment in Reinforcement Learning*. PhD thesis, University of Massachusetts, School of Computer and Information Sciences, 1984.
36. R.S. Sutton. Learning to predict by the methods of temporal differences. *Machine Learning*, 3(1):9–44, 1988.
37. R.S. Sutton. Integrated architectures for learning, planning, and reacting based on approximating dynamic programming. In Bruce W. Porter and Ray J. Mooney, editors, *Machine Learning: Proceedings of the Seventh International Conference on Machine Learning*, pages 216–224. Morgan Kaufmann, 1990.
38. R.S. Sutton, D. McAllester, S. Singh, and Y. Mansour. Policy gradient methods for reinforcement learning with function approximation. In *Advances in Neural Information Processing Systems 12*. MIT Press, 2000.

39. R.S. Sutton and S.P. Singh. On step-size and bias in temporal difference learning. In *Proceedings of the Eighth Yale Workshop on Adaptive and Learning Systems*, pages 91–96, 1994.

40. C.J.C.H Watkins. *Learning from delayed rewards*. PhD thesis, University of Cambridge, King's College, Cambridge, England, May 1989.

41. C.J.C.H. Watkins and P. Dayan. Technical note: Q-learning. *Machine Learning*, 8(3/4):279–292, 1992.

42. M. Wiering and J. Schmidhuber. Efficient model-based exploration. In R. Pfeiffer, B. Blumberg, J. Meyer, and S. W. Wilson, editors, *From Animals to Animats 5: Proceedings of the Fifth International Conference on Simulation of Adaptive Behavior*, 1998.

43. R.J. Williams and L.C. Baird. Tight performance bounds on greedy policies based on imperfect value functions. Technical Report NU-CCS-93-11, Northeastern University, College of Computer Science, November 1993.

44. Jeremy Wyatt. *Exploration and Inference in Learning from Reinforcement*. PhD thesis, University of Edinburgh, Dept. of Artificial Intelligence, Edinburgh University, May 1997.

45. Jeremy Wyatt. Exploration control in reinforcement learning using optimistic model selection. In A. Danyluk and C. Brodley, editors, *Proceedings of the Eighteenth International Conference on Machine Learning*, 2001.

A Mathematical Framework for Studying Learning in Classifier Systems

John H. Holland

The University of Michigan
Ann Arbor, MI 48109, U.S.A.

jholland@umich.edu

1. Introduction

Massively parallel, rule-based systems offer both a practical and a theoretical tool for understanding systems that act usefully in complex environments [see, for example, refs 1-4], However, these systems pose a number of problems of a high order of difficulty - problems that can be broadly characterized as problems in nonlinear dynamics. The difficulties stem from the fact that the systems are designed to act in environments with complex transition functions - environments that, in all circumstances of interest, are far from equilibrium. Interactions with the environment thus face the systems with perpetual novelty, and the usual simplifications involving fixed points, limit cycles, etc., just do not apply.

Learning procedures (adaptive algorithms) offer a way of combating these difficulties, but an understanding of the possibilities is not a simple matter. The key question is easy enough to state informally: What kinds of environmental regularity can be exploited by learning? However, if answers that are both useful and widely applicable are to be forthcoming, the question must be reformulated in a way that gives it precision without losing generality. The usual tool for this task is a mathematical framework that suitably encompasses the subject. It is the purpose of this paper to explore a framework that gives a precise definition to the notion of an environmental regularity and then treats learning procedures as procedures for revising rules in response to detected environmental regularities.

In this context, procedures for revising rules become more than a convenience, they take a central place in the design. Whether carried out by a human or a machine, rule revision requires the solution of two broad problems. First, one must rate rules as to their usefulness to the system as a whole - the *apportionment of credit* problem. Then one must devise new rules that serve the system better than the least useful of the rules already in place - the *rule discovery* problem. Though these two problems are sometimes treated separately, they are closely interrelated.

A machine learning approach is used here to illustrate the interaction of apportionment of credit and rule discovery algorithms, and then the overall system is abstracted and translated to the mathematical framework. To give the framework a concrete subject matter, section 2 introduces a particular class of highly parallel, rule-based systems called *classifier systems*. The next section,

Reprinted from Physica D, Vol 2, No 1–3, 1986, pp. 307–317, J.H. Holland: "A Mathematical Framework for Studying Learning in Classifier Systems", with permission from Elsevier.

section 3, uses a set *derived Markov processes* to describe system-environment interaction in the absence of learning. Section 4 then defines *exploitable bias* as the formal counterpart of the notion of an environmental regularity. Section 5 proves a prototype theorem that illustrates the way in which learning algorithms can utilize an exploitable bias. Section 6 presents a set of conjectures that should be provable along the lines of the prototype and then goes on to a brief discussion of potential uses of the framework.

2. Classifier Systems

Classifier systems are parallel, rule-based systems designed to be amenable to modification by learning algorithms. Individual rules in the system, called *classifiers* have affinities with the rules (productions) used in expert systems (see, for example, Davis and King [5] or Waterman and Hayes-Roth [6]), but there are major differences in the way in which the overall system operates. The systems have been tested in a variety of applications including adaptive gas pipeline operation [7], implementation of the KL-ONE knowledge representation system [8], machine learning using the game of poker [9], image processing [10, 11], and cognitive adaptation [12, 13]. Classifier systems come in many variants, but the definition here will be restricted to the simplest "kernel" class.

The system has three layers: a *performance system* that interacts with the environment via a set of rules, an *apportionment* of credit algorithm that rates the rules as to usefulness, and a *rule discovery* algorithm that generates plausible new rules to replace less useful rules. Each of these layers will be defined, briefly, in turn. The reader is referred to [4] for details.

The performance system is organized around a set of rules that process messages. All messages are binary strings of length k; that is, they are elements of the set $\{1,0\}^k$. All interactions are mediated by messages: All internal interaction is via messages posted to a global "blackboard" called a *message list;* all input from the environment comes via an input interface that translates environmental conditions into messages, typically by means of a set of threshold detectors; all output to the environment takes place via an output interface that translates messages into effector actions.

Each *classifier* (rule) in the performance system is made up of one or more *conditions,* an *action,* and a *strength.* Each condition specifies a particular kind of message that satisfies it; that is, conditions classify messages. In order to accommodate learning algorithms, the specification of a condition is kept quite simple: Each condition has the same length k as the messages, but its specification is based upon the use of an additional symbol "#" that serves as a "don't care" or "wild card". Formally, then, conditions are strings drawn from the set $\{0,1, \#\}^k$. A message satisfies a condition if it matches the condition string bit for bit except where the condition contains #'s. (For example, the string 10010 satisfies the condition 1##10). If the message set is treated as a k-dimensional space, then conditions name hyperplanes in that space. The action part of a classifier specifies a message to be posted to the message list when the rule is activated.

The basic execution cycle of the performance system consists of the following steps:

1. Messages are posted to the message list from the input interface.
2. Each classifier is checked against the message list to see if each of its conditions is satisfied by some message(s) on the message list.
3. All classifiers with satisfied conditions enter into a competition based upon strength and other factors (see below), and each winner (there may be many) posts, to the message list, the message specified by its action part. Winning classifiers are said to be activated.
4. The output interface checks the message list and produces effector actions accordingly. (Since the messages are uninterpreted internally, conflicts can only occur at the output interface and they are resolved there).
5. All messages from the previous time-step are erased from the message list (i.e., messages persist on the list for only a single time-step unless they are repeatedly posted).
6. Return to step (1).

It is clear that step (2), because of the definition of conditions, involves only a simple matching operation that depends not at all on meaning. This means that there can be no questions of consistency when it comes to determining which classifiers are activated. Moreover there is no problem with activating many rules simultaneously - more activated rules simply means more messages posted to the message list.

Note that one classifier can activate another by simply posting a message that satisfies the condition of the latter classifier. Such classifiers are said to be *coupled;* there are a variety of techniques for coupling classifiers in regular ways (see Forrest [8]). Action sequences can be implemented via coupling, and this is the basis for showing that classifier systems are computationally complete in the sense that any program that can be written in a standard computer language, such as LISP or C, can also be implemented in a classifier system.

The competition referred to in step (3) is implemented in terms of a *bidding* process. Indeed the bidding process is the heart of the apportionment of credit algorithm, called the *bucket-brigade* algorithm. A classifier, when its condition part is satisfied, makes a bid proportional to a product of its *strength* and its *specificity*. Strength, as adjusted over time by the bucket-brigade algorithm, is supposed to reflect the classifier's usefulness to the system *in the context of the other classifiers in the system*. Specificity is measured simply by dividing the number of specifying bits (non-#'s), a, in the condition part by the length k; specificity is supposed to reflect the amount of information required by the rule. In these terms a high-bidding rule is both useful and relevant. The competition between the bidding rules is resolved probabilistically, with the higher bidding rules being more likely to be among the set of winners.

The bucket-brigade algorithm adjusts the strengths of classifiers by treating them as middlemen in a complex economy. Each classifier that posts a message acts as a "supplier". Any classifier that is satisfied by a message acts as a "consumer". In this metaphor, then, when one rule is coupled to another, the first acts as a supplier and the second acts as a consumer. When a classifier wins the bidding process, it actually pays the bid to its suppliers (it has bought their messages). Its strength is reduced accordingly and their strengths are increased. (If there are many suppliers the bid is divided amongst them). Because the winning classifiers post their messages, they in turn become suppliers (hence the term middleman). (The equations describing this process are given in section 5). The ultimate consumer in this economy is the environment; it actually reinforces certain kinds of overt actions by increasing the strength of the rules active at that time. (This *payoff* can be thought of as the payoff attained for making the *final*, winning move in some game).

The intuitive idea behind the bucket-brigade algorithm is that a classifier only gains in strength if it makes a profit - it pays its suppliers less than it receives from its consumers. Note that each middleman (classifier) attends only to its direct suppliers and consumers; it has no complicated memory of what happens "upstream" or "downstream". Nevertheless, profit is likely to be sustained only if the middleman belongs to chains of suppliers that satisfy the ultimate consumers (the classifiers that receive payoff from the environment). The prototype theorem of section 4 gives some justification for this belief.

The final layer of the classifier system is the rule discovery process. The algorithm used for this purpose, a *genetic algorithm,* has been much studied (see, for example, refs. [14-16]). There is not room here to go into detail, but, in broadest terms, the genetic algorithm works by treating strong classifiers as parents, generating offspring by recombining parts of the strings from $\{0,1,\#\}^k$ used to define the classifiers. That is, at this level, the classifiers are treated as "chromosomes" with each gene having 3 alleles, (0,1,#), and strength is treated as a "fitness". The offspring then replace the weakest classifiers in the system *(not* the parents). Rather surprisingly, and contrary to common wisdom in genetics, it can be shown that the *crossover* operator (the major source of recombination) is the primary contributor of improved rules. Though the algorithm only manipulates classifiers, it *implicitly* biases the construction of new classifiers so that they are more likely to use *combinations* of alleles (the letters {0,1,#}) appearing in stronger rules. More importantly, it can be shown that the genetic algorithm rates and exploits $\gg M^3$ combinations (in biological terms, possible co-adapted sets of alleles) every time it executes M crossovers. (See ref. [4] for a detailed discussion of the action of genetic algorithms in the context of classifier systems).

In sum, performance is mediated by the message-passing classifiers, the bucket-brigade algorithm apportions strength to the classifiers on the basis of their interactions with other classifiers, and the genetic algorithm generates plausible new classifiers by recombining "building blocks" associated with strong classifiers.

3. Derived Markov Processes

At this point, the object is to provide a formal structure that captures the interaction of the classifier system with its environment. This part of the development applies to almost any kind of system that interacts probabilistically with its environment; to emphasize this aspect, the learning system will at times be called an *adaptive system*. Classifier systems then constitute one set of examples of interest.

We start by specifying the environment E of the adaptive system in terms of a finite state set $\{E_1,..., E_m\}$. In most cases of interest the number of states will be so large that, over feasible times, only a minuscule fraction of the states can be observed. The adaptive system's outputs (its effector settings at each time) act as inputs to the environment, influencing the environment's state transitions. These transitions will be taken to be probabilistic (noting that deterministic environments are a special case in which given transitions take place with probability 1). Accordingly, the transition probabilities corresponding to a given system output $r \in \mathbf{R}$ can be given by a Markov matrix M_r wherein the entry $p_{uu'}(r)$ in M_r gives the probability of a transition from state E_u to state $E_{u'}$ in the environment under the influence of adaptive system response r. Some, or even most, environmental transitions may be unaffected by the adaptive system's output, being largely determined by the environment's interior workings (its "laws").

Consider now a fixed classifier system $\mathbf{C}$ in which neither the strengths of the rules nor the rules themselves are allowed to change. Then the classifiers in $\mathbf{C}$ act upon $\mathbf{C}$'s current state S_y, and the messages from the environment, to determine both $\mathbf{C}$'s next state $S_{y'}$ and its response r. In general, $\mathbf{C}$ will determine $S_{y'}$ and r probabilistically, but, for the moment, assume that $S_{y'}$ and r are uniquely determined. Then $\mathbf{C}$ amounts to a fixed finite automaton with a finite state set $\{S_1,..., S_c\}$, and the current states of $\mathbf{C}$ and E, S_y and E_u, completely determine the response r. Then, knowing r and the matrix M_r, one can determine the probability that the environment goes to state $E_{u'}$; it is just entry $p_{uu'}(r)$ of M_r.

Accordingly, as a moment's thought will show, from $\mathbf{C}$ and the matrices $\{M_r | r \in \mathbf{R}\}$, one can determine a *derived Markov matrix* M that contains one column and one row for each possible state pair (S_y, E_u). The entry $p_{ii'}$ of M, corresponding to the transition from $i = (S_y, E_u)$ to $i' = (S_{y'}, E_{u'})$, is identical to entry $p_{uu'}$ of M_r. Stated another way, row i of M contains only entries from row u of M_r with zeros inserted wherever the transition from S_y to some state S_z is prohibited.

In the more general case, where $\mathbf{C}$ determines its state transitions and responses probabilistically, we can represent $\mathbf{C}$'s state transition function by a set of Markov matrices $\{M(E_u)\}$, where the entry $p_{yy'}(E_u)$ in $M(E_u)$ gives the probability of $\mathbf{C}$ going from state S_y to state $S_{y'}$ when the input messages to $\mathbf{C}$ are generated by environmental state E_u. We can represent $\mathbf{C}$'s response probabilities by a set of probabilities $\{p(r \mid (S_y, E_u)), r \in \mathbf{R}\}$, where $p(r \mid (S_y, E_u))$ is the probability of response r when the combined state of $\mathbf{C}$ and E is (S_y, E_u).

Following a line of argument similar to that in the deterministic case, we can obtain a *derived Markov matrix M* wherein the entry $p_{ii'}$ is now given by

$$p_{ii'} = p_{yy'}(E_u) \sum_{r \in \mathbf{R}} p_{uu'}(r) p(r \mid (S_y, E_u)).$$

That is, the summation gives the total probability of the environment going to state $E_{u'}$ under all possible responses to the state pair (S_y, E_u), while $p_{yy'}(E_u)$ gives the probability of **C** going to state $S_{y'}$ when the state pair is (S_y, E_u). The product of these two probabilities is just the probability of the pair $(S_{y'}, E_{u'})$ conditional on the pair (S_y, E_u).

The derived Markov matrix M is, of course, only a conceptual tool; it is much too large to be of any value computationally. It can be thought of as providing the level of greatest detail concerning the interacting systems **C** and **E**. In the systems of interest and over timespans of interest, only a minuscule fraction of the states of **E** will be observed by **C**. More importantly, the fixed points of M will have little relevance because individual states of **E** will have extremely small probabilities of recurrence (the "perpetual novelty" of the environment). Nevertheless we can use some generalizations of the notion of a model as a homomorphic image, together with general theorems about Markov matrices, to get at ideas about regularities and repeatability in an environment that never repeats states.

4. Exploitable Bias

What is it, then, that is *repeatable* for **C** when interacting with **E**? Clearly large numbers of states may share certain properties or features and these features may repeat with high frequency even though the states themselves do not. For example, a set of states in **E** may produce environmental messages to **C** that hold some set of values (defining bits) in common. Any rule in **C** that attends only to these values treats all the states in the given set as equivalent. **C**, then, may respond to aggregates of states (events) rather than to the states themselves, and these aggregates may recur with reasonable frequency. The basic question accordingly becomes: Under what conditions are the aggregates exploitable?

Let I be the set of indices of the states of a process having the transition function specified by the derived matrix M. (Hereafter we will identify the states with their indices). Let $J = \{ j_1, ..., j_n \}$ be the set of indices for a partition of the states I into a set of equivalence classes. Let $I_j = \{ i \mid i \in j \}$ be the indices of states contained in a particular equivalence class j. A lower bound on the probability of a transition from equivalence class j to equivalence class j' under M is given by

$$_*P_{jj'} \stackrel{def}{=} \min_{i \in I_j} \sum_{i' \in I_{j'}} p_{ii'}.$$

Consider the matrix M_J formed from the entries $*p_{jj'}$ for all pairs of states drawn from J^2. If, for all $i \in j$, it were the case that

$$\sum_{i' \in I_{j'}} p_{ii'} = *p_{jj'},$$

then M_J would be a homomorphic image of M. As it is, M_J is a sub-Markov matrix because the sum of each row is ≤ 1. While nontrivial homomorphic images of a complex process M are typically non-existent, the sub-Markov matrix M_J always exists for any set of equivalence classes J.

Let M_J^+ be any Markov matrix derived from M_J by adding increments to arbitrary elements of M_J so that the rows of the resulting matrix sum to 1. It is easily established that the fixed point of an appropriately chosen M_J^+ can be used to set a lower bound on the recurrence time of any equivalence class j under M.

4.1 Repeatability

Remembering that the process is actually controlled by M, we are interested in what properties of M are actually preserved, or at least bounded, by the process described by M_J. (In what follows, it will be assumed that derived matrix M and the equivalence classes J are such that the resulting process is regular; i.e., all states are recurrent and aperiodic). Let us begin by focussing on the repeatability of a transition from equivalence class j to equivalence class j'. We can think of this transition as a trial or experiment under control of the adaptive system $\mathbf{C}$. The natural question is, what is the expected frequency of repetition of the transition jj' under $\mathbf{C}$? To repeat the experiment, the system must first be reset to condition j, then the transition from j to j' must be executed. A bound on the expected reset time is set by looking at the jth component, p_j, of the eigenvector of M_j^+. p_j is just the recurrence probability of j, and hence the expected reset time is $1/p_{j:}$. (Note that, by a standard theorem, p_j, can be estimated using $\min_{h \in J} p^T{}_{hj}$, where $p^T{}_{hj}$ is an element of M_J^T.) The probability of a successful transition from j to j' is just $*p_{jj':}$ (From this point onward we will omit the left subscript "$*$" in $*p_{jj':}$.) Thus, $[p_j p_{jj'}]^{-1}$ gives an upper bound on the expected time (reset plus execution), for a successful repetition of the experiment.

Definition. A transition (experiment) jj' will be called T^*-repeatable under M_J if $[p_j p_{jj'}]^{-1} < T^*$

This definition can be extended to transitions taking (less than) T steps by using (all powers of M_J up to) M_J^T.

4.2 Discoverability

The adaptive system uses a set of detectors to detect certain properties of the environmental state. With no loss of generality these detectors can be taken to be predicates $a_q: I \rightarrow \{1,0\}$, wherein $a_q(i) = 1$ just in case state E_i has property q. Let $A = \{a_1,...,a_d\}$ be a set of distinct predicates corresponding to the adaptive system's detectors, and let $B \subset A$ be a subset of these predicates. Let B^* be the set of all clauses of the form $\{a'_j \& ... \& a'_h \& ... \& a'_b \mid a'_h$ is either a predicate $a_h \in B$ or else its negation $1-a_h\}$. Finally, let $J[B]$ be the set of indices of the equivalence classes on I induced by B^* under the requirement that $i \equiv i'$ if and only if there is a clause $D \in B^*$ such that $D(i) = D(i') = 1$. Intuitively, any two states that are indistinguishable by the detectors in B fall in the same equivalence class.

Definition. An equivalence class j is *discoverable* if $j \in J[B]$ for some $B \subset A$.

4.3 Expected Payoff

Up to this point we have not included payoff, the ultimate measure of the adaptive system's performance. At the most detailed level each environmental state has an associated payoff given by a function $u: I \rightarrow \{reals > 0\}$. The measure of the adaptive system's performance is simply the average payoff rate it receives on the trajectory defined by the derived matrix. We are now interested in how we can associate payoff with the states of M_J.

In general, the various states within the equivalence class j will have different associated payoffs. Moreover some of these states will have small probabilities of occurrence. A state with a very small probability of occurrence will not be observed within acceptable times and hence, no matter how high the associated payoff, should not be used in determining the *accessible* expected payoff.

The following procedure provides a bound on expected payoff that takes into account T^*-repeatability: First, order the states $i \in I_j$ according to their associated payoffs, going from highest payoff to lowest. Designate the states in this ordering $\{i(0),..., i(h),..., i(n(j))\}$. Each state $i(h)$ will have a probability of recurrence $p_{i(h)}$ determinable from the matrix M. Proceed through the ordered list, summing the associated probabilities $p_{i(h)}$ until the sum first exceeds $1/T^*$. That is,

1. Set $h = 0$ and set $\Sigma = p_{i(0)}$.
2. $\Sigma > 1/T^*$ then STOP, otherwise go to (3).
3. Increment h by 1 and add $p_{i(h)}$ to Σ.
4. Return to (2)

Let h_1 designate the state at which the procedure stops, let u_1 designate payoff assigned to state h_1, and let P_1 designate the sum of probabilities to that point. Starting with the next state after h_1 in the ordered list, and starting the sum anew, repeat the process to obtain (h_2, u_2, P_2). Iterate until the list is exhausted. It is clear from the procedure that the following holds:

Lemma. A lower bound on the *T*-accessible expected payoff*, $u_*(j)$, for equivalence class j, is given by

$$u_*(j) \stackrel{\text{def}}{=} \sum_i P_{h_i} u_{h_i},$$

where the P_{h_i} and the u_{h_i} are the values determined by the preceding procedure.

Define the (column) vector U_J over the equivalence classes J such that component j of U_J is $u_*(j)$. (Henceforth $u_*(j)$ will be labeled u_j, as the jth component of U_J.). Then, given an initial probability distribution P_J over the equivalence classes (treated as a row vector), and using the matrix M_J, a lower bound on the expected payoff at any time T is given by the expression $P_J(M_J)^T U_J$. The quantities used here can be estimated by sampling the process defined by M.

We are now in a position to describe the notion of a useful regularity in the environment. The idea is to define a useful regularity as some aspect of the environment that permits the adaptive system to increase its expected payoff by exploiting information about the regularity. Returning to the notion of an experiment based on some set J (see section 4.1), the environment offers such a possibility if there is an experiment based on a refinement of the original equivalence classes that offers higher expected payoff. More carefully, let K designate a set of equivalence classes that refines a set of equivalence classes J (that is, each equivalence class in J is a proper union of some set of classes from K). Let kk' be a transition included in the transition jj'; that is, $k \subset j$, $k' \subset j'$ and there is a transition from k to k'. Note that the expected payoff associated with a given experiment in J is given by the expression $p_j p_{jj'} u_{j'}$ (the probability of a reset followed by a successful transition, multiplied by the expected payoff $u_{j'}$ at the end state). This motivates the following definition of an exploitable regularity:

Definition. Let k, $k' \in K$ be *discoverable* and let $k \subset j$, $k' \subset j'$, where $j, j' \in J$. If the transitions kk' and jj' are *T*-repeatable*, with $u_{k'}$ and $u_{j'}$ being the respective *T*-accessible expected payoffs*, then kk' offers a *T*-exploitable bias* with respect to jj' if

$$p_k p_{kk'} u_{k'} > p_j p_{jj'} u_{j'}.$$

5. A Theorem about T^*-exploitable biases

Our objective now is to determine the effect of an *exploitable bias* upon the strength of a classifier that exploits it.

Let C be a classifier system with a behavior specifiable by a derived matrix M_J. Assume that there is a transition kk' with an *exploitable bias* relative to M_J. Let C be a classifier that has conditions responding only to j and let C^* be a classifier that has conditions responding only to k. (This is possible because both j and k are discoverable). Let C^* be the system C augmented by C^*.

Theorem. Given that the transition kk' offers an exploitable bias relative to transition jj' in the derived matrix M_J, then the strength that C^* attains at a local fixed point makes it a frequent winner in competition with C in the system C^*. [Though the theorem can be proved in a more general form, the following two assumptions will be added to keep the proof here brief: 1) there is no coupling between classifiers, and 2) all transitions from j to some state other than j' yield negligible payoff.]

Proof outline. In the notation already established, the payoff to C when it causes a transition to j' is $u_{j'}$; let u_0 be a T^*-bound on the expected payoff to C when it causes a transition to some state other than k. For C^* the corresponding quantities are $u_{k'}$ and u_{00}. Because neither C nor C^* is coupled to any successor, the income under the bucket brigade depends only upon payoff. The expected payoff when C wins within the system C is

$$v(C,C) = p_{jj'}u_{j'} + (1 - p_{jj'})u_0.$$

Within the system C^*, if C^* wins whenever state k' occurs (when C and C^* are simultaneously active), C has expected payoff

$$v(C,C^*) = p_{jj'}u_{j'-k'} + (1 - p_{jj'})u_0,$$

where $u_{j'-k'}$ is the T^*-bound on payoff determined for the complement of k' in j'. (The probability of the transition from j to j' when C wins is still $p_{jj'}$; even though the transition always winds up in the subset $j'-k'$ that is the complement of k' in j'. Note that the recurrence probability of C, in competition with C^*, drops to $p_j - p_k$.) The expected payoff for C^* within system C^* is

$$v(C^*,C^*) = p_{kk'}u_{k'} + (1 - p_{kk'})u_{00}.$$

For an arbitrary classifier with bidding ratio b, a transaction under the bucket-brigade involving income $v(t)$ changes the classifier's strength according to the equation

$$v(t + 1) = v(t) - bv(t) + v(t)$$

If repeated transactions produce the same (expected) income, the classifier's strength will approach the local fixed point determined by setting $v(t + 1) = v(t)$ in the above equation. The result is

$$v_{fp} = v(t)/b$$

From this we see that the strength of C in $\mathbf{C}$ tends to

$$v_{\text{fp}}(C,\mathbf{C}) = v'(C,\mathbf{C})/b$$
$$= [p_{jj'}u_{j'} + (1 - p_{jj'})u_0]/b,$$

while the strength of C in $\mathbf{C}^*$ tends to

$$v_{\text{fp}}(C,\mathbf{C}^*) = v'(C,\mathbf{C}^*)/b$$
$$= [p_{jj'}u_{j'-k'} + (1 - p_{jj'})u_0]/b,$$

Similarly, the strength of C^* in $\mathbf{C}^*$ tends to

$$v_{\text{fp}}(C^*,\mathbf{C}^*) = v'(C^*,\mathbf{C}^*)/b^*$$
$$= [p_{kk'}u_{k'} + (1 - p_{kk'})u_{00}]/b^*,$$

where b^* is the bidding ratio of C^*.

The *bid* made by any classifier at its fixed point strength v_{fp} is just its expected payoff (average "income") v'. That is, the bid from C is

$$bv_{\text{fp}}(C,\mathbf{C}^*) = [p_{jj'}u_{j'-k'} + (1 - p_{jj'})\,u_0],$$

and the bid from C^* is

$$b^*v_{\text{fp}}(C^*,\mathbf{C}^*) = [p_{kk'}u_{k'} + (1 - p_{kk'})\,u_{00}].$$

It is important to note that C sometimes wins the competition when the state is k', even if C^* outbids C, because the winner is determined as a *probabilistic* function of the difference in the sizes of the bids. When this happens the payoff to C is not strictly $u_{j'-k'}$ as assumed at the outset, but rather an average of $u_{j'-k'}$ and $u_{j'}$; call it u', that depends upon the proportion of times that C wins over C^* when state k' occurs. The more frequently C wins in this situation the closer u' approaches $u_{j'}$.

The proof of the theorem can be completed with the help of the following lemma:

Lemma. As u' approaches $u_{j'}$, the proportion of times that C wins in the competition with C^* falls to less than one-half.

Proof of lemma. Because $p_k p_{kk'} u_{k'} > p_j p_{jj'} u_{j'}$ under the exploitable bias, and because $p_k < p_j$ (since $k \subset j$), it must be the case that $p_{kk'} u_{k'} > p_{jj'} u_{j'}$. If C were to win all the time in the presence of C^*, when the system is in state j', then $u' = u_{j'}$ and $p_{jj'}u'$ $> p_{jj'}u_j$. But these values, inserted in the equations giving the size of the bid (the equations preceding the lemma) show that C^* outbids C. Under such circumstances C^* must win more often than C.

From the lemma it follows that C^* will always win at least part of the time in competition with C.

Corollary. The system $\mathbf{C}^*$ exploits the useful bias to increase the average payoff rate of $\mathbf{C}^*$ over that of $\mathbf{C}$.

Proof. Consider the derived matrices that include the transition kk' for both $\mathbf{C}$ and $\mathbf{C}^*$. The derived matrix of $\mathbf{C}$ has exactly the same expected payoff as the matrix M_J because the transition kk' is not distinguished - it takes place with exactly the same probability that it had under M_J. In contrast, because C^* wins some of the time in the system $\mathbf{C}^*$, the expected payoff under the transition kk' is larger than that expected under jj'. This added payoff increases the overall payoff rate.

6. Extensions

In looking at the implications of this theorem for classifier systems, it is important to note that, typically, a general classifier (many #'s in its condition part) has a much smaller bidding ratio than a specific classifier (few #'s in its condition part). From the fixed point equations, it follows that v_{fp} for a general classifier is large relative to v_{fp} for a specific classifier. Consequently, general classifiers take a relatively long time to achieve v_{fp}. It is often the case that a new classifier, such as C^*, enters the system, before a more general classifier, such as C, has achieved its local fixed point. As a result, the general classifier bids a smaller fraction of its *expected income* than the more specific classifier. Accordingly, even when the two have about the same expected payoff, the more specific classifier has an advantage. Because of this transient effect stemming from the bidding ratios, specific classifiers *that take advantage of regularities* tend to become established rather rapidly.

The simplest extensions of the prototype theorem can be attained by relaxing the two additional assumptions that were made to decrease the complexity of the proof. The first assumption was that all transitions from j to some state other than j' yield negligible payoff. This is essentially an assumption that payoff is "concentrated" in the target state j' and that transitions leading away from that state amount to errors. The theorem, of course, still goes through if

$$[p_{kk'}u_{k'} - p_{jj'}u_{j'k}]$$
$$> [(1 - p_{jj'})u_0 - (1 - p_{kk'}) u_{00}].$$

Because $(1 - p_{jj'}) < (1 - p_{kk'})$, this will be true if

$$[p_{kk'} u_{k'} - p_{jj'} u_{j'_k}] > [u_0 - u_{00}].$$

That is, the advantage of the *useful bias* must exceed any reverse bias in transitions *away* from the target states. Because the difference $[u_0 - u_{00}]$ depends directly upon the value of $u_{j'_k}$, the whole theorem can be rewritten in terms of the relative size of $u_{j'_k}$.

Relaxing the second assumption leads to a much deeper understanding of the bucket-brigade algorithm as an apportionment of credit algorithm. When a classifier C_1 is coupled to C^*, and is active on the preceding time-step, the bid of C^* becomes part of its income. (When several classifiers are coupled to C^*, the usual treatment is to divide the bid of C^* among those that are active on the preceding time-step). Thus, any strengthening of C^* results in a strengthening of its coupled precursors. However, it is important to note that the precursors are influencing the system's state transitions *prior to the activation* of C^*. If C_1 diverts the system away from states in which C^* is effective in attaining payoff, C_1 will suffer because C^*'s strength, and hence its bid, will decrease in response to the reduction in average income. On net, then, a classifier C_1 fares well only if it aids in setting the stage for C^*'s action. This argument can be extended inductively to coupled chains of indefinite length. Roughly, the bucket-brigade, even though it acts only locally, favours only couplings which increase the system's overall payoff rate. More formally, the effects of coupling lead to a *global* fixed point theorem for the strengths of the classifiers in the system.

From this point there are many possible directions to move in developing a deeper understanding of apportionment of credit. To give two examples: A more specific classifier, in winning the competition against a more general classifier, may actually prevent the general classifier from making a "mistake" (an inappropriate transition to lowered payoff). Under the bucket-brigade, the general classifier does *not* pay its bid if it does not win the competition, so it is protected from a net loss and is actually better off for the existence of the more specific competitor. In computer science terms, the general classifier serves as a *default* to be overridden in specific situations by *exceptions.* This *symbiotic* interaction between general and specific classifiers has been observed [7] and serves as a natural basis for the development of default hierarchies (where the exceptions in turn have exceptions, etc.).

As a second example, consider the introduction of *taxes* into transactions. The simplest tax would be a small "flat" tax charged periodically to all classifiers. The immediate effect of this tax is a steady reduction in the strength of *parasites,* classifiers that send messages satisfying only their own conditions. Under the bucket-brigade alone, a parasitic classifier would suffer no reduction in strength because it would pay its bid to itself. However, since it has no other income, the flat tax gradually decreases its strength. (There are more complicated cases of parasitism, many of which have been observed in classifier systems. They are affected by a variety of taxes including activity taxes, income taxes, etc.) A flat tax has an even more important global effect. It puts a premium on *rate* of payoff, as contrasted to simply attaining payoff. This goes back to the definition of an exploitable bias in section 4. In proving the prototype theorem, the recurrence probabilities, p_j and p_k, for the precursor states, j and k, did not enter importantly. However, when the rate of winning is important, the recurrence times enter directly, yielding a much more complicated version of the theorem.

To this point there has been little discussion of the most important part of the learning process - the discovery of new rules. Everything so far discussed just serves as a precursor to this process. The essence of constructing *plausible* new rules is the discovery of the "building blocks" that serve as components in rules already proved useful. Under the bucket-brigade algorithm, useful rules are the ones that have relatively high strengths. Under a genetic algorithm, the strengths serve as "fitnesses" that determine which *schemas* (building blocks) are favoured, under recombination, in the generation of new rules [4, 14]. In terms of the derived matrices, a recombination of schemas to generate a new condition for a classifier amounts to the definition of a new equivalence class and a transition based upon it. Any refinement of an equivalence class with a higher "density" of payoff, or a refinement that gives better definition to a precursor to that class, will increase the payoff rate. This changes the global fixed point (through redefinition of the derived matrix), and there will usually be a cascade effect in which the new strengths change probabilities of winning, yielding further changes in the derived matrix. There should be a "fixed point" matrix for this cascade (though this is yet to be proved). More interesting is the meta-process whereby the derived matrices undergo changes under the influence of the genetic algorithm. The trajectory through the space of matrices is a stochastic process influenced by the implicit values and densities of the schemas in the population of classifiers defining each derived matrix. In this space distances are measured in terms of the minimal (or average) number of applications of the genetic operators required to get from one point to another. From any given point (matrix), the likely discoveries (new rules) are the ones that are nearby under this metric. Schemas that are multi-functional (i.e., schemas that serve in rules operative in different environmental contexts) act as pivotal points in this process. An efficient search for useful schemas takes place as an implicit by-product of the very process of generating candidate rules (see chap. 6 of ref. 4). The mathematical framework proposed here holds many elements in common with the mathematics used to study other adaptive systems such as economies, ecologies, physical systems far from equilibrium, immune systems, etc. (see, for example, ref. 17). In each of these fields there are familiar topics, with mathematical treatments, that have counterparts in each of the other fields. Even an abbreviated list of such topics [adopting the name from the field where it has been more extensively studied] is impressive: 1) niche exploitation, functional convergence and enforced diversity [ecology]; 2) competitive exclusion [ecology]; 3) symbiosis, parasitism, mimicry [ecology]; 4) epistasis, linkage revision, and redefinition of "building blocks" [genetics]; 5) linkage and "hitchhiking" [genetics]; 6) multifunctionality of "building blocks" [genetics and comparative biology]; 7) polymorphism [genetics], 8) assortative recombination ("triggering" of operators) [genetics and immunology]; 9) hierarchical organization [phylogenetics, developmental biology, economics and AI]; 10) tagged clusters [biochemical genetics, immunogenesis, and adaptive systems theory]; 11) adaptive radiation and the "founder" effect of generalists [ecology and phylogenetics]; 12) feedback from coupled procedures [biochemistry and biochemical genetics]; 13) "retained earnings" as a function of past success and current purchases [economics]; 14) "taxation" as a control on efficiency [economics]; 15) "exploitation" (production) vs. "exploration" (research) [economics and adaptive systems theory]; 16) "tracking" vs. "averaging" [economics and adaptive systems theory]; 17) implicit evaluation of "building blocks" [adaptive systems theory]; 18) "basins of attraction" and behavior far from equilibrium [physics]; 19) amplification of small biases submerged in noise on "slow" passage through a critical point [physics]. Any complex system constructed from components interacting in a nonlinear fashion will, in one regime or another, exhibit *all* of these features. A general mathematical theory of such systems would explain both the pervasiveness of these features and the relations between them.

References

1. G.A. Agha, Actors: A Model of Concurrent Computation in Distributed Systems, Ph.D. Dissertation, University of Michigan (1985).
2. W.D. Hillis, The Connection Machine (MIT Press, Cambridge, 1985).
3. C.E. Hewitt, Viewing control structures as patterns of passing messages, J. Artificial Intelligence 8 (1977) 323-64.
4. J.H. Holland, Escaping brittleness: The possibilities of general purpose learning algorithms applied to parallel rule-based systems. Machine Learning 2, R.S. Michalski, J.G. Carbonell, T.M. Mitchell, eds. (Morgan Kaufmann, Los Altos, NM, 1986) chap. 20.
5. R. Davis and J. King, An overview of production systems. Machine Intelligence 8, E.W. Elcock and D. Michie. eds. (American Elsevier, New York, 1977) pp. 300-331.
6. D.A.Waterman and F.Hayes-Roth, eds. Pattern-Directed Inference Systems (Academic Press, New York. 1978).
7. D.E. Goldberg, Computer-Aided Gas Pipeline Operation Using Genetic Algorithms and Rule Learning. Ph.D. Dissertation, University of Michigan (1983).
8. S. Forrest, A Study of Parallelism in the Classifier System and Its Application to Classification in KL-One Semantic Networks, Ph.D. Dissertation, University of Michigan (1985).
9. S. Smith, A Learning System Based on Genetic Algorithms: Ph.D. Dissertation, University of Pittsburgh (1980).
10. A.M. Gillies, Machine Learning Procedures for Generating Image Domain Feature Detectors, Ph.D. Dissertation, University of Michigan (1985).
11. J.J. Grefenstette and J.M. Fitzpatrick, Genetic search with approximate function evaluations, Proc. Int. Conf. on Genetic Algorithms and Their Applications, J.J. Grefenstette, ed., Carnegie-Mellon University, Pittsburgh (1985).
12. L.B. Booker, Intelligent Behavior as an Adaptation to the Task Environment, Ph.D. Dissertation, University of Michigan (1982).
13. S.W. Wilson, Knowledge growth in an artificial animal, Proc. Int. Conf. on Genetic Algorithms and Their Applications, J.J. Grefenstette, ed., Carnegie-Mellon University, Pittsburgh (1985).
14. J.H. Holland, Adaptation in Natural and Artificial Systems. (Univ. of Michigan Press, Ann Arbor. MI, 1975).
15. K.A. DeJong, Adaptive system design: A genetic approach, IEEE Trans. on Systems, Man, and Cybernetics 10 (1980) 566-74.

16. A.D. Bethke, Genetic Algorithms as Function Optimizers. Ph.D. Dissertation, University of Michigan (1980).
17. J.D. Farmer, N.H. Packard and A.S. Perelson, The immune system, adaptation, and machine learning, Physica 22D (1986).

Rule Fitness and Pathology in Learning Classifier Systems

Tim Kovacs

Department of Computer Science
University of Bristol
kovacs@cs.bris.ac.uk
http://www.cs.bris.ac.uk/~kovacs

1 Introduction

When applied to reinforcement learning, Learning Classifier Systems (LCS) [5] evolve sets of rules in order to maximise the return they receive from their task environment. They employ a genetic algorithm to generate rules, and to do so must evaluate the fitness of existing rules. In order for the Genetic Algorithm (GA) [4] to produce rules which are better adapted to the task, rule fitness needs somehow to be connected to the rewards received by the system – a credit assignment problem. Precisely how to relate LCS performance to rule fitness has been the subject of much research, and is of great significance because adaptation of rules and LCS alike depends on it.

This work undertakes an analysis of the causes and effects of certain rule pathologies in traditional strength-based LCS (§2.3) and traces them ultimately to the relation between LCS performance and rule fitness – i.e., to the credit assignment system. We examine situations in which less desirable rules can achieve higher fitness than more desirable rules, which constitutes a mismatch between the goal of the LCS as a whole (adaptation to a task) and the goal of the GA (evolution of high-fitness rules).

To study rule pathology we undertake an analysis of what types of rules, and relationships between rules, are possible. Developing earlier work by Cliff and Ross [2] and Lettau and Uhlig [11] the notion of *strong overgeneral rules* (strong overgenerals, for short) is studied and it is shown exactly what requirements must be met for them to arise in both strength and accuracy-based LCS. In order to compare the two approaches we use accuracy-based XCS and its strength-based twin SB–XCS, which were designed to differ as little as possible, allowing us to isolate the effects of the fitness calculation on performance. The analysis is undertaken using a number of simplifying assumptions outlined in §4, and deals first with non-sequential tasks.

This chapter argues that different definitions of overgenerality and strong overgenerality are appropriate for the two types of LCS (§5.3). Minimal conditions and tasks which will support strong overgeneral rules are presented (sections §6, §7 and §8), their dependence on the reward function is demonstrated (§6.1), and certain theorems regarding their prevalence are proved under simplifying assumptions (§7.2 and §8). It is shown that XCS and SB–XCS have

T. Kovacs: *Rule Fitness and Pathology in Learning Classifier Systems*, StudFuzz **183**, 219–265 (2005)
www.springerlink.com

different kinds of tolerance for biases (see §6.1) in reward functions, and (within the context of various simplifying assumptions) to what extent we can bias them without producing strong overgenerals (§8.2). It is also shown what kinds of tasks will not produce strong overgenerals even without our simplifying assumptions (§8.1 and §11.7).

Next fit overgeneral rules are distinguished (§5.5) and it is shown how XCS and SB–XCS differ in their response to them (§9). Following this, the concept of a strong *under*general rule is introduced and it is noted that a generalisation bias in fitness is needed to avoid them (§10).

In §11 we proceed to consider the more complex case of sequential tasks, and show that sequential tasks amplify difficulties with strong overgenerals. In §11.6 we analyse the failure of SB–XCS on Woods2 and attribute it to the presence of strong and fit overgenerals. We conclude (§12) that SB–XCS is unsuitable for non-trivial sequential tasks and that it appears to have no niche (i.e., no useful domain of application), unless fitness sharing can resolve its problems (§13.1). This work concludes with consideration of the value of the approach taken and possible extensions (§13) and some final comments (§14).

2   Background

The arguments presented here require some basic knowledge of reinforcement learning, including the notions of reward functions, value functions, the Q-update, and discounting return. Each of these is introduced as needed, and the reader is referred to [17] for a more complete introduction. We make use of XCS and SB–XCS, and the details of their fitness calculations are relevant to the results presented later. We introduce them in the following sections, and more details are available in [10].

2.1   Classifier Systems for Reinforcement Learning

Reinforcement learning consists of cycles in which a learning agent is presented with an input describing the current environmental state, responds with an action and receives some reward as an indication of the value of its action. The reward received is defined by the *reward function R*, which maps state-action pairs to the real number line, and which is part of the problem definition [17]. For simplicity we initially consider only *non-sequential* tasks, in which the agent's actions do not affect which states it visits in the future. The goal of the agent is to maximise the rewards it receives, and, in non-sequential tasks, it can do so in each state independently. In other words, it need not consider sequences of actions in order to maximise reward.

When an LCS receives an input it forms the *match set* [M] of rules whose conditions match the environmental input. (The two systems we consider here are *stimulus-response* LCS, that is, they lack an internal message list.) The LCS then selects an action from among those advocated by the rules in [M]. The subset of [M] which advocates the selected action is called the *action set* [A].

Occasionally the LCS will trigger a reproductive event, in which it calls upon the GA to modify the population of rules.

We will consider LCS in which, on each cycle, only the rules in [A] are updated based on the reward received – rules not in [A] are not updated.

2.2   The Standard Ternary LCS Language

A number of representations have been used with LCS, in particular a number of variations based on binary and ternary strings. Using what we'll call the *standard ternary LCS language* each rule has a single condition and a single action. Conditions are fixed length strings from $\{0, 1, \#\}^l$, while rule actions and environmental inputs are fixed length strings from $\{0, 1\}^l$. In all problems considered here $l = 1$.

A rule's condition c matches an environmental input m if for each character m_i the character in the corresponding position c_i is identical or the wildcard (#). The wildcard is the means by which rules generalise over environmental states; the more #s a rule contains the more general it is. Since actions do not contain wildcards the system cannot generalise over them.

2.3   Strength-based and Accuracy-based Fitness

Although the fitness of a rule is determined by the rewards the LCS receives when it is used, LCS differ in how they calculate rule fitness. In traditional strength-based systems (see, e.g., [3, 18]), the fitness of a rule is called its *strength*. This value is used in both action selection and reproduction. In contrast, the more recent accuracy-based XCS [19] maintains separate estimates of rule utility for action selection and reproduction.

One of the goals of this work is to compare the way strength and accuracy-based systems handle overgeneral and strong overgeneral rules. To do so, we'll compare accuracy-based XCS with a strength-based LCS called SB–XCS which differs as little as possible from XCS, and which closely resembles Wilson's ZCS [18]. Specifically, SB–XCS updates rule strengths as follows:[1]

Strength (also called prediction):

$$p_j \leftarrow p_j + \beta(P - p_j) \tag{1}$$

where p_j is the prediction (or strength) of rule j, and $0 < \beta \leq 1$ is a constant controlling the learning rate. In non-sequential tasks, P is the immediate reward from the environment. In sequential tasks, P is analogous to the discounted maximum Q-value of the successor state in Q-learning (see equation 6). SB–XCS uses the same strength value for both action selection and reproduction. That is, the fitness of a rule in the GA is simply its strength.

---

[1] Wilson [18] refers to strength as *prediction* because he treats it as a prediction of the reward that the system will receive when the rule is used. We will use the terms interchangeably.

XCS uses this same update to calculate rule strength, and uses strength in action selection, but goes on to derive other statistics from it. In particular, from strength it derives the accuracy of a rule, which it uses as the basis of its fitness in the GA. This is achieved by updating a number of parameters as follows (see [19] for more). Following the update of a rule's strength p_j, we update its prediction error ε_j.

Prediction error:

$$\varepsilon_j \leftarrow \varepsilon_j + \beta \left(|\overbrace{P - p_j}^{Error}| - \varepsilon_j \right) \tag{2}$$

Next we calculate the rule's accuracy κ_j:

Accuracy:

$$\kappa_j = \begin{cases} 1 & \text{if } \varepsilon_j < \varepsilon_o \\ \alpha(\varepsilon_j/\varepsilon_o)^{-v} & \text{otherwise} \end{cases} \tag{3}$$

where $0 < \varepsilon_o$ is a constant controlling the tolerance for prediction error and $0 < \alpha < 1$ and $0 < v$ are constants controlling the rate of decline in accuracy when ε_o is exceeded. Once the accuracy of all rules in [A] has been updated we update each rule's relative accuracy κ_j' :

Relative Accuracy:

$$\kappa_j' = \frac{\kappa_j \cdot numerosity(j)}{\sum_{x \in [A]} \kappa_x \cdot numerosity(x)} \tag{4}$$

where $numerosity(j)$ is the number of copies of a rule represented by a single macroclassifier j (see [19]. Finally, each rule's fitness F_j is updated:

Fitness:

$$F_j \leftarrow F_j + \beta(\kappa_j' - F_j) \tag{5}$$

To summarise, the XCS updates treat the strength of a rule as a prediction of the reward to be received, and maintain an estimate of the error ε_j in each rule's prediction. An accuracy score κ_j is calculated based on the error as follows. If error is below some threshold ε_o the rule is fully accurate (has an accuracy of 1), otherwise its accuracy drops off quickly. The accuracy values in the action set [A] are then converted to relative accuracies (the κ_j' update), and finally each rule's fitness F_j is updated toward its relative accuracy. To simplify, in XCS fitness is an inverse function of the error in reward prediction, with errors below ε_o being ignored entirely.

Sequential Tasks For sequential tasks, rules in the previous time step's action set $[A]_{-1}$ are updated toward the sum of the previous time step's reward and the discounted maximum of the current time step's strengths/predictions:

$$P = r_{t-1} + \gamma \max_i P(a_i) \tag{6}$$

where r_{t-1} is the immediate reward on the previous time step, $0 \leq \gamma \leq 1$ is the *discount rate* which weights the contribution of the next time step to the value of P, and $P(a_i)$ is the system prediction for action a_i (defined in §2.4).

Summary of Strength and Accuracy In short, in XCS accuracy evaluates the utility of generalisation, while strength evaluates the utility of acting. In SB–XCS, in contrast, strength plays both roles.

2.4   Action Selection

In XCS, a rule's contribution to system prediction[2] is its prediction weighted by its fitness (so that less fit rules have less weight):

$$P(a_i) = \frac{\sum\limits_{c \in [M]_{a_i}} F_c \cdot p_c}{\sum\limits_{c \in [M]_{a_i}} F_c} \qquad (7)$$

where $[M]_{a_i}$ is the subset of the match set $[M]$ advocating action a_i, F_c is the fitness of rule c and p_c is its prediction.

In SB–XCS, however, a rule's fitness is its strength, so there is no separate fitness parameter to factor into the calculation – and low prediction rules already have less weight. However, we do need to weight SB–XCS's prediction by numerosity, so that macroclassifiers have influence equal to the equivalent number of microclassifiers. The XCS update does not include numerosity because XCS's fitness already includes numerosity, thanks to the relative accuracy update (4). Removing fitness from equation (7) and factoring in numerosity we obtain the *System Strength*:

$$S(a_i) = \sum\limits_{c \in [M]_{a_i}} p_c \cdot numerosity(c) \qquad (8)$$

In preparation for action selection, SB–XCS constructs a system strength array using (8), just as XCS constructs a system prediction array using (7). Note, however, that the two differ in that the system strength (8) for an action is *not* a prediction of the reward to be received for taking it. For example, suppose that in a given state action 1 receives a reward of 1000, and that the only matching macroclassifier advocating action 1 has strength 1000 and numerosity 2. The system strength for action 1 is $P(a_1) = 1000 \cdot 2 = 2000$, twice the actual reward since there are two copies of the rule.

In order to estimate the return for an action, we must divide the system prediction by the numerosity of the rules which advocate it. For this purpose, we define the *System Prediction* in SB–XCS as:

---

[2] The estimate of the return for taking a given action – essentially a Q-value in reinforcement learning terms.

$$P(a_i) = \frac{S(a_i)}{\sum\limits_{c \in [M]_{a_i}} numerosity(c)}$$

$$= \frac{\sum\limits_{c \in [M]_{a_i}} p_c \cdot numerosity(c)}{\sum\limits_{c \in [M]_{a_i}} numerosity(c)} \qquad (9)$$

The system prediction is needed to calculate the target for the Q-update in sequential tasks (6).

In summary, whereas XCS uses system prediction for both action selection and the Q-update, SB–XCS uses system strength for the former and system prediction only for the latter.

2.5   XCS, SB–XCS and other LCS

SB–XCS is not simply a straw man for XCS to outperform. It is a functional LCS, and is capable of solving some problems well. (For example, its performance on the 6 multiplexer task is similar to XCS's – see [10].) SB–XCS's value is that we can study when and why it fails, and we can attribute any difference between its performance and that of XCS to the difference in fitness calculation. See [10] for full details of both XCS and SB–XCS.

3   Known Problems with Strength LCS

In this section we review a number of known problems with strength LCS, which will serve as the starting point for the analysis presented later in this work.

3.1   Overgeneral Rules

Dealing with *overgeneral rules* – rules which are simply too general – is a fundamental problem for LCS. Such rules may specify the desired action in a subset of the states they match, but, by definition, not in all states, so relying on them harms performance. E.g., an overgeneral which matches 10 states may be correct in as many as 9 or as few as 1. Even overgenerals which are most often correct are (by definition) sometimes incorrect, so using them can harm the performance of the system.

3.2   Greedy Classifier Creation

Another problem faced by some LCS is what Cliff and Ross referred to as *greedy classifier creation* [2, 18]. To obtain better rules, a classifier system's GA allocates reproductive events preferentially to rules with higher fitness. This is simply the application of selective pressure in reproduction, one of the components of the evolutionary process, and of itself is not a problem. However, in many classifier systems a rule's fitness depends on the magnitude of the reward it receives. In such systems rules which match in higher-rewarding parts of the task will reproduce more than others. If the bias in reproduction of rules is strong enough there may be too few rules, or even no rules, matching low-rewarding states. (In the latter case, we say there's a gap in the rules' covering map of the input/action space.)

3.3 Strong Overgeneral Rules

Cliff and Ross [2] showed that the strength-based ZCS can have serious difficulty even with simple sequential tasks. They attributed ZCS's difficulties to the two problems above, and, in particular, to their interaction, an effect the author refers to [8, 9] as the problem of *strong overgeneral rules*. The interaction occurs when an overgeneral rule acts correctly in a high reward state and incorrectly in a low reward state. The rule is overgeneral because it acts incorrectly in one of the states, but at the same time it prospers because of greedy classifier creation and the high reward it receives in the other state.

Lettau and Uhlig [11, 12] independently discovered strong overgeneral rules using a very different approach from Cliff and Ross's. As they put it:

> "... a suboptimal rule might dominate the optimal if it is applicable only in "good" states of the world: bad decisions in good times can "feel better" than good decisions in bad times." [12] p. 153.

Sequential and Non-sequential Tasks Although both Cliff and Ross and Lettau and Uhlig dealt exclusively with sequential tasks, the problems discussed above clearly also apply in non-sequential tasks, as this work will demonstrate. In fact, examples of trivial non-sequential tasks which produce strong overgenerals will be shown, and it is in these cases that analysis is simplest.

Significance The proliferation of strong overgenerals can be disastrous for the performance of a classifier system: such rules are unreliable, but outweigh more reliable rules when it comes to action selection. Worse, they may prosper under the influence of the GA, and may even reproduce more than reliable but low-rewarding rules, possibly driving them out of the population. For these reasons, and for their prevalence, strong overgenerals (and the related fit overgenerals of §5.5) are a major difficulty – perhaps the major difficulty – for strength-based classifier systems.

4 Methodology for Rule Type Analysis

Classifier systems are complex systems and analysis of their behaviour can be quite difficult. To make our analysis more tractable we'll make a number of simplifications, perhaps the greatest of which will be to study very small tasks. Although very small, these tasks illustrate different types of rules and the effects of different fitness definitions on them – indeed, they illustrate them better for their simplicity.

Another great simplification will be to deal initially with the much simpler case of non-sequential tasks rather than sequential ones. Sequential tasks present their own difficulties, but those present in the non-sequential case persist in the more complex sequential case; after all, non-sequential tasks are just the special case of sequential task in which $\gamma = 0$. Study of non-sequential tasks can uncover fundamental features of the systems under consideration while limiting the complexity which needs to be dealt with. Consequently, the analysis of rule types considers non-sequential tasks, and sequential tasks are dealt with only later in §11.

To further simplify matters we'll remove rule discovery from the picture and enumerate all possible classifiers for each task, which is trivial given the small tasks we'll consider. This effectively leaves us with something like a tabular Q-learner, where each entry in the table corresponds to a rule in the ternary language (see [10]). In other words, some table entries aggregate elementary states, unlike in a standard tabular Q-learner. This approach simplifies matters since rule discovery is no longer an issue, and the behaviour of all possible rules is considered simultaneously. We'll restrict our considerations to the standard ternary LCS language because it is the most commonly used and because we are interested in fitness calculations and the ontology of rules, not in their representation.

At present we are concerned with rule type analysis; we would like to know, for example, the conditions under which it is possible for strong overgeneral rules to occur. We will not, however, normally consider the dynamic behaviour of rules over time. For example, we will not consider how the strength of a rule changes from an initial value to a value which reflects its utility (as determined by the credit assignment system). Instead, we'll consider the steady state values of already adapted (i.e., evaluated) rules (until §9, where we will see how fitness changes over time). In particular, we will not consider the effect of the learning rate β on rule updates: we assume it is declined appropriately so that in the limit rule strengths approach the expected values shown.

Similarly, we are not interested in fluctuations in a rule's strength due to stochastic effects, and so we will consider only the expected values of rules in our calculations, and not deviations from expectations. In fact, we will consider deterministic reward functions, although by considering expected values we could compensate for stochasticity in the reward function. For our analysis of rules types, however, deterministic reward functions suffice.

As a final simplification we'll assume that, in all tasks, states and actions occur equiprobably. That is, on any time step the LCS has the same chance of sensing any of the possible environmental states, and it chooses an action at random. This makes the calculation of steady state strengths particularly simple. For example, figure 1 defines a simple task with two states and two actions, and a reward associated with each state-action pair. Figure 2 lists all possible rules for this task, along with their expected strengths. Since all states and all actions occur with equal probabilities, a rule's expected strength is simply the average of the rewards for the state-actions it matches.

4.1 What can this Sort of Analysis Tell us?

These simplifications reduce a complex dynamic system – the interaction of an LCS with a task – to a very simple static model, in which each rule has a single fixed expected strength. Because we have simplified matters so much, there is much that such a model cannot tell us, e.g., about the dynamic behaviour of rules. In fact, removing rule discovery and choosing actions at random does not leave us with much of a classifier system and our simplifications mean that any quantitative results we obtain do not apply to any realistic applications of an LCS. However, because the model is so simple it is amenable to the analysis we will perform. In particular, this approach seems well suited to the qualitative study of rule ontology, and it will give us a qualitative sense of the behaviour of two types of LCS. §5 contains examples of this approach.

4.2 Default Hierarchies

Default hierarchies have not been included in the analysis presented here because XCS and SB–XCS do not support them. Default hierarchies are potentially significant in that they may allow strength LCS to overcome some of the difficulties with strong overgeneral rules we will show them to have. If so, this would increase both the significance of default hierarchies and the significance of the well-known difficulty of finding and maintaining them.

4.3 Fitness Sharing

Like default hierarchies, fitness sharing is a potential means for strength-based systems to escape problems with strong overgenerals. Its analysis is, unfortunately, beyond the scope of this work, and it is left as an important direction for future work. The incorporation of fitness sharing in SB–XCS would alter many, if not most, of the results in this work. It would also, however, greatly complicate the analysis presented here, and it is unlikely that the results obtained here would have been possible had fitness sharing been included in the analysis. This work is appropriately seen as a first step which identifies fundamental rule types; the effect of fitness sharing on them must await future work.

5 Analysis of Rule Types

5.1 Some Notation

Some simple notation will prove useful in our analysis of rule types.

- A Boolean target function f is a total function on a binary bit string, that is $f : \{0,1\}^n \to \{0,1\}$.
- Classifiers are constant partial functions, that is, they map some subset of the domain of f to either 0 or 1. Classifiers are constant because, using the standard ternary language, they always advocate the same action regardless of their input.

As a shorthand, and to approximate Sutton and Barto's reinforcement learning notation [17], we define $\mathcal{S}$ = domain and $\mathcal{A}$ = range when dealing with classifiers and target functions. That is, a task's state is an element of $\mathcal{S}(f)$ and a classifier system's action is an element of $\mathcal{A}(f)$, where f is a target function. The states matched by a classifier c form the set $\mathcal{S}(c)$, and the action advocated by c is $\mathcal{A}(c)$.

Note that f merely defines the state-action space. The learning task an LCS faces is defined by a reward function defined over this state-action space.

5.2   Correct and Incorrect Actions

Since the goal of a reinforcement learning agent is to maximise the rewards it receives, it's useful to have terminology which distinguishes between actions which do so and those which do not:

Correct action: In any given state the agent must choose from a set of available actions. A correct action is one which results in the maximum reward possible for the given state and set of available actions.

That is, an action c is correct, correct(s,c), for a state s with respect to a reward function R iff:
$$\forall a \ \ R(s,a) \leq R(s,c)$$

Incorrect action: One which does not maximise reward.

Figure 1 defines a simple non-sequential task, in which for state 0 the correct action is 0, while in state 1 both actions 0 and 1 are correct. Note that an action is correct or incorrect only in the context of a given state and the rewards available in it.

5.3   Overgeneral Rules

Figure 2 shows all possible rules for the task in figure 1 using the standard ternary language. Each rule's expected strength is also shown, using the simplifying assumption of equiprobable states and actions from §4. The classification shown for each rule will eventually be explained in sections §5.3 and §5.3.

We're interested in distinguishing overgeneral from non-overgeneral rules. Rules A, B, C and D are clearly not overgeneral, since they each match only one input. What about E and F? So far we haven't explicitly defined overgenerality, so let's make our implicit notion of overgenerality clear:

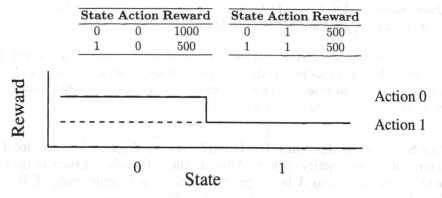

State	Action	Reward	State	Action	Reward
0	0	1000	0	1	500
1	0	500	1	1	500

Fig. 1. Reward function for a simple task.

Rule	Cond.	Action	E[Strength]	Strength Classification	Accuracy Classification
A	0	0	1000	Cons. Correct	Accurate
B	0	1	500	Cons. Incorrect	Accurate
C	1	0	500	Cons. Correct	Accurate
D	1	1	500	Cons. Correct	Accurate
E	#	0	750	Cons. Correct	Overgeneral
F	#	1	500	Overgeneral	Accurate

Fig. 2. All possible classifiers for the simple task in figure 1 and their classifications using strength-based and accuracy-based fitness. ("Cons." stands for "consistently".)

Overgeneral rule: A rule O from which a superior rule can be derived by reducing the generality of O's condition.

This definition seems clear, but relies on our ability to evaluate the superiority of rules. That is, to know whether a rule X is overgeneral, we need to know whether there is any possible Y, some more specific version of X, which is superior to X. How should we define superiority?

Are Stronger Rules Superior Rules? Can we simply use fitness itself to determine the superiority of rules? After all, this is the role of fitness in the GA. In other words, let's say X is overgeneral if some more specific version Y is fitter than X.

In SB–XCS, our strength-based system, fitter rules are those which receive higher rewards, and so have higher strength. Let's see if E and F are overgeneral using strength to define the superiority of rules.

Rule E. The condition of E can be specialised to produce A and C. C is inferior to E (it has lower strength) while A is superior (it has greater strength). Because A is superior, E is overgeneral.

This doesn't seem right – intuitively E should *not* be overgeneral, since it is correct in both states it matches. In fact all three rules (A, C and E) advocate only correct actions, and yet A is supposedly superior to the other two. This seems wrong since E subsumes A and C, which suggests that, if any of the three is more valuable, it is E.

Rule F. The condition of F can be specialised to produce B and D. Using strength as our value metric all three rules are equally valuable, since they have the same expected strength, so F is not overgeneral.

This doesn't seem right either – surely F *is* overgeneral since it is incorrect in state 0. Surely D should be superior to F since it is always correct. Clearly using strength as our value metric doesn't capture our intuitions about what the system should do. To define the value of rules let's return to the goal of the LCS, which is to maximise the reward it receives, which in turn means acting correctly in each state. It is the correctness of its actions which determines a rule's value, rather than how much reward it receives.

Recall that rule strength is derived from reward. Strength is a measure of how good – *on average* – a rule is at obtaining reward. Using strength as fitness in the GA, we will evolve rules which are – *on average* – good at obtaining reward. However, many of these rules will actually perform poorly in some states, and only achieve good average performance by doing particularly well in other states. These rules are overgeneral.

To maximise rewards, we do not want to evolve rules which obtain the highest rewards possible *in any state*, but to evolve rules which obtain the highest rewards possible *in the states in which they act*. That is, rather than rules which are *globally* good at obtaining reward, we want rules which are *locally* good at obtaining reward. In other words, we want rules whose actions are correct in all states they match. What's more, each state must be covered by a correct rule because an LCS must know how to act in each state; that is, it must have a *policy*.

To encourage the evolution of consistently correct rules, rather than rules which are good on average, we can use techniques like fitness sharing. But, while such techniques may help, there remains a fundamental mismatch between using strength as fitness and the goal of evolving rules with consistently correct actions. The effect of fitness sharing, and in particular its ability to combat overgeneral rules, deserves further study.

Strength and Best Action Maps To maximise rewards, a strength-based LCS needs a population of rules which advocates the correct action in each state. If, in each state, only the best action is advocated, the population constitutes a *best action map* [10]. While a best action map is an ideal representation in the sense that it is minimal, it is still possible to maximise rewards when incorrect actions are also advocated, as long as they are not selected. This is what we hope for in practice.

Now let's return to the question of how to define overgenerality in a strength-based system. Instead of saying X is overgeneral if some Y is fitter (stronger), let's say it is overgeneral if some Y is more consistent with the goal of forming a best action map; that is, if Y is correct in more cases than X.[3] Notice that we're now speaking of the correctness of rules (not just the correctness of actions), and of their relative correctness in specific. Let's emphasise these ideas:

Consistently Correct Rule: One which advocates a correct action in every state it matches. More formally, a classifier c is consistently correct w.r.t. a function f iff:

$$\forall s \in \mathcal{S}(c) \ \ f(s) = c(s)$$

Consistently Incorrect Rule: One which advocates an incorrect action in every state it matches. That is, a classifier c is consistently incorrect w.r.t. a function f iff:

$$\forall s \in \mathcal{S}(c) \ \ f(s) \neq c(s)$$

Correctness of a Rule: The correctness of a rule is the proportion of states in which it advocates the correct action.

The degree of correctness of a classifier c, correctness(c), w.r.t. a function f is the ratio between the states which it classifies correctly and the total number of states:

$$|\mathcal{C}| \ / \ |\mathcal{S}(c)|$$

---

[3] In reinforcement learning terms, we could say X is overgeneral if some Y is more consistent with the optimal policy.

where
$$\mathcal{C} = \{s \in \mathcal{S}(c) \mid c(s) = f(s)\}$$

Overgeneral Rule: One which advocates a correct action in some states and an incorrect action in others (i.e., a rule which is neither consistently correct nor consistently incorrect).

A classifier c is inconsistent w.r.t. a function f iff:

$$0 < \text{correctness}(c) < 1$$

That is, a classifier is inconsistent if it is neither consistently correct nor consistently incorrect.

The notion of the relative correctness of a rule allows us to say a rule Y is *more* correct (and hence *less* overgeneral) than a rule X, even if neither is consistently correct.

Now let's reevaluate E and F from figure 2 to see how consistent they are with the goal of forming a best action map. Rule E matches both states and advocates a correct action in both. This is compatible with forming a best action map, so E is not overgeneral. Rule F also matches both states, but advocates an incorrect action in state 0, making F incompatible with the goal of forming a best action map. Because a superior rule (D) can be obtained by specialising F, F is overgeneral.

Notice that we've now defined overgeneral rules twice: once in §5.3 and again above. For the tasks we're considering here the two definitions coincide, although this is not always true. For example, in the presence of perceptual aliasing (where an input to the LCS does not always describe a unique task state) a rule may be overgeneral by one definition but not by the other. That is, it may be neither consistently correct nor consistently incorrect, and yet it may be impossible to generate a more correct rule because a finer distinction of states cannot be expressed.

The above assumes the states referred to in the definition of overgenerality are task states. If we consider perceptual states rather than task states the rule is sometimes correct and sometimes incorrect *in the same state* (which is not possible in the basic tasks studied here). We could take this to mean the rule is not consistently correct, and thus overgeneral, or we might choose to do otherwise.

Accuracy and Complete Maps While all reinforcement learners seek to maximise rewards, the approach of XCS differs from that of strength-based LCS. Where strength LCS seek to form best action maps, XCS seeks to form a *complete map*: a set of rules such that each action in each state is advocated by at least one rule [19, 8]. This set of rules allows XCS to approximate the entire reward function and (hopefully) accurately predict the reward for any action in any state. XCS's fitness metric *is* consistent with this goal, and we'll use it to define the superiority of rules for XCS.

The different approaches to fitness mean that in strength-based systems we contrast consistently correct, consistently incorrect and overgeneral rules, but with accuracy-based fitness we contrast accurate and inaccurate rules.

In XCS, fitter rules are those with lower prediction errors – at least up to a point: small errors in prediction are ignored, and rules with small enough errors are considered fully accurate (specifically, those with error less than ε_0). In other words, XCS has some tolerance for prediction error, or, put another way, some tolerance for changes in a rule's strength, since changes in strength are what produce prediction error. This tolerance for prediction error is used to define overgenerality in XCS; we say that a rule is overgeneral if its prediction error exceeds the tolerance threshold, i.e., if $\varepsilon_j \geq \varepsilon_0$. In XCS 'overgeneral' is synonymous with 'not-fully-accurate'.

Although this work uses XCS as a model, we hope it will apply to other future accuracy-based LCS. To keep the discussion more general, instead of focusing on XCS and its error threshold, we'll refer to a somewhat abstract notion of tolerance called τ. Let $\tau \geq 0$ be an accuracy-based LCS's tolerance for oscillations in strength, above which a rule is judged overgeneral.

Like XCS's error threshold, τ is an adjustable parameter of the system. This means that in an accuracy-based system, whether a rule is overgeneral or not depends on how we set τ. If τ is set very high, then both E and F from figure 2 will fall within the tolerance for error and neither will be overgeneral. If we gradually decrease τ, however, we will reach a point where E is overgeneral while F is not. Notice that this last case is the reverse of the situation we had in §5.3 when using strength-based fitness. So which rule is overgeneral depends on our fitness metric.

Defining Overgenerality To match the different goals of the two systems we need two definitions of overgenerality:

Strength-based overgeneral: For strength-based fitness, an overgeneral rule is one which matches multiple states and acts incorrectly in some, but not all.[4] That is, a rule c is a strength-based overgeneral w.r.t. a function f iff:

$$|\mathcal{S}(c)| > 1$$

and $0 < \text{correctness}(c) < 1$

Accuracy-based overgeneral: For accuracy-based fitness, an overgeneral rule is one which matches multiple states, some of which return (sufficiently) different rewards, and hence has (sufficiently) oscillating strength. Here a rule is overgeneral if its oscillations exceed τ. That is, a rule c is an accuracy-based overgeneral w.r.t. a function f iff:

$$|\mathcal{S}(c)| > 1$$

and $\varepsilon_c \geq \varepsilon_0$

---

[4] This restatement of strength-based overgenerality is consistent with the two earlier definitions given in §5.3 and §5.3.

Note that the strength definition requires action on the part of the classifiers while the accuracy definition does not. Thus we can have overgenerals in a task which allows 0 actions (or, equivalently, 1 action) using accuracy (see, e.g., figure 5), but not using strength.

5.4   Strong Overgeneral Rules

Now that we've finally defined overgenerality satisfactorily let's turn to the subject of strong overgenerality. Strength is used to determine a rule's influence in action selection, and action selection is a competition between alternatives. Consequently it makes no sense to speak of the strength of a rule in isolation. Put another way, strength is a way of ordering rules. With a single rule there are no alternative orderings, and hence no need for strength. Therefore, for a rule to be a strong overgeneral, it must be stronger than another rule. In particular, a rule's strength is relevant when compared to another rule with which it competes for action selection.

Now we can define strong overgeneral rules, although to do so we need two definitions to match our two definitions of overgenerality:

Strength-based strong overgeneral: A rule which sometimes advocates an incorrect action, and yet whose *expected strength* is greater than that of some correct (i.e., not-overgeneral) competitor *for action selection*. That is, a rule c is a strength-based strong overgeneral w.r.t. a function f iff:

$$|\mathcal{S}(c)| > 1$$

and $0 < \text{correctness}(c) < 1$

and there is a consistently correct rule r with strength(r) < strength(c)

with which c competes for action selection.

Accuracy-based strong overgeneral: A rule whose strength oscillates unacceptably, and yet whose *expected strength* is greater than that of some accurate (i.e., not-overgeneral) competitor *for action selection*. That is, a rule c is an accuracy-based strong overgeneral w.r.t. a function f iff:

$$|\mathcal{S}(c)| > 1$$

and $\varepsilon_c \geq \varepsilon_o$

and there is a consistently correct rule r with strength(r) < strength(c)

with which c competes for action selection.

The intention is that competitors be possible, not that they need actually exist in a given population.

The strength-based definition refers to competition with *correct* rules because strength-based systems are not interested in maintaining incorrect rules (see §5.3). This definition suits the analysis in this work. However, situations in which more overgeneral rules have higher fitness than less overgeneral – but still overgeneral – competitors are also pathological. Parallel scenarios exist for accuracy-based fitness. Such cases resemble the well-known idea of *deception* in genetic algorithms, in which search is led away from desired solutions (see, e.g., [3]).

5.5   Fit Overgeneral Rules

In our definitions of strong overgenerals we refer to competition for action selection, but rules also compete for reproduction. To deal with the latter case we introduce the concept of *fit overgenerals* as a parallel to that of strong overgenerals. A rule can be both, or either. The definitions for strength and accuracy-based fit overgenerals are identical to those for strong overgenerals, except that we refer to fitness (not expected strength) and competition for reproduction (not action selection):

Strength-based fit overgeneral: A rule which sometimes advocates an incorrect action, and yet whose *expected fitness* is greater than that of some correct (i.e., not-overgeneral) competitor *for reproduction*.[5]

Accuracy-based fit overgeneral: A rule whose strength oscillates unacceptably, and yet whose *expected fitness* is greater than that of some accurate (i.e., not-overgeneral) competitor *for reproduction*.

We won't consider fit overgenerals as a separate case in our initial analysis since in SB–XCS fitness and strength are the same, and so strong and fit overgenerals are similar.[6] Later, in §9, we'll see how XCS handles both fit and strong overgenerals.

5.6   Parallel Definitions of Strength and Fitness

At this point we must note two terminological issues. In strength-based systems, a rule's strength is used in action selection and reproduction, and so it is with SB–XCS.[7] Although XCS uses the same update as SB–XCS, Wilson refers to this value as *prediction*, rather than strength. For simplicity this value is generally referred to simply as strength in this work.

The term "strength", however, really has two interpretations. One is the value updated by (1), and the other is the weight a rule has in action selection.

---

[5] The more formal definitions of fit overgenerals are omitted as they differ from those for strong overgenerals only in that they refer to fitness and reproduction instead of strength and action selection.

[6] Nonetheless, there is still a difference between strong and fit overgenerals in strength-based systems, since the two forms of competition may take place between different sets of rules. See §6.

[7] Strength-based systems may, however, distinguish between shared and unshared strength and use them differently.

When aggregated over the relevant rules, the latter value is referred to as system strength in SB–XCS (see below) and system prediction (equation 7) in XCS.

In SB–XCS, the two notions of strength coincide; in SB–XCS a rule's contribution to the system strength for an action is just its strength weighted by numerosity (8). In XCS, however, a rule's contribution to the system prediction for an action is a function of both its prediction and fitness (7).

Consequently, in XCS we have two notions of strength; prediction, and contribution to system prediction. In discussing strong overgenerals, it should be understood that references to strength are to the latter value.

A similar problem occurs in referring to the fitness of a rule. In XCS, fitness is a value updated by (5). Fitness, however, can also be interpreted (in fact, is normally interpreted in evolutionary computation) as the weight of a rule in reproduction. If we consider a single invocation of the niche GA, a rule's fitness parameter coincides with its weight in reproduction. However, if we consider all invocations of the GA, a rule's weight in reproduction is partly determined by its generality, thanks to the niche GA (in both XCS and SB–XCS).

In discussing fit overgenerals, it should be understood that references to fitness are to the weight of a rule in reproduction, not to the fitness parameter of a rule.

6   When are Strong and Fit Overgenerals Possible?

We've seen definitions for strong and fit overgeneral rules, but what are the exact conditions under which a task can be expected to produce them? If such rules are a serious problem for classifier systems, knowing when to expect them should be a major concern: if we know what kinds of tasks are likely to produce them (and how many) we'll know something about what kinds of tasks should be difficult for classifier systems (and how difficult).

Not surprisingly, the requirements for the production of strong and fit overgenerals depend on which definition we adopt. Looking at the accuracy-based definition of strong overgenerality we can see that we need two rules (a strong overgeneral and a not-overgeneral rule), that the two rules must compete for action selection, and that the overgeneral rule must be stronger than the not-overgeneral rule. The task conditions which make this situation possible are as follows:

1. The task must contain at least two states, in order that we can have a rule which generalises (incorrectly).[8]
2. The task may allow any number of actions in the two states, including 0 actions, or, equivalently, 1 action. (We'll see later that strength-based systems differ in this respect.)

---

[8] We assume the use of the standard LCS language in which generalisation over actions does not occur. Otherwise, it would be possible to produce an overgeneral in a task with only a single state (and multiple actions) by generalising over actions instead of states.

3. In order to be a strong overgeneral, the overgeneral must have higher expected strength than the not-overgeneral rule. For this to be the case the reward function must return different values for the two rules. More specifically, it must return more reward to the overgeneral rule.
4. The overgeneral and not-overgeneral rules must compete for action selection. This constrains which tasks will support strong overgenerals.

The conditions which will support fit overgenerals are clearly very similar: 1) and 2) are the same, while for 3) the overgeneral must have greater fitness (rather than strength) than the not-overgeneral, and for 4) they must compete for reproduction rather than action selection.

6.1   The Reward Function is Relevant

Let's look at the last two requirements for strong overgenerals in more detail. First, in order to have differences in the expectations of the strengths of rules there must be differences in the rewards returned from the task. So the values in the reward function are relevant to the formation of strong overgenerals. More specifically, it must be the rewards returned to competing classifiers which differ. So subsets of the reward function are relevant to the formation of individual strong or fit overgenerals.

Let us refer to the situation where different correct actions receive different rewards as a *bias* in the reward function. Reward functions which have no such biases are *unbiased*. More formally:

Unbiased reward function: A reward function R is unbiased iff there exists a constant r such that:

$$\forall s, a \text{ if correct}(s, a) \text{ then } R(s, a) = r$$

Figure 3 shows examples of unbiased reward functions on the left, and biased reward functions on the right. Note that the defining feature of the unbiased functions is that the highest reward in all states is the same constant value. If we plot the maximum reward for each state for any unbiased reward function we obtain a flat line.

For strong or fit overgenerals to occur, there must be a bias in the reward function at state-action pairs which map to competing classifiers. In the following section, we look at how classifiers can compete.

6.2   Competition for Action Selection

In XCS and SB–XCS, two classifiers c_1 and c_2 compete for action selection iff:

$$\mathcal{S}(c_1) \cap \mathcal{S}(c_2) \neq \emptyset \text{ and } \mathcal{A}(c_1) \neq \mathcal{A}(c_2)$$

Note that this relies on the property of the ternary language that classifiers are constant partial functions, i.e., that they advocate the same action in all states they match.

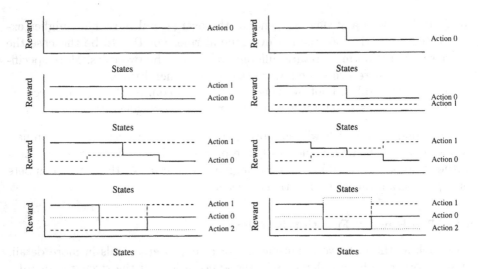

Fig. 3. Reward functions with 1, 2 or 3 actions. Those on the left are unbiased while those on the right are biased.

Figure 4 shows a reward function as a matrix with task state indexing the rows and LCS action indexing the columns. Using the standard ternary language rules can generalise over states but not actions, so a given rule will match some subset of a single column.

Rules compete for action selection when they occur in the same match set [M]. In XCS (and some other LCS) rules advocating the same action cooperate to have their action chosen, while rules advocating different actions compete. (We could say competition is between action sets [A]s, rather than individual rules.) In other LCS, rules which advocate the same action also compete against each other, and only the winner receives reward.

In either case, two rules which occur in different action sets within a match set will compete. The rules in each action set are updated toward the rewards returned for different actions in a state (rows in the matrix). Hence differences within a row of the matrix influence the strengths of competing classifiers and may result in strong overgenerals.

The rewards returned for taking the same action in different states (columns in the matrix) also affect the strengths of competing rules simply because rules can generalise over multiple states, and their strengths depend on the values in all of them. Hence differences within columns can result in strong overgenerals.

The effect, of course, of strong overgenerals on competition for action selection is that the system will tend to select an incorrect action.

6.3   Competition for Reproduction

The locus of competition for reproduction depends on the GA scheme used. With a panmictic GA all rules in the population compete for reproduction, whereas

Action

$R_{0,0}$	$R_{0,1}$	$R_{0,2}$
$R_{1,0}$	$R_{1,1}$	$R_{1,2}$
$R_{2,0}$	$R_{2,1}$	$R_{2,2}$

State

Fig. 4. A reward function as a matrix indexed by state and action.

with a niche GA only a subset of the population is eligible for reproduction. One way of running the niche GA is to restrict selection of rules to those in [M], in which case competition is between members of [M]. Similarly, with a niche GA in [A], rules in [A] compete.

More formally, using a niche GA in [M] two classifiers c_1 and c_2 compete for reproduction iff:

$$\mathcal{S}(c_1) \ \cap \ \mathcal{S}(c_2) \neq \emptyset$$

while using a niche GA in [A] c_1 and c_2 compete for reproduction iff:

$$\mathcal{S}(c_1) \ \cap \ \mathcal{S}(c_2) \neq \emptyset \text{ and } \mathcal{A}(c_1) = \mathcal{A}(c_2)$$

A niche GA limits the areas of competition within a population, and so limits how the rewards defined in the reward function interact. For example, with a GA in [A], rules belonging to different [A]s within an [M] do not compete for reproduction, so differences in the reward function between one [A] and another (i.e., differences within a row in the matrix) will not affect the reproduction of fit overgenerals in either.

Note, however, that different [A]s will overlap, as will different [M]s, and that a fully generalised rule is a member of all [M]s and all [A]s for its action. Even with a niche GA in [A] a fully general rule competes with all rules in the population which advocate the action it does. So even though a niche GA limits competition to subsets of the population, differences anywhere in a column can contribute to the fitness of two competing rules.

This means that with a GA in [A] differences in the rewards given for the same action may contribute to the fitness of rules competing for reproduction, but differences in rewards for different actions do not. With a panmictic GA all rules compete, meaning a difference between any two parts of the reward function can contribute to the fitness of rules competing for reproduction. The effect, of course, of fit overgenerals on competition for reproduction is that the fit overgenerals will tend to propagate in the population.

7   Strong Overgenerals in XCS

In §6 we saw that, using the accuracy definition, strong overgenerals require a task with at least two states, and that each state can have any number of actions. We also saw that the reward function was relevant but did not see exactly how.

Now let's look at a minimal strong overgeneral supporting task for accuracy and see exactly what is required of the reward function to produce strong overgenerals. Figure 5 shows a reward function for a task with two states and one action and all possible classifiers for it. As always, the expected strengths shown are due to the simplifying assumption that states and actions occur equiprobably (§4).

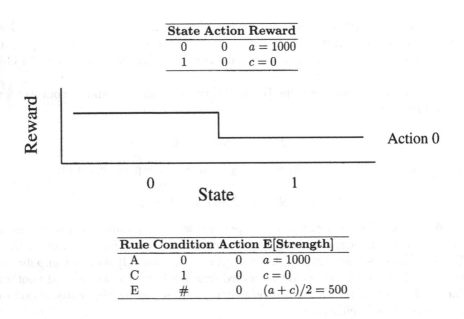

State	Action	Reward
0	0	$a = 1000$
1	0	$c = 0$

Rule	Condition	Action	E[Strength]
A	0	0	$a = 1000$
C	1	0	$c = 0$
E	#	0	$(a + c)/2 = 500$

Fig. 5. A minimal (2x1) strong overgeneral task for XCS and all its classifiers.

In §4 we made a number of simplifying assumptions, and for now let's make a further one: that there is no tolerance for oscillating strengths ($\tau = 0$), so that any rule whose strength oscillates at all is overgeneral. This means rule E in figure 5 is an overgeneral because it is updated toward different rewards. It is also a strong overgeneral because it is stronger than some not-overgeneral rule with which it competes, namely rule C. In §6 we saw that strong overgenerals depend on the reward function returning more strength to the strong overgeneral than its not-overgeneral competitor. Are there reward functions under which E will not be a strong overgeneral? Since the strength of E is an average of the two rewards returned (labelled a and c in figure 5 to correspond to the names of the fully specific rules which obtain them), and the strength of C is c, then as long as $a > c$, rule E will be a strong overgeneral. Symmetrically, if $c > a$ then E will still be a strong overgeneral, in this case stronger than not-overgeneral rule A. The only reward functions which do not cause strong overgenerals are those in which $a = c$. So in this case *any* bias in the reward function makes the formation of a strong overgeneral possible.

If we allow some tolerance for oscillations in strength without judging a rule overgeneral, then rule E is not overgeneral only if $a - c \leq \tau$ where τ is the tolerance for oscillations. In this case only reward functions in which $a - c > \tau$ will produce strong overgeneral rules.

7.1   Biases between Actions do not Produce Strong Overgenerals

State	Action	Reward	State	Action	Reward
0	0	w	0	1	y
1	0	x	1	1	z

Rule	Condition	Action	E[Strength]	Overgeneral unless
A	0	0	w	never
B	0	1	y	never
C	1	0	x	never
D	1	1	z	never
E	#	0	$(w+x)/2$	$|w-x| \leq \tau$
F	#	1	$(y+z)/2$	$|y-z| \leq \tau$

Fig. 6. A task which demonstrates that biases between actions do not result in strong overgenerals.

We've just seen an example where any bias in the reward function will produce strong overgenerals. However, this is not the case when we have more than one action available, as in figure 6. The strengths of the two fully generalised rules, E & F, are dependent only on the values associated with the actions they advocate. Differences in the rewards returned for different actions do not result in strong overgenerals – as long as we don't generalise over actions, which was one of our assumptions from §4.

7.2   Some Properties of Accuracy-Based Fitness

At this point it seems natural to ask how common strong overgenerals are and, given that the structure of the reward function is related to their occurrence, to ask what reward functions make them impossible. In this section we'll prove some simple, but perhaps surprising, theorems concerning overgeneral rules and accuracy-based fitness. However, we won't go too deeply into the subject for reasons which will become clear in §9.

Let's begin by looking at a special kind of reward function in which strong overgenerals are impossible, and proving this is so. Immediately after we'll see the general conditions which make strong overgenerals impossible, but, first, the special case:

Theorem 1. *In XCS, overgeneral rules are impossible when the reward function is constant over each action.*

Proof. The strength of a rule is a function of the values it updates toward, and these values are a subset of the rewards for the advocated action. If all such rewards are equivalent there can be no oscillations in a rule's strength, so it cannot be overgeneral. □

More generally, strong overgenerals are impossible when the reward function is sufficiently close to constancy over each action that oscillations in any rule's strength are less than τ. Now we can see when strong overgenerals are possible:

Theorem 2. *In XCS, if the task structure meets requirements 1 and 4 of §6 at least one overgeneral rule will be possible for each action for which the reward function is not within τ of being constant.*

Proof. A fully generalised rule matches all inputs and its strength is updated toward all possible rewards for the action it advocates. Unless all such rewards are within τ of equivalence it will be overgeneral. □

In other words, if the rewards for the same action differ by more than τ the fully generalised rule for that action will be overgeneral. To avoid overgeneral rules completely, we'd have to constrain the reward function to be within τ of constancy for each action. That overgeneral rules are widely possible should not be surprising. But it turns out that with accuracy-based fitness there is no distinction between overgeneral and strong overgeneral rules:

Theorem 3. *In XCS, all overgeneral rules are strong overgenerals.*

Proof. Let's consider the reward function as a vector $R = [r_1\ r_2\ r_3\ \ldots\ r_n]$, and, for simplicity, assume $\tau = 0$. An overgeneral matches at least two states, and so is updated toward two or more distinct values from the vector, whereas accurate rules are updated toward only one value (by definition, since $\tau = 0$) no matter how many states they match. For each r_i in the vector there is some fully specific (and so not overgeneral) rule which is only updated toward it. Consequently, any overgeneral rule (which must match at least two states) competes with at least two accurate rules.

Now consider the vector $X = [x_1\ x_2\ x_3\ \ldots\ x_y]$ which is composed of the subset of vector R toward which the overgeneral in question is updated. Because we've assumed states and actions occur equiprobably, the strength of a rule is just the mean of the values it is updated toward. So the strength of the overgeneral is $\bar{X}$, the mean of X.

The overgeneral will be a strong overgeneral if it is stronger than some accurate rule with which it competes. The weakest strength for such a rule is $\min x_i$. The inequality $\min x_i < \bar{X}$ is true for all reward vectors except those which are constant functions, so all overgenerals are strong overgenerals. □

Taking theorems 2 and 3 together yields:

Theorem 4. *In XCS, if the task structure meets requirements 1 and 4 of §6 at least one* **strong** *overgeneral rule will be possible for each action for which the reward function is not within τ of being constant.*

In short, using accuracy-based fitness and reasonably small τ only a highly restricted class of reward functions and tasks do not support strong overgeneral rules.

These 4 theorems are independent of the number of actions available in a task. Note that the 'for each action' part of the theorems depends on the inability of rules to generalise over actions, a syntactic limitation of the standard LCS language. If we remove this arbitrary limitation then we further restrict the class of reward functions which will not support strong overgenerals.

8 Strong Overgenerals in SB–XCS

We've seen how the reward function determines when strong overgeneral classifiers are possible in accuracy-based systems. Now let's look at the effect of the reward function using SB–XCS, our strength-based system. Recall from the strength-based definition of strong overgenerals that we need two rules (a strong overgeneral and a not-overgeneral correct rule), that the two rules must compete for action selection, and that the overgeneral rule must be stronger than the correct rule. The conditions which make this situation possible are the same as those for accuracy-based systems, except for a change to condition 2: there needs to be at least one state in which at least two actions are possible, so that the overgeneral rule can act incorrectly. (It doesn't make sense to speak of overgeneral rules in a strength-based system unless there is more than one action available.)

A second difference is that in strength-based systems there is no tolerance for oscillations in a rule's strength built into the update rules. This tolerance is simply not needed in SB–XCS where all that matters is that a rule advocate the correct action, not that its strength be consistent.

A complication to the analysis done earlier for accuracy-based systems is that strength-based systems tend toward best action maps (§5.3). Simply put, SB–XCS is not interested in maintaining incorrect rules, so we are interested in overgenerals only when they are stronger than some correct rule. For example, consider the binary state binary action task of figure 7. Using this unbiased reward function, rules E & F are overgenerals (since they are sometimes incorrect), but not strong overgenerals because the rules they are stronger than (B & C) are incorrect. (Recall from the definition of a strong overgeneral in a strength LCS in §5.4 that the strong overgeneral must be stronger than a *correct* rule.) This demonstrates that in strength-based systems (unlike accuracy-based systems) not all overgeneral rules are strong overgenerals.

What consequence does this disinterest in incorrect rules have on the dependence of strong overgenerals on the reward function? The reward function in this example is not constant over either action, and the accuracy-based concept of tolerance does not apply. In an accuracy-based system there must be strong overgenerals under such conditions, and yet there are none here.

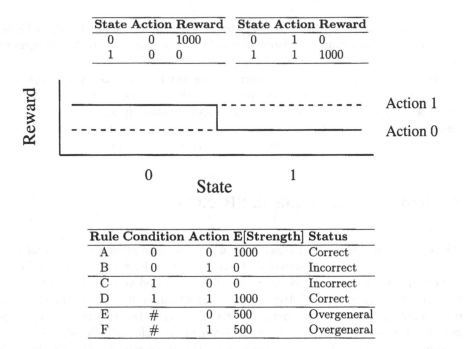

State	Action	Reward	State	Action	Reward
0	0	1000	0	1	0
1	0	0	1	1	1000

Rule	Condition	Action	E[Strength]	Status
A	0	0	1000	Correct
B	0	1	0	Incorrect
C	1	0	0	Incorrect
D	1	1	1000	Correct
E	#	0	500	Overgeneral
F	#	1	500	Overgeneral

Fig. 7. An unbiased reward function and all its classifiers. Unbiased functions will not cause strong overgenerals.

8.1   When are Strong Overgenerals Impossible in SB–XCS?

Let's begin with a first approximation to when strong overgenerals are impossible. Later, in §8.2, we'll ask when strong overgenerals *are* possible, and we'll get a more precise answer to the question of when they are impossible.

Theorem 5. *In SB–XCS, strong overgenerals are impossible when the reward function is unbiased (i.e., constant over correct actions).*

Proof. A correct action is one which receives the highest reward possible in its state. If all correct actions receive the same reward, this reward is higher than that for acting incorrectly in *any* state. Consequently no overgeneral rule can have higher strength than a correct rule, so no overgeneral can be a strong overgeneral. $\square$

To make theorem 5 more concrete, reconsider the reward values in figure 6. By definition, a correct action in a state is one which returns the highest reward for that state, so if we want the actions associated with w and z to be the only correct actions then $w > y, z > x$. If the reward function returns the same value for all correct actions then $w = z$. Then the strengths of the overgeneral rules are less than those of the correct accurate rules: E's expected strength is $(w + x)/2$ which is less than A's expected strength of w and F's expected strength is $(y + z)/2$ which is less than D's z, so the overgenerals

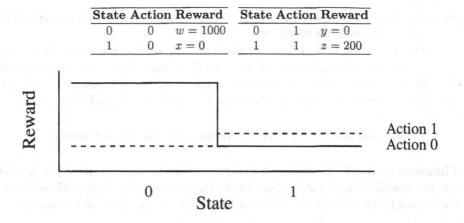

State	Action	Reward	State	Action	Reward
0	0	$w = 1000$	0	1	$y = 0$
1	0	$x = 0$	1	1	$z = 200$

Rule	Condition	Action	E[Strength]	Strong overgeneral if
A	0	0	$w = 1000$	never
B	0	1	$y = 0$	never
C	1	0	$x = 0$	never
D	1	1	$z = 200$	never
E	#	0	$(w + x)/2 = 500$	$(w + x)/2 > z$
F	#	1	$(y + z)/2 = 100$	$(y + z)/2 > z$

Fig. 8. A 2x2 biased reward function which is a minimal strong overgeneral task for strength-based LCS, and all its classifiers.

cannot be strong overgenerals. (If $w < y$ and $z < x$ then we have a symmetrical situation in which the correct action is different, but strong overgenerals are still impossible.)

8.2 What Makes Strong Overgenerals Possible in SB–XCS?

It is possible to obtain strong overgenerals in SB–XCS by defining a reward function which returns different values for correct actions. An example of a minimal strong overgeneral supporting task for SB–XCS is given in figure 8. Using this reward function, E *is* a strong overgeneral, as it is stronger than the correct rule D with which it competes for action selection (and for reproduction if the GA runs in the match set or panmictically).

However, not all differences in rewards are sufficient to produce strong overgenerals. How much tolerance does SB–XCS have before biases in the reward function produce strong overgenerals? Suppose the rewards are such that the actions associated with w and z are correct (i.e., $w > y, z > x$) and the reward function is biased such that $w > z$. How much of a bias is needed to produce a strong overgeneral? That is, how much greater than z must w be? Rule E competes with D for action selection, and will be a strong overgeneral if its expected strength exceeds D's, i.e., if $(w + x)/2 > z$, which is equivalent to $w > 2z - x$.

So a bias of $w > 2z - x$ means E will be a strong overgeneral with respect to D, while a lesser bias means it will not.

E also competes with A for reproduction, and will be fitter than A if $(w + x)/2 > w$, which is equivalent to $x > w$. So a bias of $x > w$ means E will be a fit overgeneral with respect to A, while a lesser bias means it will not. (Symmetrical competitions occur between F & A (for action selection) and F & D (for reproduction).)

We'll take the last two examples as proof of the following theorem:

Theorem 6. *In SB–XCS, if the task structure meets requirements 1 and 4 of §6 and the modified requirement 2 from §8, a strong overgeneral is possible whenever the reward function is biased such that $(w + x)/2 > z$ for any given rewards w, x & z.*

8.3 SB–XCS's Tolerance for Reward Biases

The examples in the previous section show there is a certain tolerance for biases (differences) in rewards within which overgenerals are not strong enough to outcompete correct rules. Knowing what tolerance there is is important as it allows us to design reward functions which will not produce strong overgenerals. Unfortunately, because of the simplifying assumptions we've made (see §4) these results do not apply to more realistic tasks. However, they do tell us how biases in the reward function affect the formation of strong overgenerals, and give us a sense of the magnitudes involved. An extension of this work would be to find limits to tolerable reward function bias empirically. Two results which do transfer to more realistic cases are theorems 1 and 5, which tell us under what conditions strong overgenerals are impossible for the two types of LCS. These results hold even when our simplifying assumptions do not.

9 Fit Overgenerals and the Survival of Rules under the GA

We've examined the conditions under which strong overgenerals are possible under both types of fitness. The whole notion of a strong overgeneral is that of an overgeneral rule which can outcompete other, preferable, rules. But, as noted earlier, there are two forms of competition: action selection and reproduction. Our two systems handle the first in the same way, but handle reproduction differently. In this section we examine the effect of the fitness metric on the survival of strong overgenerals.

XCS and SB–XCS were compared empirically on the tasks in figures 7 and 8. The GA was disabled and all possible rules inserted in the LCS at the outset. Settings were $\beta = 0.2$, and $\varepsilon_o = 0.01$. Wilson's pure explore/exploit scheme [19] was used.

9.1   Comparison on an Unbiased Reward Function

First we compared XCS and SB–XCS on the reward function from figure 7. Figure 9 shows the fitness of each rule using strength (left) and accuracy (right), with results averaged over 100 runs. The first thing to note is that we are now considering the development of a rule's strength and fitness over time (admittedly with the GA turned off), whereas until this section we had only considered steady state strengths (as pointed out in §4). We can see that the actual strengths indeed converge toward the expected strengths shown in figure 7. We can also see that the strengths of the overgeneral rules (E & F) oscillate as they are updated toward different values.

Using strength (figure 9, left), the correct rules A & D have highest fitness, so if the GA was operating we'd expect SB–XCS to reproduce them preferentially and learn to act correctly in this task.

Using accuracy (figure 9, right), all accurate rules (A, B, C & D) have high fitness, while the overgenerals (E & F) have low fitness. Note that even though the incorrect rules (B & C) have high fitness and will survive with the GA operational, they have low strength, so they will not have much influence in action selection. Consequently we can expect XCS to learn to act correctly in this task.

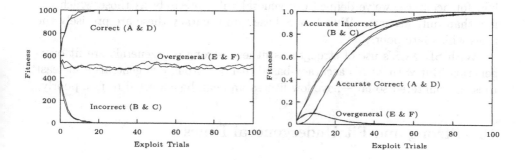

Fig. 9. Rule fitness using strength-based SB–XCS (left) and accuracy-based XCS (right) on the unbiased function from figure 7.

9.2   Comparison on a Biased Reward Function

While both systems seem to be able to handle the unbiased reward function, compare them on the same task when the reward function is biased as in figure 8. Consider the results shown in figure 10 (again, averaged over 100 runs). Although XCS (right) treats the rules in the same way now that the reward function is biased, SB–XCS (left) treats them differently. In particular, rule E, which is overgeneral, has higher expected strength than rule D, which is correct, and with which it competes for action selection. Consequently E is a strong overgeneral

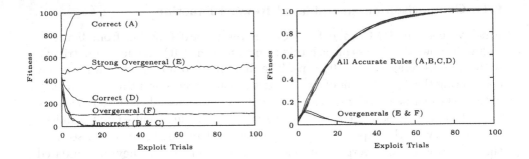

Fig. 10. SB–XCS (left) and XCS (right) on the biased function from figure 8.

(and a fit overgeneral if E and D also compete for reproduction). (Further notes on figures 9 and 10 are available in [10].)

9.3   Discussion

In these trivial tasks XCS's accuracy-based fitness is effective at penalising overgeneral, strong overgeneral, and fit overgeneral rules. This shouldn't be surprising: for accuracy, we've defined overgeneral rules precisely as those which are less than fully accurate. With fitness based on accuracy these are precisely the rules which fare poorly.

With SB–XCS's use of strength as fitness, strong overgenerals are fit overgenerals. But with XCS's accuracy-based fitness, strong overgenerals – at least those encountered so far – have low fitness and can be expected to fare poorly.

10   Strong and Fit Undergeneral Rules

We have seen strong and fit overgeneral rules, which are overgenerals that are stronger or fitter than some correct competitor, and that they result from biases in the reward or variance function. Now we will see that it is also possible for biases in the reward function to produce undergeneral rules which are stronger and fitter than a more-general-yet-correct competitor. The problem with this is that it interferes with the evolution of (accurate) general rules, and so accurate, general representations of the task at hand.

The task in figure 11 demonstrates such a case. At the bottom of the figure are three rules of particular interest, their expected strengths and an evaluation of their generality. Although rule C is the most general-yet-accurate, rules B and A have higher strength. (The principle can be demonstrated with only 2 states, but the example seemed clearer with 4.) This is not a problem in action selection since they all advocate the same action, but it is in reproduction. That is, strong undergeneral rules are not a problem, but fit undergeneral rules are. Let's consider SB–XCS and XCS in turn.

10.1 SB–XCS

With SB–XCS's strength/fitness, the less general rules are fitter, which is clearly undesirable. This effect of fit undergenerals depends on the reward function being biased; with an unbiased reward function, fit undergenerals cannot occur. However, even with unbiased reward functions there must still be some bias toward generality, otherwise SB–XCS will not prefer the more general of two consistently correct rules. This is a real problem, since without a generality preference the population will swell up with an enormous number of correct but overly specific rules.

These problems demonstrate that SB–XCS must factor generality into rule fitness if it is to evolve accurate, general rules. Happily, the niche GA provides an effective fitness bonus which does just this.

10.2 XCS

In XCS, the fitness of A depends on the setting of the accuracy criterion. If it is strict enough, A is, by definition, an overgeneral, and the low fitness it receives is appropriate.

If, however, A is not overgeneral according to the accuracy criterion, it is only as fit as B and C. In other words, XCS will have equal preference for the three, on the basis of their fitness. Thus, although XCS does not suffer from fit overgenerals it still needs some bias toward generality, which, happily, it has thanks to the niche GA.

11 Sequential Tasks

In non-sequential tasks a reinforcement learner approximates the reward function defined by the experimenter, which is in principle enough to maximise the reward it receives. In sequential tasks, however, consideration of the reward function alone is not sufficient, as it defines immediate reward (the reward on a given time step) only. In sequential tasks the learner's actions influence the state of the task and hence which rewards it may receive in the future. Consequently the learner must take into account future consequences of current actions if it is to maximise the total amount of reward it receives over multiple time steps. Many reinforcement learning systems do so by learning a *value function*, which maps each state to an estimate of its long-term value. The value function for a task is implicitly defined by the task definition.

11.1 Q-learning

We'll examine the learning of sequential tasks by Q-learning agents, since XCS and SB–XCS both use the Q-update to estimate rule strength. Q-learners take long-term consequences into account by learning a *Q-function* (also called an *action-value* function) which maps state-action pairs to an estimate of their long

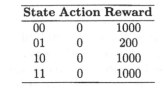

State	Action	Reward
00	0	1000
01	0	200
10	0	1000
11	0	1000

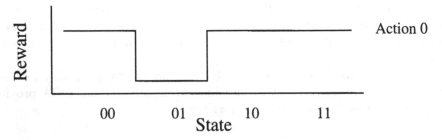

Rule	Condition	Action	E[Strength]	Generality
A	00	0	1000	undergeneral
B	1#	0	1000	undergeneral
C	##	0	800	optimally general

Fig. 11. A task which supports strong undergeneral rules.

term value called their *Q-value*. (Q stands for quality of the state-action pair.) A Q-function is really just a special kind of value function, in which we estimate the long-term value of state-action pairs rather than states. In fact, we can define a value function $V(s)$ as:

$$V(s) = \max_a Q(s, a) \qquad (10)$$

That is, the value of a state s is the value of its highest Q-value.

In non-sequential tasks XCS and SB–XCS both update rules strengths toward their immediate reward, and the population of rules estimates the reward function. In sequential tasks, when using the sequential definition of P (equation 6), the strength of a rule is an estimate of long-term value, and the population of rules and their strengths approximate the value function. Just as in non-sequential tasks, strong and fit overgenerals depend on the reward function, so in sequential tasks they depend on the value function (or Q-function). Consequently, it is of interest to define a unbiased value functions, under which strong and fit overgenerals are impossible following theorems 1 and 5.

Unbiased value function: A value function V is unbiased iff there exists a constant c such that:

$$V(s) = c \qquad (11)$$

for all $s \in \mathcal{S}$. That is, a value function is unbiased if it is a constant function.

Recall from §6.1 that a reward function is unbiased if it is constant *over correct actions*, not simply constant. The difference occurs because the value function is not parameterised by actions; by (10) the value of a state is the value of its highest-valued state-action.

11.2   The Need to Pass Values Back

Q-values are estimates of the long-term value of taking a given action in a given state, not just the immediate reward for doing so. To effect this, each Q-value is updated toward a fraction of the value of its successor. In this way, the estimate of the value of acting now comes to reflect some of the value of what happens later. To see how this works, let's look at the Q-update:

$$
Q(s_t, a_t) \leftarrow Q(s_t, a_t) + \beta \left[\overbrace{r_{t+1} + \gamma \max_a Q(s_{t+1}, a)}^{Target} - \overbrace{Q(s_t, a_t)}^{OldEstimate} \right] \tag{12}
$$

$$\underbrace{\qquad\qquad\qquad\qquad\qquad\qquad\qquad}_{Error}$$

where $Q(s, a)$ is the quality (Q-value) of state-action pair s, a, t is the time step, β is the learning rate, and r is the immediate reward.

Q-values are updated toward the target component of (12), that is, toward the immediate reward r plus some fraction γ of the value of the state which follows it ($\max_a Q(s_{t+1}, a)$). (We say the value of the following state s_{t+1} is discounted and passed back to its predecessor s_t.) In this way a state-action which receives an immediate reward of 0 will have a Q-value greater than 0 if the state it leads to has a Q-value greater than 0 (assuming $\gamma > 0$). In other words, by passing value back a state takes on some of the value of its successor. In this way, Q-updates drive Q-values toward the long-term value of a state-action.

A simple task is shown in figure 12 to illustrate Q-learning. The immediate reward r and true Q-value Q for each state-action, assuming a discount rate of $\gamma = 0.9$, are shown. We define the value of the terminal state to be 0 so that the value of the transition labelled c is due entirely to the immediate reward there.

We can initialise estimates of each Q-value arbitrarily, and the Q-updates will move them toward the true Q-values shown. (In fact, given infinite revisits to each state-action, and appropriately declined α, the estimated Q-values are guaranteed to converge to their true values.)

Knowledge of the true (as opposed to estimated) Q-function for a task is sufficient to act optimally by simply taking the action in the current state with the highest associated Q-value. In figure 12 this means taking a' rather than i' and a rather than i.

11.3   The Need for Discounting

We've just seen that we must pass value back in order to take long-term consequences into account. But how much value should we pass back? In the Q-update,

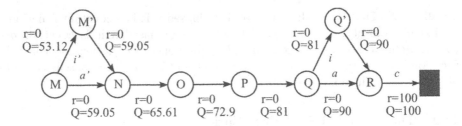

Fig. 12. A simple sequential task showing immediate rewards r and Q-values Q for state transitions using $\gamma = 0.9$. M is the start state and the square state is terminal.

this is parameterised by the discount rate γ. We can think of γ controlling how much consideration the system gives to future rewards in making decisions. At $\gamma = 1.0$ no discounting occurs, and the system will learn the path through the task's states which results in the most reward, regardless of how long the path is. This is often not what we want. For example, being paid £5 a year from now is not as desirable as being paid £5 today, for a number of reasons, e.g., the £5 cannot be spent until it has been paid a year from now, over the course of a year inflation will reduce its value, and, worse, either party might die within a year, preventing payment. If we used $\gamma = 1.0$ in figure 12 then both the i and a transitions would have Q-values of 100 and the system would be unable to choose between them. At the other extreme, if we set γ to 0.0 the system will be shortsighted and take no interest in the future consequences of its actions. This too is often undesirable, as it would lead the system to choose £5 today rather than £1000 tomorrow. In figure 12, $\gamma = 0.0$ would give i and a Q-values of 0, and again the system would be unable to choose between them. Typically we will want to set γ to some value between 0 and 1 in order to give possible future rewards a suitable weighting.

11.4   How Q-functions become Biased

Passing values back from one state to another is necessary for the system to take future consequences of its actions into account. Discounting of these values is necessary for the system to find shorter paths through the state space. However, passing values back and discounting tend to produce one of the two criteria for the production of strong overgenerals: a biased Q-function. Figure 12 demonstrates this effect. Notice that even though there are only two different immediate rewards (r is either 0 or 100), there are many Q-values. Even though the reward function has only two different rewards for correct actions (and so is a little biased) the Q function has many different rewards for correct actions (and so has more biases than the reward function).

The other criterion for the formation of strong overgenerals, that it be possible to act incorrectly, is met in all non-trivial tasks, specifically, in all tasks in which we have a choice of alternative actions in at least one state.

11.5   Examples

Let's look at some examples of strong overgenerals based on figure 12.

Short Sequences can Produce Strong Overgenerals Imagine the situation where an overgeneral matches the state-actions labelled c and i. Now imagine a second rule which is correct and which advocates only the state-action labelled a. This rule competes with the overgeneral for action selection in state Q, and its strength is 90.

Using the strength definition of strong overgenerality (§5.4), the overgeneral rule is a strong overgeneral if it is stronger than a correct competitor. If we assume the overgeneral experiences both the transitions it advocates with equal frequency, we can use the inequality from theorem 6 to tell us whether it is a strong overgeneral:

$$(c + i)/2 > a \qquad (13)$$

which evaluates to:

$$(100 + 81)/2 > 90$$
$$90.5 > 90$$

This example demonstrates that strong overgenerals can be obtained even with very short sequences of states, even the minimal sequence shown here.

Any $0 < \gamma < 1$ can Produce Strong Overgenerals If the value of a state s is $V(s)$, discounting results in $\gamma V(s)$ being passed to the state preceding s. More generally, a state n steps ahead of s receives $\gamma^n V(s)$ from s. Let n be the number of steps between c and both i and a. Then we can rewrite the strong overgeneral inequality (13) as:

$$(c + \gamma^n c)/2 > \gamma^n c \qquad (14)$$

which is true for any $c > 0$, $n \geq 1$, and, significantly, any $0 < \gamma < 1$. In other words, according to our approximate expression, passing back *any* fraction of the value of a state (other than all of it, i.e., $\gamma = 1$) will produce a Q-function capable of supporting strong overgeneral rules in this task. This does not occur if $\gamma = 1$, but we've already seen in §11.3 that passing back the entire value of a state often does not produce the results we want. Of course our calculations have been greatly oversimplified, but it should be clear that all but the simplest sequential tasks can support at least some strong overgeneral rules.

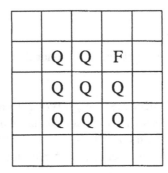

Fig. 13. The basic Woods2 pattern.

Longer Sequences Increase the Bias Now let's look at another example, in which the overgeneral matches in states R and M, and the correct rule matches only in state M. We now use the Q-values labelled a' and i' for a and i in $(c + i)/2 > a$ and obtain $76.56 > 59.05$. Notice that in this example the overgeneral acts incorrectly farther from the goal than in the first example, but its strength exceeds the threshold required of a strong overgeneral by a greater amount. The farther i and a are from the goal, the stronger the strong overgeneral will be, compared to the correct classifier. Notice also that the farther i and a are from the goal, the easier it is to produce strong overgenerals because there are more state transitions in which c can occur and gain enough strength to produce a strong overgeneral. (We can show the same thing by increasing n in equation (14).)

11.6   Woods2 Revisited

In [10] tabular Q-learning, XCS and SB–XCS were evaluated on the sequential Woods2 task (a simple Markov Decision Process), and it was found that although the first two achieved optimal performance of an average of 1.7 steps to food, SB–XCS only reached approximately 3 steps to food. Why is this? This section has argued that sequential tasks have biased value functions, and that this tends strongly to cause strong and fit overgenerals in SB–XCS. Let's look at Woods2 in more detail.

Woods2 is very regular and consists of a basic 5x5 pattern (shown in Figure 13) repeated many times. In some copies of the pattern, G is substituted for F, and Os may be substituted for Qs. Since both kinds of rock and both kinds of food behave identically, the only effect of these substitutions is to increase the number of inputs which the animat may experience, and to create equivalence classes among the inputs, over which the animat can generalise.

Figure 14 shows the basic Woods2 pattern, but with the number of steps needed to reach a goal state shown in each blank cell, and the food and rock states crossed out for legibility. Of the 25 cells in the basic pattern, 16 are blank. Of these, 5 are 1 step from a goal state and 11 are 2 steps.

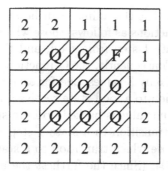

Fig. 14. The basic Woods2 pattern with steps to food from each blank cell.

Given random initial placement of the animat in a blank cell, the average number of steps to food is $(5 \times 1) + (11 \times 2)/16 = 1.6875 \approx 1.7$. That at most 2 actions – i.e., a *minimal* sequence – are needed to reach a goal state suggests Woods2 is not a particularly difficult sequential decision task, even if random behaviour averages 27 steps to food.

The Reward Function Although Woods2 contains 5 types of cell (2 types of food, 2 types of rock, and blanks), the animat can only occupy blank cells. Given the sensory coding specified in Woods2 [19] there are 70 distinct states (input strings) the animat can encounter. These 70 states occur in the 16 blank cells in the basic pattern which the animat can occupy, and we can group states according to the cell in the basic pattern in which they occur. That is, cells can be thought of as macro-states, or equivalence classes among states. Transitions to food states result in a reward of 1000, while all other actions result in a reward of 0. This produces a biased reward function, as in some states the correct action results in 1000 reward and in others 0 reward. The following figure shows the reward for state-actions grouped according to which of the 16 empty cells they occur in, numbering the cells from the top left of the basic pattern and working around its edges clockwise. For each state $\max_a R(s, a)$ is shown as a solid line, while the reward for all others state-actions is shown with a dashed line.

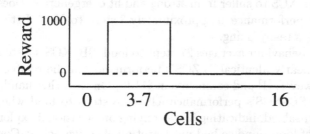

Given the analysis earlier in this chapter, a biased reward function immediately suggests SB–XCS may suffer from strong and fit overgenerals. Let us look next at the Q-function function.

The Q-function Given that the animat can be at most 2 steps from a food state, we can classify all actions as follows: those which transition to food, those which transition to a state 1 step from food, and those which transition to a state 2 steps from food. Consequently, the Q-function for Woods2 is particularly simple, having a range of only 3 values, one for each case above.

Actions which transition to a food state receive a reward of 1000, and, since food states are terminal, no value is backed up from successor states. Assuming $\gamma = 0.71$, actions which transition to a state 1 step from food have a Q-value of $\gamma 1000 = 710$, while those which transition to a state 2 steps from food have a Q-value of $\gamma^2 1000 \approx 504$. (Since the furthest the animat can start from the food is 2 steps, transitions *to* a state 2 steps from food only occur either when it moves from a state 1 step from food (i.e., moves the wrong way), or when it attempts to move into a wall from a state 2 steps from food.)

The following figure shows the Q-function for state-actions belonging to the 16 empty cells, numbering them clockwise from the top left as before. The solid line indicates the Q-value of the optimal action in each state, while the dashed lines show the values of suboptimal actions.

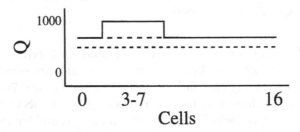

Notice that the Q-function is less biased than the reward function, in that the Q-values are closer together than the rewards. It is more biased, however, in that it may take on 3 values, whereas the reward function has a range of only 2 values. This means there are more points at which the Q-function is biased than the reward function. That is, there are more pairs of state-action pairs whose Q-values differ than there are pairs whose rewards differ.

How Hard is Woods2? Given that the Q-function for Woods2 is biased, we can expect SB–XCS to suffer from strong and fit overgenerals. Does this account for SB–XCS's performance of approximately 3 steps to food? Let's consider how well SB–XCS is really doing.

If random behaviour averages 27 steps to food, SB–XCS's 3 steps seems quite good, and is nearly identical to ZCS's performance on the related Woods1 task [18]. (See, however, [1] and comments in §11.6.) On the other hand, in the experiment in [10] SB–XCS's performance fell to 28 steps to food when exploratory actions were disabled, indicating it was relying on occasional exploratory actions to get it out of loops, and so had not learnt an effective policy. Considering that the animat is only ever at most 2 steps from food, we have to ask how well SB–XCS (and, incidentally, ZCS) are really adapting to this task when they achieve 3 steps to food.

One way to approach this question is to ask how well an agent which cannot learn about sequences of actions would do. That is, how well would an agent do if it learnt how to act in cells adjacent to food, but could not propagate information from these cells to cells further from the food? To find out, tabular Q-learning was run in Woods2 with $\gamma = 0$, so that it would not adapt except in those cells immediately adjacent to food. Other parameters were as in [10].

Averaged over 10 runs, this system converged to approximately 4.3 steps to food, despite the fact that it adapted in only 5 of 16 blank cells and behaved randomly in the others. (In all runs it adapted in those 5 cells which were adjacent to food.) In other words, the performance of SB–XCS and ZCS at 3 steps to food is not much better than a system which is incapable of learning sequences of actions!

How is the relatively good performance from the non-sequential learner possible? To begin, in 5/16 of the trials the animat starts in a cell adjacent to food. In these states it is able to learn the optimal policy, thanks to the immediate reward it receives when it takes the correct action. In the other 11/16 of the trials, it behaves randomly, and usually stumbles by chance across a state adjacent to food – in which it knows how to act – within a few steps. (The system did, however, occasionally time out after 100 time steps.) Inspection of the learnt Q-values confirms that this is how the system operates; state-actions which lead to food have Q-values of 1000, and all others have Q-values of 0. Essentially, Woods2 is not a very difficult sequential decision task. A system which can learn the optimal policy in states adjacent to food (which requires no sequential learning), and which acts randomly elsewhere performs reasonably well.

Inspection of rules evolved by SB–XCS suggests that it adopts the same technique of learning to act optimally in states adjacent to food. (Note that if we consider only these states, the reward function is unbiased.) Given this, its performance in other states must be particularly bad. Rules with numerosity greater than 3 are shown in figure 15, sorted first by numerosity and then by strength. The most numerous rules are unsurprisingly those with strength 1000, i.e., those which map the states adjacent to food. There are, however, a number of high-numerosity overgeneral rules present. Such rules are easily detected in the list as their strengths differ considerably from the true Q-values towards which a tabular Q-learner will converge in this task (1000, 710 and 504).[9] And yet, these rules have considerable experience, suggesting that they would by now have converged to stable values were they not overgeneral. Additionally, the prediction error ε is high for these rules. (SB–XCS calculates prediction error to provide additional statistics on rules, but does not use it itself.)

These rules are in fact strong and fit overgenerals, as they compete with less strong and fit accurate rules. The less strong competitors do not appear in figure 15 simply because they have numerosity less than 4, suggesting that the strong and fit overgenerals are effectively depriving them of numerosity, and, in many cases, actually driving them from the population.

---

[9] The true Q-value of a state-action is defined mathematically by the task definition. The Q-values learnt by an LCS or tabular Q-learner are estimates.

The only correct rule with strength less than 1000 is the last shown. Notice that despite having numerosity 4, it only has experience 6, indicating that it was created only recently. There are very few other correct rules with strength less than 1000 in the population, all with numerosity less than 4, and all with low experience. This suggests that although SB–XCS is able to find some correct rules, they do not tend to survive long, nor do they accumulate large numerosities. (The numerosity of 4 for the last rule shown is probably unusually high for such a rule.) This makes sense; although desirable, such rules have lower fitness than their fit overgeneral competitors. SB–XCS's fitness calculation does not value many of the rules it needs to adapt to this task.

Condition			Act.	Str.	ε	Num.	Exp.
0##00#00	##0##000	##0##1##	NW	1000	0.0	26	1517
##0#0#00	000#00##	1#00#1#0#	SW	1000	0.0	17	2253
0###0#00	#11#####	##000#0#	SE	1000	0.0	16	387
#000#00#	#0#0##0#	1##1##0#	W	1000	0.0	9	376
0######0	0#00#0#0	#00000##	SE	656	0.08	9	318
00000#00	#11#####	##0#0#0#	SE	1000	0.0	8	295
#000#00#	##0#0###	###1##0#	W	1000	0.0	7	292
##00##00	000##0##	1#00##0#	SW	1000	0.0	7	1026
0##00#00	##0##000	1####1##	NW	1000	0.0	6	153
00000000	000#1###	########	S	1000	0.0	6	197
#00##0##	#0#00###	###1##0#	W	1000	0.0	6	324
0#000#00	##0##000	##0##1##	NW	1000	0.0	6	464
0##00#00	000#00##	1#00##0#	SW	1000	0.0	6	1809
#000#00#	#0#00###	###1##0#	W	1000	0.0	6	1577
##000###	#0#000#	##00#0#	SW	772	0.13	6	217
#00#0#00	########	#####0##	SW	701	0.14	6	499
0###0#00	#11#####	##0#0#0	SE	1000	0.0	5	314
##0#0#00	000#00##	1000##0#	SW	1000	0.0	5	461
000###0#	##0##1##	##0###0	S	1000	0.0	5	317
#0######	##0#0###	###1##0#	W	1000	0.0	5	367
0###0000	011#####	##0#####	SE	1000	0.0	5	477
##0#0#00	000#00##	1#000#0#	SW	1000	0.0	5	1335
00##00#0	0##0##00	1##1#0#	W	1000	0.0	5	499
##00##0#	000##0##	1#00##0#	SW	1000	0.0	5	1467
#0000000	000#1###	##0#####	S	1000	0.0	5	963
00##000#	0####0##	#####00#	W	914	0.15	5	263
#00#0#00	########	####00#0	SW	743	0.15	5	446
#0000000	000#1###	##0####0	S	1000	0.0	4	154
000#0#00	##0#1###	##0####0	S	1000	0.0	4	184
0###0000	011#####	##0#0#00	SE	1000	0.0	4	286*
#0#0000#	##0#1###	###000#0	S	1000	0.0	4	330
##0#0#00	000#0##1	#######	SW	1000	0.0	4	1303
#0####0#	##00####	###1##0#	W	1000	0.0	4	530
00000#00	#11#####	##00##0#	SE	1000	0.0	4	1367
##0#0#00	000#00##	1#00##0#	SW	1000	0.0	4	2326
#000#00#	##0#0###	###1#0#0	W	941	0.16	4	229
00###00#	##0#0###	###1####	W	8930	0.22	4	260
0##01#0#	0#0#####	##00###1	SE	709	0.0	4	6

Fig. 15. The most numerous rules evolved by SB–XCS for Woods2.

That SB–XCS and ZCS perform slightly better than the $\gamma = 0$ Q-learner indicates that they *are* adapting somewhat in the states 2 steps from food, which is likely to result from the higher strengths of rules which move towards food. It is clear, however, that even in this very simple task, with a simple,

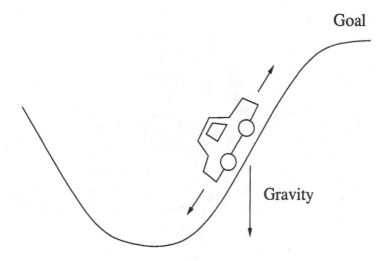

Fig. 16. The mountain car task.

not very biased Q-function, that SB–XCS has not learnt the optimal policy in states 2 steps from the goal. That is, its ability to learn sequences of actions in this task appears to be extremely limited. This suggests that strong and fit overgeneral rules are a considerable problem even when the Q-function is not heavily biased. This is emphasised by the fact that strong overgenerals actually make up a small proportion of the total numerosity in figure 15, and yet appear to prevent SB–XCS from adapting.

ZCS Redux Recently, Bull and Hurst [1] have shown that different parameter settings, and turning the GA off after the system has adapted, allow ZCS to reach near-optimal performance on Woods1. Clearly, in this case it is able to adapt in states which are not adjacent to food. It is possible that SB–XCS could reach similar performance if it were better parameterised. However, it may be that ZCS's fitness sharing is responsible for its near-optimal performance in [1], in which case SB–XCS (lacking fitness sharing) should be unable to match ZCS. This matter deserves further investigation.

The Mountain Car Task The Q-function for Woods2 is rather simple. As a more interesting example, consider the value function for the mountain car task, in which an under-powered car is trapped between two mountains. To escape, the car must learn to rock back and forth in the valley between the two in order to gain sufficient momentum to reach the peak (see figure 16).

The reward function returns -1 on each time step of the task, and so is unbiased. Figure 17 shows the value function for the mountain car task with $\gamma = 1$, using the discretisation found by Reynolds's adaptive resolution reinforcement learning system [13–15]. That is, for each state $V(s) = \max_a Q(s, a)$ is shown. (Figures 16 and 17 appear courtesy of Stuart I. Reynolds.)

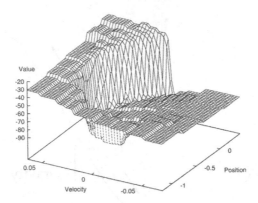

Fig. 17. Value function $V(s) = \max_a Q(s,a)$ for the mountain car task.

Although the value function contains many plateaus (thanks to the state aggregation performed by this learning system), it is nonetheless decidedly very highly biased! Given the great biases in this function, we can expect SB–XCS to produce an overwhelming number of strong and fit overgeneral rules and so be unable to adapt. This task and its complex, highly irregular value function are more typical of tasks studied in the mainstream reinforcement learning literature than Woods2. This task has has been included here simply to illustrate the complexity of a typical value function, and to suggest the difficulty which SB–XCS should have on such tasks.

11.7   When Will the Value Function be Unbiased?

SB–XCS should be able to adapt to tasks with unbiased value functions, since this makes strong and fit overgenerals impossible (§8.1). Under what conditions will a value function be unbiased?

To address this, let's pose the question more carefully. In the task in figure 12, under what reward functions and values of γ will we obtain a value function which is unbiased over non-terminal states? (We do not ask that the terminal state have the same value as the other states since we defined its value to be 0.)

In terms of the reward function, the value function will be unbiased over non-terminal states only when:

$$\max_a R(s,a) = (1 - \gamma)V(s') \tag{15}$$

for all $s, s' \in S$, that is, when the reward function exactly makes up for the value lost from the successor state due to discounting. Two cases where this occurs are:

1. It occurs in non-sequential tasks when the reward function is unbiased.
2. It occurs when the reward function is constant 0 over correct actions (γ can take any value).

In other words, the value function is only unbiased either in non-sequential tasks with unbiased reward functions (case 1), or in uninteresting sequential tasks with a degenerate reward function (case 2).

Note that to assert that the value function is unbiased is to assert that all states have equal value. That is, $V(s) = V(t)$ for all $s, t \in \mathcal{S}$. If this is the case, the task is effectively non-sequential, since there are no sequential decisions to make; being in any state is as good as being in any other. The only issue is what action to take, and, for a classifier system, how to generalise over states and actions.

12   What Tasks can we Solve with SB–XCS?

The circumstances under which a value function will be unbiased, and so under which SB–XCS can be expected to adapt, are limited, consisting of non-sequential tasks with relatively unbiased reward functions, and sequential tasks which are effectively non-sequential.

Furthermore, many of the unbiased non-sequential tasks for which SB–XCS is suitable are probably often better modelled as supervised learning tasks. If we can specify the correct action in each state, we have enough information to do supervised learning. Since the supervised learning paradigm provides the learner with more information (allowing it to avoid the explore/exploit dilemma), agents should be able to adapt more quickly when a task is formulated as supervised learning. Unpublished work has shown a supervised-learning-like XCS to outperform the standard XCS on the 6 multiplexer [7].

SB–XCS's prospects are particularly poor for sequential tasks. Recall that in the task in figure 12 any discounting, and any length of action sequence were sufficient to produce strong overgenerals, under our simplifying assumptions (§11.5).

This analysis suggests SB–XCS will suffer from strong and fit overgenerals in a very wide range of tasks. How much of a problem are strong and fit overgenerals? Experiments with Woods2 show that SB–XCS adapts reasonably well, but largely because much of the task is non-sequential [10]. In the sequential aspects of the task, the relatively few strong overgenerals prevent it from learning an effective policy, meaning it must rely on occasional random actions to break it out of loops. In the unbiased 6 multiplexer, SB–XCS is able to adapt well but is outperformed by XCS [10], and should be outperformed by supervised learners.

This analysis would seem to leave SB–XCS, as it is, with a rather small niche. However, as we'll see in the next section, the addition of fitness sharing might make it a more useful system.

13 Extensions

This section briefly considers some extensions to the work presented earlier.

13.1 Fitness Sharing

We claim XCS avoids strong and fit overgenerals because its accuracy-based fitness penalises overgeneral rules (§9). We claim SB–XCS cannot adapt to tasks with (sufficiently) biased reward functions, because it suffers from strong and fit overgenerals. We have not, however, considered the addition of fitness sharing to SB–XCS. Fitness sharing is known to counter the propagation of overgeneral rules [16, 6, 1], at least in some cases. The addition of fitness sharing to SB–XCS, and its use in other systems, may allow successful adaptation to tasks with biased reward functions, although this has yet to be demonstrated conclusively. Clearly, this is an important direction for future work, and the analysis of rule types in this work and of representations in [10] are two possible starting points for such work.

13.2 Other Factors Contributing to Strong Overgenerals

This work has emphasised the role of the reward and value functions, and of fitness calculation, in the formation of strong and fit overgenerals. Clearly these are major factors, but there are others. Unfortunately, the analysis in this work is a gross oversimplification of more realistic learning tasks, in which it can be very difficult to determine how much of a problem strong and fit overgenerals are likely to be. Additional factors are:

the classifiers - they often apply in many states, not only two which in isolation make strong or fit overgenerals possible.

the explore/exploit policy - The strategy adopted affects how often classifiers are
updated toward their different rewards.

the frequency with which given states are seen - in the non-sequential case this
depends on the training scheme, and on the learner and the task itself in the sequential case.

the selection mechanisms - how high selective pressure is in reproduction and deletion.

the fitness landscape - to what extent strong and fit overgenerals compete with stronger and fitter correct rules.

As a simple example of these factors, an overgeneral might act correctly in 10 states with reward c and incorrectly in only 1 with reward i. Using the strength-based strong overgeneral inequality (theorem 6 from §6), its expected strength would be $(10c+i)/11$, and it would be a strong overgeneral if this value exceeded the strength of some accurate competitor. Similarly, the overgeneral might match in 10 states with reward i and only 1 with reward c.

Although the complexity of the issue makes a more complete analysis difficult it should be clear that the nature of the reward and value functions affect the prevalence of strong and fit overgenerals, and that they are not uncommon.

In the mainstream reinforcement learning literature strength-like values are often stored using look-up tables with an entry for each state-action pair. Such tabular systems are relatively insensitive to the form of the reward and value functions, which may account for the lack of attention this subject has received in the mainstream reinforcement learning literature. SB–XCS, however, is clearly sensitive to the form of the reward and value functions. Other strength-based LCS, even with fitness sharing, must still be influenced by the form of the reward and value functions. That is, even if fitness sharing is able to completely overcome strong and fit overgenerals, and allow strength-based LCS to adapt regardless of the form of the value function, complex value functions are still likely to be more difficult for strength-based LCS and require greater effort to learn. Fitness sharing may overcome strong overgenerals, but with some effort. This constitutes an important difference between strength-based LCS and tabular reinforcement learners. It is curious that the form of these functions has not received more attention in the LCS literature given their sensitivity to them.

13.3   Qualitative and Quantitative Approaches

We could extend the approach taken in this work by removing some of the simplifying assumptions made in §4 and dealing with the resultant additional complexity, and by including the factors in §13.2. For example, we could put aside the assumption of equiprobable states and actions, and extend the inequalities showing the requirements of the reward function for the emergence of strong overgenerals to include the frequencies with which states and actions occur. Taken far enough such extensions might allow quantitative analysis of non-trivial tasks. Unfortunately, while some extensions would be fairly simple, others would be rather more difficult.

At the same time, the most significant results from this approach may be qualitative, and some have been obtained: we have refined the concept of over-generality and argued that strength and accuracy-based LCS have different goals (§5.3), and introduced the concepts of fit overgenerals (§5.5), and strong and fit undergenerals (§10).

We've seen that, qualitatively, strong and fit overgenerals in SB–XCS depend on biases in the reward or value function, and that they are very common. We've also seen that the newer XCS has, so far, dealt with reward function biases much better than SB–XCS (although we have not considered fitness sharing or default hierarchies). This is in keeping with the analysis in §5.3 which suggests that using strength as fitness results in a mismatch between the goals of the LCS and its GA.

In addition to these qualitative and empirical results, some interesting quantitative results have been obtained, despite our simplifications. We've seen that unbiased reward and value functions will not support strong overgenerals (sections §1 and §8.1), and we've seen the conditions under which a value function will be unbiased (§11.7).

Rather than pursue further quantitative results it would be preferable to extend the qualitative approach used here to consider the effects of default hierarchies and mechanisms to promote them, and fitness sharing.

14   Conclusion

We've analysed and extended the concept of overgeneral rules under different fitness schemes. Dealing with such rules is a major issue for Michigan-style evolutionary rule-based systems in general, not just for the two classifier systems considered here. For example, use of alternative representations (e.g., fuzzy classifiers), rule discovery systems (e.g., evolution strategies) or addition of internal memory should not alter the fundamental types of rules which are possible. In all these cases, the system would still be confronted by the problems of greedy classifier creation, overgeneral, strong overgeneral, and fit overgeneral rules. Only by modifying the way in which fitness is calculated (or by restricting ourselves to benign reward functions, if they are suitable), can we influence which types of rules are possible.

Although we have not described it as such, we have examined the fitness landscapes defined by the reward function, γ, task structure, rule representation and fitness scheme used. To avoid pathological landscapes we need appropriate fitness schemes.

15   Acknowledgements

Many thanks to Manfred Kerber, Riccardo Poli, Robert Smith, Stewart Wilson and Stuart Reynolds.

References

1. Larry Bull and Jacob Hurst. ZCS Redux. Evolutionary Computation, EC 10(2): 185-205, 2002.
2. Dave Cliff and Susi Ross. Adding Temporary Memory to ZCS. *Adaptive Behavior*, 3(2):101–150, 1995.
3. David E. Goldberg. *Genetic Algorithms in Search, Optimization, and Machine Learning*. Addison-Wesley, Reading, MA, 1989.
4. John H. Holland. *Adaptation in Natural and Artificial Systems*. University of Michigan Press, 1975.
5. John H. Holland. *Adaptation*. In Rosen and Snell, editors, Progress in Theoretical Biology 4. Plenum, 1976.
6. Jeffrey Horn and David E. Goldberg. Towards a Control Map for Niching. In *Foundations of Genetic Algorithms (FOGA)*, pages 287–310, 1998.

7. Tim Kovacs. Steady State Deletion Techniques in a Classifier System. Unpublished document. School of Computer Science, University of Birmingham, 1997.
8. Tim Kovacs. Strength or Accuracy? Fitness Calculation in Learning Classifier Systems. In Pier Luca Lanzi, Wolfgang Stolzmann, and Stewart W. Wilson, editors, *Learning Classifier Systems. From Foundations to Applications*, volume 1813 of *LNAI*, pages 143–160. Springer-Verlag, Berlin, 2000.
9. Tim Kovacs. Towards a theory of strong overgeneral classifiers. In Worthy Martin and William M. Spears, editors, *Foundations of Genetic Algorithms Volume 6*, pages 165–184. Morgan Kaufmann, 2001.
10. Tim Kovacs. *Strength or Accuracy: Credit Assignment in Learning Classifier Systems.* Springer, 2004.
11. Martin Lettau and Harald Uhlig. Rules of Thumb and Dynamic Programming. Technical report, Department of Economics, Princeton University, 1994.
12. Martin Lettau and Harald Uhlig. Rules of thumb versus dynamic programming. *American Economic Review*, 89:148–174, 1999.
13. Stuart I. Reynolds. Decision boundary partitioning: Variable resolution model-free reinforcement learning. In *Proceedings of the 17th International Conference on Machine Learning*, San Fransisco, 2000. Morgan Kaufmann.
14. Stuart I. Reynolds. Adaptive representation methods for reinforcement learning. In *Advances in Artificial Intelligence, Proceeding of AI-2001, Ottawa, Canada*, Lecture Notes in Artificial Intelligence (LNAI 2056), pages 345–348. Spring Verlag, June 2001.
15. Stuart I. Reynolds. *Reinforcement Learning with Exploration.* PhD thesis, School of Computer Science, University of Birmingham, 2002.
16. Robert E. Smith and Manuel Valenzuela-Rendón. A Study of Rule Set Development in a Learning Classifier System. In J. David Schaffer, editor, *Proceedings of the 3rd International Conference on Genetic Algorithms (ICGA-89)*, pages 340–346, George Mason University, June 1989. Morgan Kaufmann.
17. Richard S. Sutton and Andrew G. Barto. *Reinforcement Learning: An Introduction.* MIT Press, Cambridge, MA, 1998.
18. Stewart W. Wilson. ZCS: A Zeroth Level Classifier System. *Evolutionary Computation*, 2(1):1–18, 1994.
19. Stewart W. Wilson. Classifier Fitness Based on Accuracy. *Evolutionary Computation*, 3(2):149–175, 1995.

Learning Classifier Systems:
A Reinforcement Learning Perspective

Pier Luca Lanzi

Dipartimento di Elettronica e Informazione
Politecnico di Milano
pierluca.lanzi@polimi.it

1 Introduction

Reinforcement learning is defined as the problem of an *agent* that learns to perform a certain task through *trial and error interactions* with an unknown *environment* [27]. Most of the research in reinforcement learning focuses on algorithms that are inspired, in a way or another, by methods of Dynamic Programming (e.g., Watkins' Q-learning [29]). These algorithms have a strong theoretical framework but assume a tabular representation of the value function; thus, their applicability is limited to problems involving few input states and few actions. Alternatively, these methods can be extended for large applications by using function approximators (e.g., neural networks) to represent the value function [27]. In these cases, the general theoretical framework remains but convergence theorems no longer apply.

Another way to tackle reinforcement learning problems is that introduced by Holland [12] with learning classifier systems. These are a learning paradigm in which an agent learns to perform a task by *evolving* a population of condition-action rules (i.e., the classifiers) through *temporal difference learning* [26] and *genetic algorithms* [11]. In particular, they employ a finite population of *classifiers* to represent the current knowledge of the system; *temporal difference learning* to distribute the incoming *reward* to the classifiers accountable for the rewards obtained; *genetic algorithms* to improve the current solution (i.e., the population) through the discovery of "better" classifiers [12].

In the literature, various researchers have compared learning classifier systems and tabular reinforcement learning to highlight the differences and similarities of these two approaches (e.g., [26, 10, 27, 19]). But these works usually leave some open questions concerning how the two approaches relate each other. For instance, both learning classifier systems and reinforcement learning techniques use credit assignment procedures which, in many cases, are based on similar algorithms (e.g., Wilson's XCS [30] uses a modification of Watkin's Q-learning [29]). Thus we can consider genetic algorithms as the most distinctive difference between these two approaches. But *why is there a genetic algorithm in learning classifier systems?* In other words, since genetic algorithms are a search heuristic, *what do genetic algorithms search for in classifier systems?* Moreover, if we

P.L. Lanzi: *Learning Classifier Systems: A Reinforcement Learning Perspective*, StudFuzz
183, 267–284 (2005)
www.springerlink.com

could define what is the goal of evolution in learning classifier systems (e.g., generalization as in Wilson's XCS [30]), *are genetic algorithms the only option?* And finally, in case there are other techniques that might replace genetic algorithms, *is there any advantage in using genetic algorithms?*

One well accepted answer to the first question (*"why is there a genetic algorithm in learning classifier systems?"*) is that the genetic algorithm discovers *better* classifiers and helps the development of better behaviors, that is, better performance (see for instance [12, 11]). However, the word *better* has many different meanings in learning classifier systems. Sometimes *better* means that the rule predicts an higher payoff [12], sometimes *better* means that the rule is more *accurate* [30], while in other cases *better* has no well defined meaning in terms of performance [3, 22]. Moreover, in the literature we have a number of systems which develop satisfactory behaviors *without* an evolutionary component [8, 9, 23] and others which are strongly based on the evolutionary component [3, 22, 30]. So, what are the feasible answers to previous questions?

We believe it is difficult to compare or to *explain* learning classifier systems in terms or tabular techniques by starting from specific learning classifier systems models. In fact, this approach would involve a process of *reverse engineering* aimed at explaining (i) the role of the different classifier systems components and (ii) the different decisions that were taken by the designers of specific classifier system models.

In this paper we present a novel approach for comparing tabular reinforcement learning with classifier systems. Our approach is *constructive* in that it starts from scratch and builds up learning classifier systems on the basis of tabular Q-learning. Accordingly, it does *not* assume any a priori knowledge about classifier systems. We start from the fundamentals of reinforcement learning: a problem modeled as a Markov Decision Process and an algorithm (Q-learning) which is proved to converge to an optimal solution for that problem under adequate hypotheses. Then we ask ourselves: *What do we need to develop a rule-based implementation of Q-learning with generalization capabilities?* To answer this question we develop a formal framework in which we introduce a rule-based representation that we use to implement Q-learning. We do not focus on a specific representation (e.g., the usual ternary representation [12]) to keep the approach general. Instead, we define *rules* by specifying a minimal set of requirements which guarantees that rules are *"adequate"* for implementing *tabular* Q-learning with a rule representation. Then we consider the most important problem of reinforcement learning, i.e., generalization. We formally define generalization with respect to our framework and show that, in this context, generalization can be restated in terms of a *Concept Learning* problem [18]. We shortly discuss the different techniques that can be used to tackle concept learning tasks. We argue that among the available methods for adding generalization to our framework, genetic algorithms are the most general solution since they do not require any assumption on the underlying representation of rules. We add a genetic algorithm to our rule-based implementation of Q-learning and find out that the overall

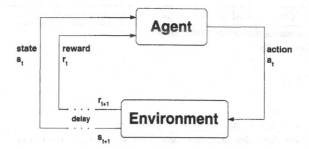

Fig. 1. The agent-environment interaction in reinforcement learning.

framework results in a classifier system model very similar to Wilson's XCS [30]. Finally, we propose some answer to the questions posed at the beginning of this paper.

2   Reinforcement Learning

Reinforcement learning is defined as the problem of an *agent* that learns to perform a task through *trial and error interactions* with an unknown *environment* which provides feedback in terms of numerical *reward* [27]. The agent and the environment interact continually (Figure 1). At time t the agent senses the environment to be in state s_t; based on its current sensory input s_t the agent selects an action a_t in the set A of the possible actions; then action a_t is performed in the environment. Depending on the state s_t, on the action a_t performed, and on the effects of a_t in the environment, the agent receives a *scalar reward* r_{t+1} and a new state s_{t+1}. The agent's goal is to learn how to maximize the amount of reward received. More precisely, the agent usually learns to maximize the *discounted* expected payoff (or *return* [27]) which at time t is defined as:

$$E\left[\sum_{k=0}^{\infty}\gamma^k r_{t+1+k}\right]$$

The term γ is the *discount factor* ($0 \le \gamma \le 1$) which effects how much future rewards are valued at present. To maximize the expected payoff, the agent develops either an action-value function that maps state-action pairs into the expected payoff (as for instance in Q-learning), either a value function that maps states into the maximum payoff that the agent expects starting from that state (as for instance in TD(λ)). The sketch of the typical reinforcement learning algorithm is reported as Algorithm 1: *episodes* represent problem instances, the agent starts an episode in a certain state and continues until a terminal state is entered so that the episode ends; t is the time step; s_t is the state at time t; a_t is the action taken at time t; r_{t+1} is the immediate reward received as a result of performing action a_t in state s_t; function π ($\pi : S \to A$) is the agent's *policy* that specifies

---

Algorithm 1 The typical reinforcement learning algorithm.

1: Initialize the value function arbitrarily
2: **for all** episodes **do**
3:    $t \leftarrow 0$
4:    Initialize s_t
5:    **repeat**
6:        $a_t \leftarrow \pi(s_t)$
7:        perform action a_t; observe r_{t+1} and s_{t+1}
8:        update the value function based on s_t, a_t, r_{t+1}, and s_{t+1}
9:        $t \leftarrow t + 1$
10:   **until** s_t is terminal
11: **end for**

---

how the agent selects an action in a certain state. Note that, π depends on different factors, such as the value of actions in the state, the problem to be solved, and the learning algorithm involved [27].

Most of the research in reinforcement learning focuses on problems which can be modeled with a finite Markov Decision Process (MDP). This is formally defined by: a finite set S of states; a finite set A of actions; a transition function T ($T : S \times A \rightarrow \Pi(S)$) which assigns to each state-action pair a probability distribution over the set S, and a reward function R ($R : S \times A \rightarrow \mathbb{R}$). Given a reinforcement learning problem modeled as an MDP, under adequate hypotheses, the Q-learning algorithm converges with probability one to the optimal action-value function Q^* which maps state-action pairs to the associated expected payoff [28]. More precisely, Q-learning computes by successive approximations the table of all values $Q(s, a)$, named Q-table. $Q(s, a)$ is defined as the payoff predicted under the hypothesis that the agent performs action a in state s, and then it carries on always selecting the actions which predict the highest payoff. The Q-learning algorithm is reported as Algorithm 2. For each state-action pair, $Q(s, a)$ is initialized with a random values, and updated at each time step t that action a_t is performed in state s_t, obtaining a reward r_{t+1} and a new input s_{t+1}, according to the formula:

$$Q(s_t, a_t) \leftarrow Q(s_t, a_t) + \alpha(r_{t+1} + \gamma \max_a Q(s_{t+1}, a) - Q(s_t, a_t))$$

The term α is the *learning rate* ($0 \leq \alpha \leq 1$); γ is the *discount factor*; r_{t+1} is the reward received for performing a_t in state s_t; and s_{t+1} is the state the agent encounters after performing a_t in s_t.

Tabular Q-learning is simple and easy to implement but it is infeasible for problems of interest because the size of the Q-table (which is $|S| \times |A|$) generally grows exponentially in the problem dimensions. This is a major drawback in real applications since the bigger the Q-table: (i) the more the experiences required to converge to a good estimate of the optimal value function; and (ii) the more the memory required to store the table [27]. To cope with the complexity of the tabular representation the agent must be able to *generalize* over its experiences,

---

Algorithm 2 The Q-learning algorithm.

1: Initialize $Q(\cdot, \cdot)$ arbitrarily
2: **for all** episodes **do**
3:    $t \leftarrow 0$
4:    Initialize s_t
5:    **repeat**
6:       $a_t \leftarrow \pi(s_t)$
7:       perform action a_t; observe r_{t+1} and s_{t+1}
8:       $Q(s_t, a_t) \leftarrow Q(s_t, a_t) + \alpha(r_{t+1} + \gamma \max_{a \in A} Q(s_{t+1}, a) - Q(s_t, a_t))$
9:       $t \leftarrow t + 1$
10:    **until** s_t is terminal
11: **end for**

---

i.e., to produce a good approximation of the Q-table from a limited number of experiences, using a small amount of storage. In reinforcement learning generalization is usually realized by using function approximation techniques (see [27] for a review).

3 Rule-based Representation for Q-learning

Consider a reinforcement learning problem modeled as an MDP M defined over a set of states S and a set of actions A. Suppose that we use Q-learning to find an optimal Q-table Q^* for M. For each state-action pair $\langle s, a \rangle$ the value $Q(s, a)$ is an estimate of the payoff p that the agent expects if in s, it performs a, and then it goes on following the best policy. We can represent any Q-table Q as a relation $R_Q \subseteq S \times A \times \mathbb{R}$ defined as:

$$\langle s, a, p \rangle \in R_Q \Leftrightarrow Q(s, a) = p$$

we can interpret a triple $\langle s, a, p \rangle$ in R_Q as the following elementary rule:

$$\text{"if } s \text{ then } a \text{ predicts } p\text{"}$$

which explicits the role of s, a, and p with respect to the reinforcement learning problem. The rule states that if in state s the agent performs action a then the agent should expect a payoff p in the future.

Relation R_Q is of the same size as the Q-table since one triple in R_Q represents exactly one cell of Q. To reduce the size of R_Q we have to increase the representation capabilities, i.e., we need *general rules* which can represent more cells of the Q-table. To define general rules we need a way to specify *general conditions* over the set of states S, i.e., we need a language *adequate* for *building* conditions over S and a way to *interpret* them.

What are the mandatory features of such a language? First, we want to be sure that our language can represent any action-value function, otherwise we would not be guaranteed that the language can represent the optimal Q-table. In addition, we need a (computable) procedure that allows us to build conditions from set of states, otherwise we would not be able to build a rule-based representation starting from a particular action-value function. Finally, we wish to be able to compose simple conditions into more complex ones and to define the semantics of complex conditions by composition of the semantics of their subparts.

Definition 1. Language and Interpretation. *Let S be a set of states, C a grammar, and $L(C)$ the language generated by C. We say that C is representationally adequate for S (or adequate for short) if and only if there exists a computable interpretation function I ($I : L(C) \to 2^S$, where 2^S indicates the powerset of S) and a computable covering function named "cover" (cover $: S \to L(C)$) such that:*

   i. $\forall s \in S \, (\exists c \in L(C) : I(c) = \{s\})$
  ii. $cover(s) = c \Leftrightarrow I(c) = \{s\}$
 iii. *Let $\otimes$ be a compositional operator in C which allows the building of conditions from other conditions; let f be a function defined as $f : 2^S \times 2^S \to 2^S$. Then $\forall c, d \in L(C) : I(c \otimes d) = f(I(c), I(d))$.*

Given a set of states S we denote with C_S a grammar adequate for S.

A grammar C_S is representationally adequate for S if it allows the building of *conditions* which can be interpreted over the set of states S. In particular we require that (i) C_S is able to represent conditions which match single states; that (ii) given a state s it is possible to generate the condition c which matches exactly that state; and that (iii) the *meaning* of conditions depends on the meaning of their subparts, or equivalently, that the semantic of a compound condition depends on the semantic of its subparts.

Observe that in Definition 1, C_S is defined in terms of its representational capabilities rather than its actual syntax. We are not interested in what conditions look like. Instead our main concern is to define an appropriate *semantic* for conditions by means of the interpretation function I. The interpretation function I hides all the details concerning the interpretation of a specific representation while the covering function (*cover*) guarantees that from any state s we can build a condition c which matches exactly s hiding all the details concerning the building of conditions.

Definition 2. Rules. *Consider an MDP M defined over the set of states S and the set of actions A. Let C_S be a grammar adequate for S and let us denote with $\mathcal{R}(C_S, A)$ the set $L(C_S) \times A \times \mathbb{R}$. We define rules as triples $\langle c, a, p \rangle \in \mathcal{R}(C_S, A)$.*

Definition 3. Projections. *Over the set $\mathcal{R}(C_S, A)$, we define three projection functions:*

$$cond : \mathcal{R}(C_S, A) \to C_S \text{ defined as } cond(\langle c, a, p \rangle) = c \tag{1}$$

$$act : \mathcal{R}(C_S, A) \to A \text{ defined as } act(\langle c, a, p \rangle) = a \tag{2}$$

$$pred : \mathcal{R}(C_S, A) \to \mathbb{R} \text{ defined as } pred(\langle c, a, p \rangle) = p \tag{3}$$

Notation: in the following, given a rule $r \in \mathcal{R}(C_S, A)$, we denote with r.c the result of cond(r), with r.a the result of act(r), and with r.p the result of pred(r).

Definition 4. Match operator. *Given an MDP M defined over the set of states S and the set of actions A, and a grammar C_S adequate for S, we say that a rule $r \in \mathcal{R}(C_S, A)$ matches a state $s \in S$ (written as match(r, s)) if and only if $s \in I(r.c)$. More formally:*

$$\forall s, a \; \exists r : \; match(r, s) \Leftrightarrow s \in I(r.c)$$

Rules can be used to represent Q-tables.

Definition 5. *Let S be a set of states, A the set of actions, C_S a grammar adequate for S, P a set of rules ($P \subseteq \mathcal{R}(C_S, A)$), and Q a Q-table. We say that: P represents Q (shortly $P \equiv Q$) if and only if:*

$$\forall \langle c, a, p \rangle \in P : \; (\forall s \in I(c) : Q(s, a) = p)$$

The definition of "$\equiv$" states that a set of rules P *represents* a Q-table Q if it associates the same payoff as Q to each state-action pair. Note that relation "$\equiv$" is quite simple since if the Q-table has all distinct values, there will be exactly one rule for each state-action pair, that is, relation "$\equiv$" will not allow any generalization. It is straightforward to prove the following theorem.

Theorem 1. *Let M be an MDP which defines a reinforcement learning problem over a set of states S and a set of actions A. Let C_S be a grammar adequate for S. For every Q-table Q defined over S and A there exists a set of rules P ($P \subseteq \mathcal{R}(C_S, A)$) which represents Q.*

Proof. Given the set $S = \{s_0 \ldots s_n\}$, since C_S is adequate for S there exists a set of conditions $c_0, \ldots, c_n$ such that $\forall i(cover(s_i) = c_i)$. We can define the following set of classifiers $P = \{\langle c, a, p \rangle \mid cover(s) = c \wedge Q(s, a) = p\}$ which represents Q, i.e., $P \equiv Q$.

Extension. Definition 5 states that $P \equiv Q$ if P is an *exact* representation of Q. This is however a major limitation. In fact, when solving reinforcement learning problems in complex environments we are generally available to sacrifice some predictive accuracy in favor of more compact representations. For instance, function approximators can replace tabular representation when Q-learning is applied to large problems [27]. Accordingly, we can improve the definition of $\equiv$ by including an error threshold ϵ on the representation accuracy provided by P.

Definition 6. Relation $\equiv_\epsilon$. *Let S be a set of states, A the set of actions, and C_S a grammar adequate for S. Let P be a set of rules ($P \subseteq \mathcal{R}(C_S, A)$); Q a Q-table; and $\epsilon \in \mathbb{R}$. We say that P ϵ-represents Q (shortly $P \equiv_\epsilon Q$) if and only if $\forall \langle c, a, p \rangle \in P (\forall s \in I(c) : |Q(s, a) - p| \leq \epsilon)$, or equivalently:*

$$(\forall s, a \; \exists r : \; match(r, s)) \wedge \left(\max_{r \in P \wedge s \in I(r.c)} |Q(s, a) - r.p| \leq \epsilon \right)$$

According to this definition, a set of rules P represents a Q-table Q with an error ϵ if the payoff predicted by the rules in P differs for at most ϵ from the payoff predicted by Q. In practice, ϵ is a similarity threshold which indicates the amount of information we are willing to sacrifice when representing Q with P. Although relation "$\equiv_\epsilon$" is a more general approach than "$\equiv$" to define equivalence between rules and tables, for the sake of simplicity, we prefer to use "$\equiv$". The relation "$\equiv_\epsilon$" will be considered again at the end of the paper in the discussion section.

4   Rule-based Implementation of Q-learning

We now introduce a rule-based implementation of Q-learning which we enrich with generalization in the next sections. First, we implement the basic steps of the agent-environment interaction with rules; second, we define how the expected payoff is associated to actions; then we add the Q-learning update; finally, we outline the overall algorithm.

4.1   The Agent-Environment Interaction

Consider a reinforcement learning problem modeled with an MDP M over the set of states S and actions A. Let C_S be a grammar adequate for S and P a set of rules $(P \subseteq \mathcal{R}(C_S, A))$ which represents a Q-table Q (i.e., $P \equiv Q$). According to the typical reinforcement learning cycle (Algorithm 1), at time t, the agent senses the environment to be in state s_t; it selects an action a_t (according to the policy π); it performs a_t; and finally, it uses the incoming reward r_{t+1} and the new input state s_{t+1} to update the value function. We now implement the first three steps with rules.

We represent the current value function with a set of rules P. To select a_t the agent first needs to determine which rules apply in s_t, i.e., which rules in P match s_t. Given the current state s_t and the set of rules P we define the *match set* $M(P, s_t)$ as:

$$M(P, s_t) = \{r \mid r \in P \land match(r, s_t)\}$$

The set $M(P, s_t)$ contains the rules in P which can be applied in s_t. Therefore, $M(P, s_t)$ can be considered a representation of the row of the Q-table corresponding to s_t. Note that, if no rules in P match s_t, that is $M(P, s_t)$ is empty, new rules are added to P by means of the cover function. Finally, an action to be performed can be selected based on the agent's current policy.

4.2   Evaluating Actions' Value

To perform the Q-learning update, the agent needs an evaluation of the payoff that the rules in $M(P, s_t)$ predict for each possible action. For every action $a \in A$ we define an *action set* $A(P, s_t, a)$ as:

$$A(P, s_t, a) = \{r \mid r \in M(P, s_t) \land a = act(r)\}$$

The action set $A(P, s_t, a)$ contains the rules in P that apply in s_t and advocate action a. Thus $A(P, s_t, a)$ may be considered a representation of a single entry $Q(s_t, a)$ of the Q-table. Note however that in general, $A(P, s_t, a)$ contains more than one rule while $Q(s_t, a)$ is a *unique* value. Accordingly, we introduce a computable *Expected payoff* function E_p that associates an estimate of the expected payoff to a set of rules. At this time it is not important how E_p is computed; we only require that the expected payoff associated to *one* rule is exactly the prediction of the rule. More formally,

$$\forall r \in \mathcal{R}(C_S, A) : E_p(\{r\}) = r.p$$

With the sets $A(P, s_t, a)$ and the function E_p the agent can perform the Q-learning update over the rule-based representation.

4.3   Adding Q-Learning

In Q-learning, the Q-table Q is updated, at time step t, according to the formula:

$$Q(s_t, a_t) \leftarrow (1 - \alpha)Q(s_t, a_t) + \alpha \Delta_{t+1} \tag{4}$$

where

$$\Delta_{t+1} = r_{t+1} + \gamma \max_{a \in A} Q(s_{t+1}, a)$$

We previously argued that the set $A(P, s_t, a_t)$ represents $Q(s_t, a_t)$ and we introduced the function E_p to associate an estimate of the expected payoff to a set of rules. We now use E_p to calculate the value Δ_t for rules (denoted with ΔR_t) as follows:

$$\Delta R_t = r_t + \gamma \max_{a \in A} E_p(A(P, s_t, a))$$

To implement the Q-learning update (4) with rules we introduce a function *update* which, given a rule r, the learning rate α, and a value δ, returns a rule with the same condition, the same action as r, and the prediction parameter updated according to (4). More precisely:

$$update(r, \alpha, \delta) = \langle r.c, r.a, (1 - \alpha)r.p + \alpha\delta \rangle$$

We use the function *update* to implement the Q-learning update on the rules in $A(P, s_{t-1}, a_{t-1})$ so as to form the new action set A:

$$A_{-1} \leftarrow \{update(r_{t+1}, \alpha, \Delta R_{t+1}) \mid r \in A(P, s_t, a_t)\}$$

Finally, the updated rules in A_{-1} replace those in $A(P, s_t, a_t)$ into the current representation P of the Q-table, i.e.:

$$P \leftarrow (P \backslash A(P, s_t, a_t)) \cup A$$

4.4   Rule-Based Implementation of Q-learning

The rule-based implementation of tabular Q-learning is outlined as Algorithm 3. The variable P is the set of rules that represents the Q-table; t is the time step. Initially (line 1) P is empty and (line 3) t is zero. At time t, the rules that can be applied in s_t, i.e., that match s_t, are put in M_t (line 6). If M_t is empty (line 7) then: (i) a condition c matching only s_t is generated through covering (line 8); (ii) for all possible actions $a \in A$ a rule with condition c and an initial prediction value p_I is added to P (lines 9-11); finally (iii) the match set M_t is computed again (line 12). Next, an action a_t is selected according to the agent's policy and performed in the environment; as a result the agent receives a reward r_{t+1} and a new input s_{t+1} (line 15). Then (lines 16-18) Q-learning is used to update the prediction of rules used at time t. First the value of ΔR is computed (line 16); then (line 17) the predictions of rules which acted at the previous time step $A(P, s_t, a_t)$ are updated and the new rules are put in A_{-1}. Finally (line 18) the rules that have been updated (i.e., those in $A(P, s_{t-1}, a_{t-1})$) are replaced with the new ones (i.e., those in A_{-1}) in P. This cycle is repeated until the agent enters a terminal state.

It is easy to prove that Algorithm 3 is equivalent to tabular Q-learning. Basically, Algorithm 3 can be viewed as a Q-learning in which initially the table is empty, in that it has no cells. When the agent encounters a new state s_t the corresponding row is added (lines 7-13). In particular, in state s_t, for each possible state-action pair, exactly one rule is added to P, that is, each position in the Q-table is represented by exactly one rule. Therefore, at each time step the rule update statements are applied just to one rule, making the update strictly equivalent to Q-learning.

5   Rule-based Generalization

The rule-based implementation we presented in the previous section has no generalization and represents each cell of the Q-table with a distinct rule. To get rid of this limitation, we must exploit the generalization capability of the rule representation we introduced. In reinforcement learning generalization is commonly introduced by means of function approximators. The action-value function is viewed as a function parametrized by a vector θ that maps state-action pairs into the expected payoff; experience tuples $\langle s_t, a_t, r_{t+1} \rangle$ collected during the interactions with the environment are used to build the input examples to a supervised learning method (e.g., neural networks) which approximates the action-value function. In most of these approaches, the action-value function is represented *implicitly* by the parameter vector θ; also generalization is defined *implicitly* by the type of function approximator used. For instance, when using neural networks to implement Q-learning, the Q-table is implicitly represented by the network weights and the Q-learning update process is mapped into an update of the weight vector; the type of generalization developed is implicitly

---

Algorithm 3 Rule-based implementation of tabular Q-learning.

1: $P \leftarrow \{\}$
2: **for all** episodes **do**
3:      $t \leftarrow 0$
4:      Initialize s_t
5:      **repeat**
6:          $M_t \leftarrow M(P, s_t)$
7:          **if** M_t is empty **then**
8:              $c \leftarrow cover(s_t)$
9:              **for all** a **do**
10:                 $P \leftarrow P \cup \{\langle c, a, p_I \rangle\}$
11:             **end for**
12:             $M_t \leftarrow M(P, s_t)$
13:         **end if**
14:         $a_t \leftarrow \pi(s_t)$
15:         perform action a_t; observe r_{t+1} and s_{t+1}
16:         $\Delta R \leftarrow r_{t+1} + \gamma \max_a E_p(A(P, s_{t+1}, a))$
17:         $A_{-1} \leftarrow \{update(r, \alpha, \Delta R) \mid r \in A(P, s_t, a_t)\}$
18:         $P \leftarrow (P \backslash A(P, s_t, a_t)) \cup A_{-1}$
19:         $t \leftarrow t + 1$
20:     **until** s_t is not terminal
21: **end for**

---

defined by the network topology. In contrast, rules provide an explicit representation of solutions which allows an explicit definition of generalization. In fact, within a rule-based representation, generalization can be defined as the problem of finding the the smallest set of (maximally general) rules P^* which represent Q^*, that is:

$$P^* = \arg \min_{P \in \{T \mid T \equiv Q^*\}} |P| \qquad (5)$$

Because relation "$\equiv$" requires that P^* is an exact representation of Q^*, the above minimization task can be partitioned into a series of *separate* minimization problems as follows. For all $p \in \mathbb{R}$ and $a \in A$ we denote with $P^*_{a,p}$ the sets of rules in P^* with action a and prediction p. Since the sets $P^*_{a,p}$ form a partition of P^*, the problem of finding P^* is equivalent to the problem of finding the sets $P^*_{a,p}$. Let us denote with $S_{a,p}$ the set of states s such that $Q^*(s, a) = p$; with $C^*_{a,p}$ the smallest set of classifier conditions ($C^*_{a,p} \subseteq L(C_S)$) that *covers* $S_{a,p}$, i.e., such that the union of all the interpretation of the conditions in $C^*_{a,p}$ is $S_{a,p}$.[1] It is easy to show that:

$$P^* = \bigcup_{(p \in \mathbb{R}) \wedge (a \in A)} \{\langle c, a, p \rangle \mid c \in C^*_{a,p}\}$$

Generalization with rule-based representation can be thus decomposed into a series of *set covering* problems: the computing of P^* is equivalent to finding the

---

[1] Formally, $C^*_{a,p}$ covers $S_{a,p} \Leftrightarrow [\cup_{c \in C^*_{a,p}} I(c)] = S_{a,p}$; where I is the interpretation function that must be defined in every grammar adequate for S.

smallest set $C_{a,p}^*$ of conditions which covers the states in $S_{a,p}$. Unfortunately, set covering is an NP-complete problems likewise the minimization problem in (5); therefore exact solutions are infeasible in practice. However, in the field of *concept learning* a number of techniques have been developed that can tackle these types of problems which could be used to add generalization capabilities to our rule-based implementation of Q-learning.

Before we proceed any further we wish to point out that this definition of generalization has two major limitations. First, it can be applied only off-line since P^* can be computed only *after* Q^* was already computed; thus it is infeasible for reinforcement learning applications. Second, if Q^* has all distinct values there will be exactly one classifier for each condition-action pair, i.e., there will be no generalization. On the positive side, our definition relates rule-based generalization to *concept learning* which might provide some useful hints for adding generalization to our rule-based Q-learning.

6   Adding Generalization

We showed that the problem of computing the minimal representation P^* of Q^* is equivalent to finding the smallest sets of classifier conditions $C_{a,p}^*$ which cover the sets of states $S_{a,p}$ corresponding to action a and payoff p in Q^*. Although this formalization suggests off-line solutions, which are infeasible in our case, yet we may extend our implementation of Q-learning by adapting known (either on-line or off-line) concept learning techniques to our problem.

6.1   Concept Learning

Given a set of examples which describe a Boolean concept, concept learning algorithms search for the *most general* set of hypotheses which *explain* the examples. A concept learning algorithm can be viewed as having three components: (i) a *representation* (i.e., a formal language) used to describe the hypotheses; (ii) an *evaluation function* used to evaluate the quality of hypotheses; (iii) a *search procedure* used to search for the hypotheses space.

While there are many representations that can be used to describe hypotheses [18, 7, 21], most of the search strategies proposed fall into two categories i.e.: *specific to general* or *general to specific* [18]. Algorithms employing a *specific to general* search strategy start with very specific hypotheses, which cover few examples, and explore the hypotheses space by generalizing current hypotheses. The search stops when the evaluation function shows that further generalizations deteriorate the current solution.

The vast majority of concept learning algorithms work top down employing a *general to specific* search strategy. They start from the most general hypothesis, which covers all the examples, and explore the hypotheses space top down by specializing current hypotheses. As in the former case, the search stops when the evaluation function shows that further specializations do not improve the current solution. In both cases, the quality of a candidate solution is usually measured as its accuracy with respect to examples. These strategies explore the hypotheses space locally and tend to be stuck in local optima when the hypotheses space is large [18]. Accordingly, other search strategies have been developed that are less sensible to local optima. For instance, some algorithms employ a mix of *general to specific* and *specific to general* [18], others employ completely different search heuristics such as *genetic algorithms* [21], or no explicit search at all as for instance with neural networks [1].

6.2 Concept Learning for Rule-Based Generalization

If we do not consider algorithms based on representations infeasible for rule-based representations (e.g., neural networks), we note that most concept learning algorithms assumes (i) a (partial) *general to specific* ordering relation over the hypotheses space, (ii) some ways to specialize or to generalize an hypothesis, and (iii) a measure of the quality of hypotheses. The ordering relation (i) and the methods (ii) are used to compare hypotheses and to navigate the hypotheses space in search of the most general set of hypotheses which adequately explains the examples with respect to measure (iii).

In the framework presented here, it is possible to define such an ordering relation over the hypotheses space. In fact, we defined conditions by means of a grammar that is *adequate* (Definition 1), thus, a partial ordering ($\preceq$) is induced over conditions by the interpretation function I as follows:

$$\forall c_i, c_j \in C_S : c_i \preceq c_j \Leftrightarrow I(c_i) \subseteq I(c_j)$$

It would also be straightforward to define a function to evaluate the quality of conditions (i.e., our hypotheses). Accordingly, we *might* be tempted to adapt concept learning methods to our framework.

On the other hand, it is worth noting that our effort would be feasible only if the ordering relation ($\preceq$) is easy to evaluate, for instance if it can be *computed in polynomial time*. Otherwise, the comparison of hypotheses might become NP-complete. Let us illustrate this with a example.

Suppose we want to solve a problem involving states defined over bitstrings of size n, i.e., $S = \{0,1\}^n$. Suppose that our conditions are strings of size n in the ternary alphabet $\{0,1,\#\}$, interpreted as usual, i.e.: the $\#$ symbol matches both 0 and 1, all the others match the same input value. In this case, it is very easy to sort hypotheses from the more specific ones, containing only 1s and 0s, to the most general one, containing only $\#$s; in addition, the complexity of "$\preceq$" is linear. Accordingly, concept learning techniques might be a feasible approach to add generalization in our rule-based Q-learning. Obviously, the effectiveness of the approach will depend on the chosen method and on the problem structure. Alternatively, suppose we want to have a richer representation involving conditions defined as *generic expressions* over a set of n Boolean variables. Without any limitation on the condition structure, the only way to determine whether one condition (i.e., one hypothesis) is more general than another one (i.e., to evaluate "$\preceq$") is to perform an input-by-input comparison. In this case, the complexity of "$\preceq$" is 2^n and therefore the navigation of the hypotheses space is infeasible. As a consequence, if we want to use generic expressions, concept learning methods becomes infeasible.

6.3   Generalization with Genetic Algorithms

We want to keep our approach general, thus we must not make any assumptions on the representation of conditions. We should be able to implement generalization without the need of an explicit ordering of the hypotheses space, i.e., without computing "$\preceq$" explicitly. What we need is an on-line definition of generalization.

If condition c_j is more general than condition c_i, then we can hypothesize that c_j will apply more often than c_i, i.e., c_j will match more states than than c_i. More simply, we argue that *general rules will be used more often than specific ones*. Thus, the frequency of rule activation might replace the ordering relation "$\preceq$". This frequency might be estimated explicitly with a new parameter and used by a heuristic to implement the ordering relation. Alternatively, we may use an *implicit* estimate of activation frequency by using a heuristic which acts on the rules in [A]$_{-1}$ at fixed intervals. In fact general classifiers appear more often in A_{-1} (Algorithm 3, line 17) since they are used more often; therefore, a heuristic applied at fixed intervals on A_{-1} will act more often on general classifiers. In addition, to evaluate the quality of hypotheses, we adopt the approach usually followed in concept learning algorithms: we introduce a function ε^{-1} which estimates the quality of an hypothesis as its *accuracy* with respect to the set of examples.[2] To keep the framework general we do not define ε^{-1} here; although, within a more specified framework it could be easily defined (e.g., [30]).

We can finally introduce the heuristic to be applied in A_{-1} with the function ε^{-1} in order to add generalization capabilities to our algorithm. For the sake of generality, this heuristic should be independent from a specific representation, otherwise we would be obliged to specify more details about rules before introducing such a heuristic. Genetic algorithms fit our need perfectly. In fact, they provide a good *representation-independent* paradigm to implement the heuristic we are looking for. Although their original definition considers a binary representation of individuals, there are many successful extensions (e.g., Genetic Programming [13]) based on generic representations. Accordingly, we can add generalization capabilities to our algorithm by applying a genetic algorithm to the classifiers in A_{-1}; as the fitness of the genetic algorithm we use ε^{-1} which estimates the quality of classifiers as their accuracy. The overall result is that rules which represent accurate and general hypotheses will be reproduced and recombined preferentially.

---

[2]In concept learning the quality of an hypothesis is usually estimated as its accuracy or its predicted error; here we prefer to use accuracy since it leads to a simpler definition of fitness for the genetic algorithm.

Algorithm 4 Rule-based Q-learning with Generalization.

```
 1: P ← {}
 2: for all episodes do
 3:     t ← 0
 4:     Initialize s_t
 5:     repeat
 6:         M_t ← M(P, s_t)
 7:         if M_t is empty then
 8:             c ← cover(s_t)
 9:             for all a do
10:                 P ← P ∪ {⟨c, a, p_I⟩}
11:             end for
12:             M_t ← M(P, s_t)
13:         end if
14:         a_t ← π(s_t)
15:         perform action a_t; observe r_{t+1} and s_{t+1}
16:         ΔR ← r_{t+1} + γ max_a E_p(A(P, s_{t+1}, a))
17:         A_{-1} ← {update(r, α, ΔR) | r ∈ A(P, s_t, a_t)}
18:         P ← (P\A(P, s_t, a_t)) ∪ A_{-1}
19:         genetic_algorithm(P, A_{-1})
20:         t ← t + 1
21:     until s_t is not terminal
22: end for
```

Note that we cannot define the genetic operators, here since they depend on the syntax of conditions, i.e., on C_S. On the other hand, because of condition (iii) in Definition 1 we are guaranteed that rule conditions can be arbitrarily split and recombined by exploiting the compositional operators of the grammar C_S.

6.4 The Overall Picture

The algorithm obtained from the rule-based version of Q-learning (Algorithm 3) by adding generalization capabilities is reported as Algorithm 4. The two algorithms differ only for one statement, $genetic\_algorithm(P, A_{-1})$ in line 19, which invokes the genetic algorithm on the sets P and A_{-1}.[3] The readers familiar with Wilson's XCS basic structure will recognize the many analogies between Algorithm 4 and XCS. In particular, by specifying more details of Algorithm 4, we can easily derive all the class of XCS classifier systems [30, 14, 15, 31, 32]. While we refer the readers unfamiliar with XCS structure to Butz and Wilson algorithmic description for a comparison [20].

---

[3]We included P to point out that by acting on A_{-1} the genetic algorithm also modifies the overall solution P.

7 Summary

In this paper we have suggested some answers to what we believe are two important questions for learning classifier system research: Why is there a genetic algorithm in learning classifier systems? Or are genetic algorithms the only option?

To propose an answer to these questions we *did not* start from existing frameworks, as done in former works reported in the literature [26, 10, 27, 19]. Instead we *"started from scratch"*, that is, from the most known reinforcement learning algorithm, Q-learning, and developed a rule-based Q-learning with generalization capabilities. First, we have formally defined *rules* by specifying a minimal set of characteristics which allow us to implement a rule-based (tabular) Q-learning. We discussed *generalization* in the context of rule representation and formally defined *generalization* as the problem of finding the minimal set of rules which represent the optimal Q-table. We showed that, under certain hypotheses, generalization can be viewed as a concept learning problem. We noted that concept learning techniques (adequate for rule-based representation) might be feasible when the representation of rule conditions is computationally convenient. Accordingly, we suggested that, unless we make more assumptions on the representation employed, genetic algorithms are likely to be the best way to approach generalization in the framework we proposed. We added a genetic algorithm to the former rule-based version of Q-learning and found out that the resulting algorithm is a "generalization" of Wilson's XCS and that all the XCS models developed so far can be derived from that algorithm by specifying more details. To end the paper, we wish to suggest some answers to the questions we introduced at the beginning of this paper.

Why is there a genetic algorithm in learning classifier systems? The answer to this question is somewhat obvious and not much novel [30]. If we look at learning classifier systems as reinforcement learning techniques we believe that *generalization* is the most plausible answer to this question. Of course, by focusing on reinforcement learning we implicitly leave out that important branch of research that looks at learning classifier systems as a way of modeling complex adaptive systems [16]. On the other hand, we are also convinced that it would be difficult to discuss learning classifier systems without a formal description of the environment, and Markov Decision Processes serve such purpose quite effectively.

Are genetic algorithms the only option? The answer to this question is much trickier and not so obvious. Our way of looking at learning classifier systems suggests that genetic algorithms are probably the most general approach to add generalization to a rule-based implementation of Q-learning. On the other hand, if we allow a representation of rule conditions that is computationally convenient (for instance those used in concept learning [18]) it could be the case that there are *many* concept learning techniques which might prove as effective as (or even more than) genetic algorithms. This is somehow supported by the results reported with Anticipatory Classifier Systems which usually do not use any evolutionary component [23–25] (see [4,5] for results involving Anticipatory Classifier Systems with genetic algorithms). Note however that because of the assumptions made on the rule syntax such concept learning techniques might become inapplicable to the learning classifier systems models that have been recently developed. Accordingly, we believe that a plausible answer to this question is that genetic algorithms *are not* the only option for generalization but they are likely to be the most *general* way of approaching generalization. Indeed, concept learning are an option if the rule syntax allows. However, when using general purpose representations concept learning techniques could result to be infeasible or just perform worse than genetic algorithms.

Acknowledgements

Pier Luca wishes to thank Stewart Wilson and Marco Colombetti for many discussions and inspirations.

References

1. J.A. Anderson and E. Rosenfeld, editors. *Neurocomputing: Foundations of Research*. MIT Press, 1988.
2. Wolfgang Banzhaf, Jason Daida, Agoston E. Eiben, Max H. Garzon, Vasant Honavar, Mark Jakiela, and Robert E. Smith, editors. *Proceedings of the Genetic and Evolutionary Computation Conference (GECCO-99)*. Morgan Kaufmann: San Francisco, CA, 1999.
3. Lashon B. Booker. Do We Really Need to Estimate Rule Utilities in Classifier Systems? In Lanzi et al. [17], pages 125–142.
4. Butz, M. and Goldberg, D.E. and Stolzmann, W. Introducing a Genetic Generalization Pressure to the Anticipatory Classifier System Part 1: Theoretical Approach. In *Proceedings of the 2000 Genetic and Evolutionary Computation Conference (GECCO-2000)*. Morgan Kaufmann, 2000.
5. Butz, M. and Goldberg, D.E. and Stolzmann, W. Introducing a Genetic Generalization Pressure to the Anticipatory Classifier System Part 2: Performance Analysis. In *Proceedings of the 2000 Genetic and Evolutionary Computation Conference (GECCO-2000)*. Morgan Kaufmann, 2000.
6. Dave Cliff, Philip Husbands, Jean-Arcady Meyer, and Stewart W. Wilson, editors. *From Animals to Animats 3. Proceedings of the Third International Conference on Simulation of Adaptive Behavior (SAB94)*. A Bradford Book. MIT Press, 1994.
7. Pedro Domingos. A unified approach to concept learning. 1997.
8. Jean-Yves Donnart and Jean-Arcady Meyer. A hierarchical classifier system implementing a motivationally autonomous animat. In Cliff et al. [6], pages 144–153.
9. Jean-Yves Donnart and Jean-Arcady Meyer. Spatial Exploration, Map Learning, and Self-Positioning with MonaLysa. In Pattie Maes, Maja J. Mataric, Jean-Arcady Meyer, Jordan Pollack, and Stewart W. Wilson, editors, *From Animals to Animats 4. Proceedings of the Fourth International Conference on Simulation of Adaptive Behavior (SAB96)*, pages 204–213. A Bradford Book. MIT Press, 1996.
10. Marco Dorigo and Hugues Bersini. A Comparison of Q-Learning and Classifier Systems. In Cliff et al. [6], pages 248–255.

11. David E. Goldberg. *Genetic Algorithms in Search, Optimization, and Machine Learning*. Addison-Wesley, Reading, Mass., 1989.
12. John H. Holland. Escaping Brittleness: The possibilities of General-Purpose Learning Algorithms Applied to Parallel Rule-Based Systems. In Mitchell, Michalski, and Carbonell, editors, *Machine learning, an artificial intelligence approach. Volume II*, chapter 20, pages 593–623. Morgan Kaufmann, 1986.
13. John Koza. *Genetic Programming*. MIT Press, 1992.
14. Pier Luca Lanzi. Extending the Representation of Classifier Conditions Part I: From Binary to Messy Coding. In Banzhaf et al. [2], pages 337–344.
15. Pier Luca Lanzi. Extending the Representation of Classifier Conditions Part II: From Messy Coding to S-Expressions. In Banzhaf et al. [2], pages 345–352.
16. Pier Luca Lanzi and Rick L. Riolo. A Roadmap to the Last Decade of Learning Classifier System Research (from 1989 to 1999). In Lanzi et al. [17], pages 33–62.
17. Pier Luca Lanzi, Wolfgang Stolzmann, and Stewart W. Wilson, editors. *Learning Classifier Systems: From Foundations to Applications*, volume 1813 of *LNAI*. Springer-Verlag, Berlin, 2000.
18. Tom Mitchell. *Machine Learning*. McGraw Hill, 1997.
19. D. E. Moriarty, Alan C. Schultz, and John J. Grefenstette. Evolutionary algorithms for reinforcement learning. *Journal of Artificial Intelligence Research*, 11:199–229, 1999. http://www.ib3.gmu.edu/gref/papers/moriarty-jair99.html.
20. M.V. Butz and Stewart W. Wilson. An algorithmic description of XCS. Technical Report 2000017, Illinois Genetic Algorithms Laboratory, University of Illinois at Urbana-Champaign, April 2000.
21. Filippo Neri. *First Order Logic Concept Learning by means of a Distributed Genetic Algorithm*. PhD thesis, University of Milano, Italy, 1997.
22. R. E. Smith, B. A. Dike, B. Ravichandran, A. El-Fallah, and R. K. Mehra. The Fighter Aircraft LCS: A Case of Different LCS Goals and Techniques. In Lanzi et al. [17], pages 285–302.
23. Wolfgang Stolzmann. Anticipatory classifier systems. In *Proceedings of the Third Annual Genetic Programming Conference*, pages 658–664, San Francisco, CA, 1998. Morgan Kaufmann.   http://www.psychologie.uni-wuerzburg.de/stolzmann/gp-98.ps.gz.
24. Wolfgang Stolzmann. An Introduction to Anticipatory Classifier Systems. In Lanzi et al. [17], pages 175–194.
25. Wolfgang Stolzmann and Martin Butz. Latent Learning and Action-Planning in Robots with Anticipatory Classifier Systems. In Lanzi et al. [17], pages 303–320.
26. Richard S. Sutton. Learning to predict by the methods of temporal differences. In *Machine Learning 3*, pages 9–44. Boston: Kluwer, 1988.
27. Richard S. Sutton and Andrew G. Barto. *Reinforcement Learning – An Introduction*. MIT Press, 1998.
28. Christopher Watkins and P. Dayan. Technical note: Q-learning. *Machine Learning*, 8:279–292, 1992.
29. C.J.C.H. Watkins. Learning from delayed reward. PhD Thesis, Cambridge University, Cambridge, England, 1989.
30. Stewart W. Wilson. Classifier Fitness Based on Accuracy. *Evolutionary Computation*, 3(2):149–175, 1995. http://prediction-dynamics.com/.
31. Stewart W. Wilson. Get Real! XCS with Continuous-Valued Inputs. In Lanzi et al. [17], pages 209–220.
32. Stewart W. Wilson. Mining Oblique Data with XCS. volume 1996 of *LNAI*, Berlin, 2001. Springer-Verlag.

Learning Classifier System with
Convergence and Generalization

Atsushi Wada[1,2], Keiki Takadama[1,3],
Katsunori Shimohara[1,2], and Osamu Katai[2]

[1] ATR Network Informatics Laboratories, 2-2-2 Hikaridai, "Keihanna Science City"
Kyoto 619-0288, JAPAN
[2] Kyoto University, Graduate School of Informatics, Kyoto University,
Yoshida-Honmachi, Sakyo-ku, Kyoto 606-8501, Japan
[3] Tokyo Institute of Technology, Interdisciplinary Graduate School of Science and
Engineering, 4259 Nagatsuta-cho, Midori-ku, Yokohama, 226-8503, Japan

1 Introduction

Learning Classifier Systems (LCSs) are rule-based systems whose rules are named
classifiers. The original LCS was introduced by Holland [1, 2], and was intended
to be a framework to study learning in condition-action rules. It included the
distinctive features of a *generalization* mechanism in rule conditions and a *rule
discovery* mechanism using genetic algorithms (GAs) [3]. Later, this original LCS
was revised to its "standard form"[4], which produced many variants [5–8].

Although LCSs were mainly developed in the field of evolutionary computation, they include the concept of *credit assignment*, which actually has an essential connection with reinforcement learning (RL) methods, especially temporal
difference (TD) methods [9]. The *bucket brigade* algorithm [10], a well-known
credit assignment mechanism for LCS, is quite similar to the Sarsa algorithm[11].
Wilson proposed two types of classifier systems, ZCS and XCS having reinforcement processes similar to Q-learning[12], which is called the *Q-bucket brigade*
algorithm.

Despite these essential similarities, when focusing on theoretical aspects,
LCSs do not have such a strong basis as RL methods. While RL methods are
more amenable to mathematical analysis including convergence proofs for several RL methods [13, 14], analysis of LCSs is more difficult and no such proof
exist. This weakness comes from the rule discovery process in LCSs using GA,
which distinguishes LCSs from RL methods but also complicates mathematical
analysis.

Consequently, is it possible to state that an LCS is equal to a RL method
when the LCS's rule discovery process is suppressed but possesses the other
essential properties of reinforcement and generalization? Can we partially introduce a convergence theorem to such an LCS by applying RL theories? Our
objective is to answer these questions in two stages: (1) to compare LCS's reinforcement process with Q-learning enhanced by a common generalization technique, *function approximation* (FA) method; and (2) to propose an LCS whose
reinforcement process has the convergence proof of a RL method.

A. Wada et al.: *Learning Classifier System with Convergence and Generalization*, StudFuzz
183, 285–304 (2005)
www.springerlink.com   © Springer-Verlag Berlin Heidelberg 2005

The rest of this chapter is organized as follows. Section 2 introduces related research. Section 3 describes the reinforcement processes of ZCS, XCS, and Q-learning with FA. Sections 3 and 4 analyses the reinforcement processes of ZCS and XCS from the viewpoint of Q-learning with FA, which reveals an equivalence of ZCS and an inconsistency of XCS compared with Q-learning with FA. Section 5 proposes a new LCS whose reinforcement process is applicable to a convergence proof. Finally, Section 6 gives our conclusions.

2   Related Research

Since LCSs and RL methods share an essential similarity regarding reinforcement, some work comparing LCSs and RL methods, especially Q-learning has been carried out. Dorigo et al. introduced Very Simple Classifier System (VSCS) and compared with Q-learning and showed the equivalence of the reinforcement process under the limitation of VSCS having neither generalization ability nor creation and deletion of classifiers [15]. In a general study that compared LCS and Q-learning, Lanzi implemented LCS by starting from simple Q-learning, extended it to a rule-based model, added a reinforcement mechanism, and finally added generalization ability [16]. However, the objective of this work is neither to show equivalence with Q-learning nor to prove learning convergence in particular. Consequently, the final description of LCS with generalization is in a form incompatible with Q-learning.

The above works achieved our objective to a minor extent by showing the consistency of LCSs with Q-learning when such LCSs have neither the ability to generalize nor invoke the rule discovery mechanism. However, these results are quite limited because most existing LCSs do allow the generalization of classifier conditions.

Therefore, we propose a fair comparison between LCSs and RL methods, which deals not only with the similarity of the reinforcement process, but also with the ability to generalize. For this comparison, we focus on a common generalization technique for RL methods: *function approximation* (FA) method [11]. To clarify the relation between LCS and RL methods with FA, we compare Q-learning enhanced by FA method with two well-known LCSs, Zeroth-level Classifier System (ZCS) [17] and eXtended Classifier System (XCS) [8]. Next, based on these comparisons, we extend a convergence proof for a RL method with FA to a LCS under the condition that its rule discovery process is suppressed, but its generalization ability is retained.

3   Reinforcement Processes of ZCS, XCS, and Q-learning with Function Approximation

In this section, we describe the reinforcement processes of ZCS, XCS, and Q-learning with FA which are necessary for comparisons.

3.1   Reinforcement process of ZCS

Zeroth-level Classifier System (ZCS) is a simple LCS that retains the distinctive features of LCSs: the reinforcement process, rule condition generalization, and a rule discovery process using GA. Due to its simplicity, ZCS has been studied from a minimalist approach to explore LCS's essence [18], which mainly focuses on the rule discovery process. Here, we focus on the other essential property of ZCS, the reinforcement process, and do not describe the rule discovery process. See [17] for the entire algorithm.

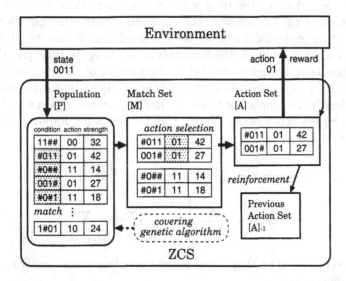

Fig. 1. System architecture of ZCS, which shows process flow from ZCS receiving an input state from environment to ZCS obtaining a reward from environment by performing an action.

Figure 1 shows how ZCS operates by following the flow of processes. The basic component of ZCS is a set of rules named classifiers. Each classifier is comprised of three parts: condition, action, and strength. ZCS maintains a set of classifiers named population [P], and when an input state arrives, the condition part of each classifier in [P] is matched with the state. All classifiers whose conditions match the specified state are collected to organize a *match set* [M]. For example, in Fig. 1, four classifiers whose conditions match input state "0011" are collected to organize a match set. Here, we do not describe in detail the representation of the classifier condition and its matching definition, although each condition of the classifiers in the match set matches at least one state, including "0011." Next, an action is selected from among those advocated by the members of [M]. Many action selection schemes are possible, for example roulette-wheel selection, which chooses an action stochastically by proportional probability based on the total strength of the classifiers whose action part is the same. After action selection, an *action set* [A] is formed from classifiers in the match set [M] that have the same action as the selected action. For example, in Fig. 1, two classifiers whose action is "01" are collected to organize an action set [A], which is preserved to the next time step but renamed as $[A]_{-1}$ while the new action set [A] is formed. The reinforcement process occurs when ZCS obtains an immediate reward, which has the following result:

$$S_{[A]_{-1}} \longleftarrow S_{[A]_{-1}} + \beta \left[r + \gamma S_{[A]} - S_{[A]_{-1}} \right]. \tag{1}$$

Expression $S_{[A]}$ denotes the total value of the strength of all classifiers in action set [A], and the left arrow in the equation denotes the operation to assign the value of the right-hand side to the left-hand side. Parameter β denotes the *learning rate*, which controls the flexibility of learning. In this case, the value of the right-hand side is equally divided by the number of classifiers in the previous action set $[A]_{-1}$, and added to each classifier in $[A]_{-1}$. By introducing notation $|A_{-1}|$ to denote the number of classifiers in the previous action set $[A]_{-1}$, Equation 1 can be rewritten as the following update equation for strength s_j of each classifier $cl_j \in [A]_{-1}$:

$$s_j \longleftarrow s_j + \frac{\beta}{|A_{-1}|} \left[r + \gamma S_{[A]} - S_{[A]_{-1}} \right] \tag{2}$$

$$= s_j + \frac{\beta}{|A_{-1}|} \left[r + \gamma \sum_{cl_k \in [A]} s_k - \sum_{cl_k \in [A]_{-1}} s_k \right], \tag{3}$$

Wilson also proposed an alternative update equation quite similar to the update of Q-learning:

$$S_{[A]_{-1}} \longleftarrow S_{[A]_{-1}} + \beta \left[r + \gamma \max_a S_{[M]|_a} - S_{[A]_{-1}} \right], \tag{4}$$

where $[M]|_a$ denotes the set of classifiers included in match set [M] having action a. This equation can also be rewritten as an update equation for strength s_j for each classifier $cl_j \in [A]_{-1}$:

$$s_j \longleftarrow s_j + \frac{\beta}{|A_{-1}|} \left[r + \gamma \max_a S_{[M]|_a} - S_{[A]_{-1}} \right] \tag{5}$$

$$= s_j + \frac{\beta}{|A_{-1}|} \left[r + \gamma \max_a \sum_{cl_k \in [M]|_a} s_k - \sum_{cl_k \in [A]_{-1}} s_k \right]. \tag{6}$$

In [17], Wilson discussed the relevance between ZCS and Q-learning based on the equation above, which is named the Q-bucket brigade algorithm. However, such a macroscopic viewpoint only deals with the aggregated value of the classifier strengths. It lacks the ability of a microscopic view to relate ZCS to Q-learning at the level of each classifier strength reinforcement, which is the topic discussed in Section 4.

3.2 Reinforcement Process of XCS

EXtended Classifier System (XCS) [8] is a LCS based on ZCS extended to use a distinctive fitness criterion, the *accuracy* for its rule discovery process. Due to this modification, XCS is known to be capable of acquiring appropriately generalized classifier populations [19, 20] and became a mainstream model in LCS fields. Although several theoretical analyses concerning its rule discovery process [21, 22] have been proposed, an analysis regarding both the reinforcement process and the ability of classifier condition generalization has not been addressed. To focus on the relation between XCS and RL methods, here we contrast XCS's reinforcement process with ZCS. See [23] for a detailed description of XCS that includes the rule discovery process.

Classifier. In XCS, a classifier is extended to have three main attributes concerning the reinforcement process: (1) *prediction p*, which estimates the expected payoff when the classifier is used; (2) *error ϵ*, which estimates the error of classifier prediction; and (3) *fitness F*, which estimates the accuracy of the classifier calculated from ϵ. XCS also adopts a *macro-classifier* concept, where classifiers having the same condition and action are aggregated as a single macro-classifier with an additional attribute, *numerosity num*, which denotes the number of aggregated classifiers.

Payoff. For action selection, XCS calculates *payoff $P(a_i)$* for each action a_i in a match set [M] defined as:

$$P(a_i) = \frac{\sum_{cl_k \in [M]|a_i} p_k \times F_k}{\sum_{cl_k \in [M]|a_i} F_k}, \tag{7}$$

instead of using an aggregated value of the classifier strength in ZCS. Here, p_k and F_k denote the prediction and the fitness for the k-th classifier cl_k.

Update. As XCS derives most of its basic framework from ZCS, it also adopts the idea of the Q-bucket brigade algorithm. Instead of using the maximum total strength for updating classifier strength, XCS uses maximum payoff value $\max_a P(a)$. Thus, the update equation of prediction p_j of each classifier cl_j in the previous action set $[A]_{-1}$ is defined as:

$$p_j \leftarrow p_j + \beta(P - p_j), \tag{8}$$

where target payoff value P is defined as:

$$P \leftarrow r + \gamma \max_a P(a). \tag{9}$$

The update equation for error ϵ_j of each classifier $cl_j \in [A]_{-1}$ is defined as:

$$\epsilon_j \leftarrow \epsilon_j + \beta(|P - p_j| - \epsilon_j). \tag{10}$$

The update procedure of fitness F_j is slightly complicated, and is described as the following set of update equations:

$$\kappa_j = \begin{cases} 1 & \text{if } \epsilon_j \leq \epsilon_0 \\ \beta(\epsilon_j/\epsilon_0)^{-\nu} & \text{otherwise.} \end{cases} \tag{11}$$

$$\kappa'_j = \frac{(\kappa_j \times num_j)}{\sum_{cl_k \in [A]_{-1}} (\kappa_k \times num_k)} \tag{12}$$

$$F_j \leftarrow F_j + \beta(\kappa'_j - F_j) \tag{13}$$

where κ_j and κ'_j are values representing the *absolute accuracy* and the *relative accuracy* calculated from error ϵ_j to update fitness F_j. The fitness calculation parameters ϵ_0 and ν are expected to be adjusted depending on a problem to be solved.

3.3   Q-learning with a function approximation method

Q-learning is probably the most well-known RL method capable of online learning, in which an explicit model of the target problem or environment is not required in advance. The name Q denotes *action value function* $Q(x, a)$, which estimates the *action value* for a state-action pair (x, a)[4]. An action value for a state-action pair (x, a) is defined as an expected value of the *return*, a total of future rewards when taking an action a in state x. By denoting a reward given in time step t as r_t, the return at time step t is defined as $R_t = \sum \gamma^k r_{t+k+1}$. Parameter γ denotes a *discount factor* for determining the present value of future rewards, which is also important for avoiding the divergence of action values.

The essence of Q-learning estimates action value $Q(x, a)$ for each state-action pair (x, a) by interacting with a given environment, which results in finding an optimal *policy*, an appropriate action selection for each state which will maximize the return. By denoting an agent's state, action, received reward, and Q values at time step t as x_t, a_t, r_t and Q_t respectively, the update equation of action values is defined[5] as:

$$Q_t(x_{t-1}, a_{t-1}) = Q_{t-1}(x_{t-1}, a_{t-1}) + \alpha \left[v_{t-1} - Q_{t-1}(x_{t-1}, a_{t-1})\right], \tag{14}$$

where v_{t-1} is a target value for the update of $Q_{t-1}(x_{t-1}, a_{t-1})$ defined as:

$$v_{t-1} = r_t + \gamma \max_a Q_{t-1}(x_t, a). \tag{15}$$

Parameter α denotes the *learning rate* which controls the flexibility of learning. Action value function $Q(x, a)$ is often called a Q-table, since it holds action values for all combinations of states and actions represented as $\mathcal{X} \times \mathcal{A}$, where $\mathcal{X}$ and

---

[4] To avoid confusion, in this chapter, the symbol x is used to denote a state instead of using common symbol s, which is already used to denote classifier strength.

[5] To maintain the consistency of the expression with LCSs, here the time steps in the update equation are replaced from $(t + 1)$ and t, which is common in RL literature, with t and $(t - 1)$.

$\mathcal{A}$ are sets of all possible states and actions. This causes the serious state-space explosion problem when the number of dimensions of the states becomes too large.

To avoid this problem, a function approximation method can be applied to compress a Q-table with a large number of states by approximating it with a small number of parameters. Instead of updating a single cell in a Q-table, these parameters are updated using a gradient-descent method described as follows. Let $\boldsymbol{\theta}_t = (\theta_t(1), \theta_t(2), \cdots, \theta_t(n))^T$ ("T" here denotes transposition) approximate the action value function, where $Q_t(x, a)$ is a smooth differentiable function of $\boldsymbol{\theta}_t$ for all $x \in \mathcal{X}$ and $a \in \mathcal{A}$. Gradient-descent methods update $Q_t(x, a)$ by adjusting the parameter vector as in the following equation,

$$\boldsymbol{\theta}_t = \boldsymbol{\theta}_{t-1} + \Delta\boldsymbol{\theta}, \tag{16}$$

where delta value $\Delta\boldsymbol{\theta}$ is defined as:

$$\Delta\boldsymbol{\theta} = \alpha \left[v_{t-1} - Q_{t-1}(x_{t-1}, a_{t-1}) \right] \nabla_{\boldsymbol{\theta}_{t-1}} Q_{t-1}(x_{t-1}, a_{t-1}). \tag{17}$$

Here, gradient $\nabla_{\boldsymbol{\theta}}$ for function f is defined as follows.

$$\nabla_{\boldsymbol{\theta}} f(\boldsymbol{\theta}) = \left(\frac{\partial f(\boldsymbol{\theta})}{\partial \theta(1)}, \frac{\partial f(\boldsymbol{\theta})}{\partial \theta(2)}, \cdots, \frac{\partial f(\boldsymbol{\theta})}{\partial \theta(n)} \right)^T, \tag{18}$$

where $\partial f(\boldsymbol{\theta}) / \partial \theta(k)$ is a partial differential of function f on parameter $\theta(k)$. Especially if Q_t is linear to each parameter in parameter vector $\boldsymbol{\theta}_t$, Q_t can be expressed as a product of parameter vector $\boldsymbol{\theta}_t$ and feature vector $\boldsymbol{\phi}_{xa}$ independent of $\boldsymbol{\theta}_t$. Such an FA is called a linear FA.

$$Q_t(x, a) = \sum_i \theta_t(i) \phi_{xa}(i) = \boldsymbol{\theta}_t^T \boldsymbol{\phi}_{xa}. \tag{19}$$

Note that simple Q-learning using a Q-table can be described as a special case of linear FA in which the parameter vector is composed by listing all of the action values in the Q-table in a row.

4 Comparing ZCS with Q-learning with FA

In this section, we clarify the relationship between the reinforcement process of ZCS and Q-learning with FA[6] by taking the following three steps: (1) introducing required notations for the comparison; (2) comparing representations between the classifier population in ZCS and the approximated action-value function in Q-learning with FA, which leads to our idea that the classifier population in ZCS can be represented as an approximated action-value function in Q-learning with FA; and (3) comparing the update processes of Q-learning with FA and ZCS by applying both update processes to the same ZCS representation.

---

[6] We first proposed the contributions of this section in [24, 25], which focuses on the FA method in RL for the analysis of LCS.

4.1 Notation

For subsequent analysis, some notations are introduced here. Sets $[P_t]$, $[M_t]$, and $[A_t]$ denote classifier population, match set, and action set at time step t, respectively, which contain classifiers as their elements. Let cl_j be a classifier, and then each of the three parts composing the classifier are labeled $condition_j \in C$, $action_j \in A$, and $s_j \in \mathcal{R}$, where set C denotes the set of all possible condition expressions allowed under the classifier representation. Function $equal(x, x')$ returns 1 when $x = x'$, or otherwise 0. Function $match(x, c)$ for $x \in \mathcal{X}$, $c \in C$ returns 1 when condition c matches state s, or otherwise it returns 0[7].

4.2 Comparing representations

To compare the representations of ZCS and Q-learning with FA, first the type of representation used in each model should be stated. In Q-learning with FA, the action-value of state-action pair (x_t, a_t) is directly used for action selection, and calculated from the approximated action-value function denoted as $Q_t(x_t, a_t)$. In ZCS, action selection is done by using the total value of each strength in a set of classifiers matching state x_t and having action a_t.

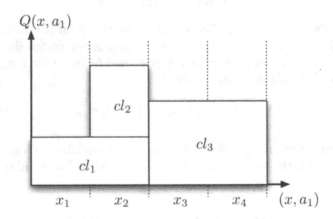

Fig. 2. Relation between action-values and the values of classifier strength.

Because both the values of the approximated action-value function and the aggregated values are used for action selection, we can regard them as corresponding with each other. This idea is described in a graph in Fig. 2, where

---

[7] To keep the discussion general, here we omit detailed definitions of C and $match(x, c)$. By defining C and $match(x, c)$ appropriately, the following discussions can be applied to several classifier representations, such as the real-valued representations used in the XCSR classifier system [26].

the horizontal axis specifies the state-action pair and the vertical axis denotes the corresponding action-values. Here, the three classifiers cl_1, cl_2, and cl_3 match states $\{x_1, x_2\}$, $\{x_2\}$, and $\{x_3, x_4\}$, respectively, all of which have the same action a_1. In the graph in Fig. 2, each classifier is described as a box whose classifier strength contributes to the total strength of each state-action pair. For example, two classifiers cl_1 and cl_2 match state x_2 having action a_1, so the total value of classifier strength $(s_1 + s_2)$ is used for action selection, whose value corresponds to the approximated function denoted as $Q(x_2, a_1)$.

To extend this idea to all possible state-action pairs, all classifiers in population $[P_t]$ should be considered; however, the influence of classifiers unrelated to the state-action pair in question must be excluded. This problem can be solved by using the match function previously defined, which results in the following equation:

$$Q_t(x_t, a_t) = \sum_{cl_j \in [P_t]} s_j \times match_j(x_t, a_t), \qquad (20)$$

where action-value function $Q_t(x_t, a_t)$ is represented by classifier population $[P_t]$ with each classifier cl_j having strength value s_j weighted with match function $match_j(x_t, a_t)$ defined as:

$$match_j(x_t, a_t) = match(x_t, condition_j) \times equal(a_t, action_j). \qquad (21)$$

Equation 20 shows that the classifier population in ZCS can be represented as an approximated action-value function whose approximation design depends on the composition of classifiers in the population. Here the strength of each classifier corresponds to a parameter to approximate the action-value function.

4.3   Comparing update processes

The previous analysis revealed the exchangeability of ZCS representation with that of Q-learning with FA. However, discussing the relationship between both reinforcement processes is insufficient since update processes might be different between ZCS and Q-learning with FA.

To clarify the relation between the update processes, we apply the update equation of Q-learning with FA defined in Equations 16 and 17 to ZCS, whose representation is translated into the form of the approximated action-value function described in Equation 20. This permits a comparison of the two update processes.

Let $\boldsymbol{\theta}_t$ be a parameter vector composed of the set of strength values s_i for all classifiers in classifier population $[P_t] = \{cl_1, cl_2, ..., cl_n\}$, where n denotes the total number of classifiers in $[P_t]$. Then let $\boldsymbol{\phi}_{x_t a_t}$ be a feature vector defined as $(match_1(x_t, a_t), ..., match_n(x_t, a_t))^T$. These two definitions transform Equation 20 into the form of linear FA as Equation 19.

$$Q_t(x_t, a) = \sum_{cl_j \in [P_t]} s_j \times match_j(x_t, a_t) \qquad (22)$$

$$= \begin{bmatrix} s_1 \\ \vdots \\ s_n \end{bmatrix}^T \cdot \begin{bmatrix} match_1(x_t, a_t) \\ \vdots \\ match_n(x_t, a_t) \end{bmatrix} \qquad (23)$$

$$= \boldsymbol{\theta}_t^T \cdot \boldsymbol{\phi}_{x_t a_t}. \qquad (24)$$

By calculating $\nabla_{\boldsymbol{\theta}_{t-1}} Q_{t-1}(x_{t-1}, a_{t-1})$, we get

$$\nabla_{\boldsymbol{\theta}_{t-1}} Q_{t-1}(x_{t-1}, a_{t-1})$$

$$= \left(\frac{\partial Q_{t-1}(x_{t-1}, a_{t-1})}{\partial \theta_{t-1}(1)}, \cdots, \frac{\partial Q_{t-1}(x_{t-1}, a_{t-1})}{\partial \theta_{t-1}(n)} \right)^T \qquad (25)$$

$$= \left(\frac{\partial \sum_{cl_j} s_j \cdot match_j(x_{t-1}, a_{t-1})}{\partial s_1}, \cdots, \frac{\partial \sum_{cl_j} s_j \cdot match_j(x_{t-1}, a_{t-1})}{\partial s_n} \right)^T \qquad (26)$$

$$= (match_1(x_{t-1}, a_{t-1}), \cdots, match_n(x_{t-1}, a_{t-1}))^T \qquad (27)$$

$$= \phi_{x_{t-1} a_{t-1}}^T. \qquad (28)$$

Using these results, we finally get the update equation,

$$\boldsymbol{\theta}_t = \boldsymbol{\theta}_{t-1} + \alpha \left[\left(r_t + \gamma \max_a Q_{t-1}(x_t, a) \right) - Q_{t-1}(x_{t-1}, a_{t-1}) \right] \phi_{x_{t-1} a_{t-1}}^T. \quad (29)$$

This equation can be interpreted from a ZCS viewpoint by: (i) expanding parameter vector $\boldsymbol{\theta}_t$ as a set of classifier strength s_j; (ii) expanding feature vector $\boldsymbol{\phi}_{xa}$ with a set of match functions $match_j(x, a)$; and, (iii) representing action value function $Q_t(x, a)$ with strength s_j and match function $match_j(x, a)$ regarding classifier $cl_j \in [P_t]$ as defined in Equation 20. Applying these replacements, Equation 29 is transformed into an update equation:

$$s_j \longleftarrow s_j + \alpha \left[\left(r_t + \gamma \max_a \sum_{cl_j \in [P_t]} s_j \cdot match_j(x_t, a) \right) \right.$$

$$\left. - \sum_{cl_j \in [P_{t-1}]} s_j \cdot match_j(x_{t-1}, a_{t-1}) \right] \times match_j(x_{t-1}, a_{t-1}). \quad (30)$$

Regarding the definition of match function $match_j(x, a)$, here the aggregated value $\sum_{cl_j \in [P_t]} s_j \cdot match_j(x_t, a^*)$ can be simplified:

$$\sum_{cl_j \in [P_t]} s_j \cdot match_j(x_t, a^*)$$

$$= \sum_{cl_j \in [M_t]|_{a^*}} s_j \cdot match_j(x_t, a^*) + \sum_{a \neq a^*} \sum_{cl_j \in [M_t]|_a} s_j \cdot match_j(x_t, a^*)$$

$$= \sum_{cl_j \in [M_t]|_{a^*}} s_j \cdot 1 + \sum_{a \neq a^*} \sum_{cl_j \in [M_t]|_a} s_j \cdot 0$$

$$= \sum_{cl_j \in [M_t]|_{a^*}} s_j, \qquad (31)$$

where a^* is a greedy action to maximize this aggregated value $\sum_{cl_j \in [M_t]|_{a^*}} s_j$. The other aggregated value $\sum_{cl_j \in [P_{t-1}]} s_j \cdot match_j(x_{t-1}, a_{t-1})$ can also be converted as:

$$\sum_{cl_j \in [P_{t-1}]} s_j \cdot match_j(x_{t-1}, a_{t-1})$$

$$= \sum_{cl_j \in [M_{t-1}]|_{a_{t-1}}} s_j \cdot match_j(x_{t-1}, a_{t-1}) + \sum_{a \neq a_{t-1}} \sum_{cl_j \in [M_{t-1}]|_a} s_j \cdot match_j(x_{t-1}, a_{t-1})$$

$$= \sum_{cl_j \in [M_{t-1}]|_{a_{t-1}}} s_j \cdot 1 \quad + \sum_{a \neq a_{t-1}} \sum_{cl_j \in [M_{t-1}]|_a} s_j \cdot 0$$

$$= \sum_{cl_j \in [A_{t-1}]} s_j. \tag{32}$$

Applying Equations 31 and 32 to Equation 30 results in:

$$s_j \longleftarrow s_j + \alpha \left[r + \gamma \max_a \sum_{cl_k \in [M]|_a} s_k - \sum_{cl_k \in [A]_{-1}} s_k \right] \times match_j(x_{t-1}, a_{t-1}). \tag{33}$$

This equation represents the updating strength process of the classifiers matching the state-action pair (x_{t-1}, a_{t-1}), that is, the classifiers in the previous action set $[A]_{-1}$, while the other classifiers' strengths are not modified. Here, if we set the learning rate of Q-learning with FA α to the value $\beta/|A|_{-1}$ regarding ZCS's learning rate, this update process would be equivalent to that of ZCS described as Equation 5.

Finally, from the equivalence regarding both the representation and the update process, we conclude that the reinforcement process of ZCS and Q-learning with FA are equivalent.[8]

5   Comparing XCS with Q-learning with FA

In this section, XCS's reinforcement process is compared with Q-learning with FA by following the same steps performed in the previous section that compared ZCS's reinforcement process: (1) introducing required notations for the comparison; (2) comparing representations between classifier population in XCS and the approximated action-value function in Q-learning with FA, to represent the classifier population in XCS as an approximated action-value function in Q-learning with FA; and (3) comparing the update processes of Q-learning with FA and XCS.

---

[8] Note that this equivalence assumes that ZCS's rule discovery process is suppressed, that is, the design of the action value function $Q(x, a)$ is not changed through learning. Thus, any operation to change the composition of classifier population [P] will break the equivalence of the reinforcement processes between Q-learning and ZCS. This limitation is quite strong for ZCS, since it cuts off the most distinctive feature of dynamic classifier creation and deletion. However, it still assures equivalence in the time periods between the invocations of such operations.

5.1   Notation

For subsequent analysis, some additional notations regarding XCS are introduced. For classifier cl_j, prediction, error, fitness, and numerosity are denoted as p_j, ϵ_j, F_j, and num_j respectively. The match function $match_j(x, a)$ is defined in the same way as described in Section 4.1, where $match_j(x, a)$ returns 1 if the classifier cl_j matches the state x and has action a, otherwise it returns 0.

5.2   Comparing representations

To compare the representations of XCS and Q-learning with FA, payoff $P(a)$ in XCS defined as Equation 7 is translated into the form of approximated action-value function $Q(x, a)$ as:

$$Q(x, a) = \frac{\sum_{cl_k \in [P]} p_k \cdot F_k \cdot match_k(x, a)}{\sum_{cl_k \in [P]} F_k \cdot match_k(x, a)}, \tag{34}$$

by using match function $match_j(x, a)$. Here prediction p_k of a classifier $cl_j \in [P]$ is regarded as the parameter to approximate action-value function $Q(x, a)$, which can be defined as $\theta = \{p_1, p_2, \cdots, p_n\}$, where n is the number of classifiers in population $[P]$.

Note that Equation 34 is non-linear with parameter p_k due to fitness F_k. Since fitness F_k is calculated indirectly from prediction p_k by sequentially updating Equations 8 to 13, fitness F_k should be regarded as a function of prediction p_k, which makes the equation non-linear with prediction p_k.

5.3   Comparing update processes

To compare the update processes between XCS and Q-learning with FA, we focus on XCS's prediction update defined as Equation 8, which can be rewritten as:

$$p_j \longleftarrow \begin{cases} p_j + \beta \left(r + \gamma \max_a \dfrac{\sum_{cl_k \in [M]|a_i} p_k \times F_k}{\sum_{cl_k \in [M]|a} F_k} - p_j \right) & \text{if } cl_j \in [A]_{-1} \\ p_j & \text{if } cl_j \notin [A]_{-1}, \end{cases} \tag{35}$$

by explicitly describing the range of classifiers involved in a single update process, that is, the previous action set $[A]_{-1}$. Using match function $match_j(x_{t-1}, a_{t-1})$, these two exclusive cases $cl_j \in [A]_{-1}$ or $cl_j \in [A]_{-1}$ can be unified as:

$$p_j \longleftarrow p_j + \beta \left(r + \gamma \max_a \frac{\sum_{cl_k \in [M]|a_i} p_k \times F_k}{\sum_{cl_k \in [M]|a} F_k} - p_j \right) \cdot match_j(x_{t-1}, a_{t-1}). \tag{36}$$

Furthermore, regarding Equation 34, the fitness weighted sum of prediction p_k for classifiers $cl_k \in [M]|_a$ can be transformed to action-value function $Q(x_t, a)$ as:

$$\frac{\sum_{cl_k \in [M]|a} p_k \times F_k}{\sum_{cl_k \in [M]|a} F_k} = \frac{\sum_{cl_k \in [P]} p_k \cdot F_k \cdot match_k(x_t, a)}{\sum_{cl_k \in [P]} F_k \cdot match_k(x_t, a)} = Q(x_t, a). \tag{37}$$

Assigning Equation 37 to Equation 36 and denoting prediction p_j as a parameter $\theta_t(j)$ regarding time step t, we finally get the update equation of XCS represented in Q-learning with FA style:

$$\theta_t(j) \; = \; \theta_{t-1}(j) + \beta \left[v_{t-1} - \theta_{t-1}(j)\right] \cdot match_j(x_{t-1}, a_{t-1}), \qquad (38)$$

where target value v_{t-1} is defined as Equation 15.

For comparisons between XCS and Q-learning with FA, the update equations of Q-learning with FA, Equations 16 and 17, are combined and rewritten in a form regarding each parameter $\theta_t(j)$:

$$\theta_t(j) \; = \; \theta_{t-1}(j) + \alpha \left[v_{t-1} - Q_{t-1}(x_{t-1}, a_{t-1})\right] \cdot \frac{\partial Q_{t-1}(x_{t-1}, a_{t-1})}{\partial \theta_{t-1}(j)}. \qquad (39)$$

Thus far, the update equations of XCS and Q-learning with FA are transformed into Equations 38 and 39, which are in a form comparable with each other. By comparing these equations, two different terms emerge: (1) the difference between the middle terms $[v_{t-1} - \theta_{t-1}(j)]$ in Equation 38 and $[v_{t-1} - Q_{t-1}(x_{t-1}, a_{t-1})]$ in Equation 39; and (2) the difference between the last terms $match_j(x_{t-1}, a_{t-1})$ in Equation 38 and $\partial Q_{t-1}(x_{t-1}, a_{t-1})/\partial \theta_{t-1}(j)$ in Equation 39. Here, for convenience, we name the former the *residual term* and the latter the *gradient term* and discuss in detail.

Residual term. The residual term calculates the difference between the target and the current values that drives the update. In RL literature, this term is called *Bellman residual*, where the name "residual term" originates. In both XCS and Q-learning with FA, the target values are defined as the same value, v_{t-1}. However, the current values which are used to measure the errors are different: parameter $\theta_{t-1}(j)$ in XCS and action value $Q_{t-1}(x_{t-1}, a_{t-1})$ in Q-learning with FA. This inconsistency reveals the essential difference of the update strategy between the two equations. In XCS, parameter $\theta_{t-1}(j)$, that is, prediction p_j is updated based on the local error measured between target values v_{t-1} and p_j. On the other hand, in Q-learning with FA, all the parameter updates use a single global error between target value v_{t-1} and previous action value $Q_{t-1}(x_{t-1}, a_{t-1})$.

Gradient term. In Q-learning with FA, gradient term $\partial Q_{t-1}(x_{t-1}, a_{t-1}) \, / \, \partial \theta_{t-1}(j)$ represents the influence of parameter $\theta_{t-1}(j)$ on action value $Q_{t-1}(x_{t-1}, a_{t-1})$, which is weighted to the corresponding residual term to decide the delta value of the parameter update. The corresponding term in XCS is defined as $match_j(x_{t-1}, a_{t-1})$, which limits the update of the classifiers to the range of previous action set $[A_{-1}]$. However, if following the update equation of Q-learning with FA defined as Equation 39, the gradient term would be the partial gradient of Equation 34 with respect to parameter $\theta_{t-1}(j)$, that is, prediction p_j, which would be inconsistent[9] with $match_j(x_{t-1}, a_{t-1})$, the original XCS update.

---

[9] This inconsistency regarding XCS's gradient term was originally mentioned in [27] to propose a variant of XCS adopting an update equation with a gradient term,

Consequently, from the essential differences regarding the residual and the gradient terms discussed above, the XCS's reinforcement process is shown to be inconsistent with Q-learning with FA regarding both update equations.

6   Applying residual gradient algorithm to LCS

So far, two common LCSs, ZCS and XCS, were compared with Q-learning with FA in Sections 4 and 5, which revealed the equivalence of ZCS and the inconsistency of XCS compared with Q-learning with FA. Based on these results, in this section, we focus on the next objective to introduce a convergence proof of RL methods to the reinforcement process of a LCS by the following three steps: (1) locating ZCS's reinforcement process in the FA classes for RL regarding their convergence; (2) focusing on a residual gradient algorithm, which is a general technique to introduce convergence proof to RL methods with linear FA; and (3) applying a residual gradient algorithm to ZCS to obtain an LCS applicable of convergence proof.

6.1   ZCS and Convergence proof for RL method with FA

In general, simple Q-learning using a Q-table receives the benefit of the convergence theorem proving convergence with a probability 1 under the condition that learning rate α decreases appropriately [12]. However, in the case of Q-learning with FA, the applicability of the convergence theorem depends on the class of FA. For some special cases in the class of linear FA, such as state aggregation, the convergence theorem has already been proved [28–30]. However, in the case of linear FA in general, Baird presented some counterexamples that show that the value function diverged as learning proceeded.

As shown in Section 4, ZCS's reinforcement process can also be viewed as Q-learning with linear FA, which uses Equation 20 for its approximated action value function. Accordingly, the stability of ZCS cannot be proved, even under equivalence conditions (e.g. the rule discovery process is being suppressed.)

However, this limitation can be avoided by applying *residual gradient algorithms* [31], an alternative algorithm for updating values proposed by Baird. Residual gradient algorithm introduce proof of convergence to several classes of FA for many reinforcement learning methods including Q-learning, which is introduced in the following subsection.

---

namely, XCSG. In XCSG, gradient term $F_j / \sum_{cl.k \in [A]_{-1}} F_k$ is used, which is derived as a partial differential of the payoff defined as Equation 7 on the prediction p_j by regarding fitness F_j as constant to p_j. Although XCSG outperforms XCS in multi-step maze problems, this approximation regards F_j as independent of p_j along with the inconsistency of the residual term; XCSG is also inconsistent with Q-learning with FA.

6.2   Residual gradient algorithms for Q-learning

Residual gradient algorithms are modifications of gradient descent algorithms for updating parameters used with FA methods for reinforcement learning. They were originally proposed by Baird [32] to avoid the limitations of gradient descent algorithms shown by several counter examples showing the instability in FA methods applied to such off-policy TD control methods as Q-learning with linear and non-linear approximation classes.

By using target value v_t defined as Equation 15, the residual gradient algorithm for Q-learning is described by the alternative delta value for the parameter vector update:

$$\Delta\theta = \alpha \left[v_{t-1} - Q_{t-1}(x_{t-1}, a_{t-1})\right] \left[\phi_{x_{t-1}a_{t-1}} - \gamma\phi_{x'a^*}\right], \tag{40}$$

where a^* is a greedy action to maximize $Q_{t-1}(x', a)$. State x' denotes a following state after visiting state x_{t-1} and taking action a_{t-1}, which must be generated independently against x_t. When state transition is deterministic, x' becomes equal to x_t, which is the case we assume in this section.

This delta value is divided into two delta values as $\Delta\theta = \Delta_1\theta + \Delta_2\theta$, where $\Delta_1\theta$ and $\Delta_2\theta$ are defined as:

$$\Delta_1\theta = \alpha \left[v_{t-1} - Q_{t-1}(x_{t-1}, a_{t-1})\right] \phi_{x_{t-1}a_{t-1}}, \tag{41}$$

$$\Delta_2\theta = -\alpha\gamma \left[v_{t-1} - Q_{t-1}(x_{t-1}, a_{t-1})\right] \phi_{x_t a^*}. \tag{42}$$

The difference between this update and the original update described as Equation 17 is the additional delta value $\Delta_2\theta$, which modifies the values of parameters concerned with $Q_{t-1}(x_t, a^*)$. Thus, a residual gradient algorithm can be obtained by inserting an additional update term $\Delta_2\theta$ into the original update.

6.3   ZCS with residual gradient algorithms

To obtain ZCS with residual gradient algorithm, an additional term $\Delta_2\theta$ defined as Equation 42 must be introduced to ZCS's reinforcement process. This can be done by translating Equation 42 in the manner of ZCS as follows:

$$\Delta_2\theta = -\alpha\gamma \left[v_{t-1} - Q_{t-1}(x_{t-1}, a_{t-1})\right] \phi_{x_t a^*} \tag{43}$$

$$= -\frac{\beta\gamma}{|A_{-1}|} \left[r + \gamma \sum_{cl_k \in [M]|_{a^*}} s_k - \sum_{cl_k \in [A]_{-1}} s_k\right] \phi_{x_t a^*}, \tag{44}$$

which can be decomposed to the delta value for each parameter $\theta(j)$, that is, the delta value for each strength s_j:

$$\Delta_2 s_j = -\frac{\beta\gamma}{|A_{-1}|} \left[r + \gamma \sum_{cl_k \in [M]|_{a^*}} s_k - \sum_{cl_k \in [A]_{-1}} s_k\right] match_j(x_t, a^*). \tag{45}$$

Since a^* is defined as an action a that maximizes $Q_{t-1}(x_t, a)$, that is, the aggregated value $\sum_{cl_k \in [M]|_a} s_k$, Equation 45 can be further expanded as:

$$\Delta_2 s_j = \begin{cases} -\dfrac{\beta \gamma}{|A_{-1}|} \left[r + \gamma \displaystyle\sum_{cl_k \in [M]|_{a^*}} s_k - \sum_{cl_k \in [A]_{-1}} s_k \right] & \text{if } cl_j \in [M]|_a^* \\ 0 & \text{if } cl_j \notin [M]|_a^*. \end{cases} \tag{46}$$

By inserting the update process regarding this Equation 46 to ZCS's reinforcement process, a ZCS with a residual gradient algorithm is obtained whose algorithmic description is shown in Figure 3. It is almost the same as the original ZCS; however, it differs where the set of classifiers $[M]|_{a^*}$ are updated from lines 23 to 25 in Fig. 3. By suppressing the rule discovery process of lines 5, 15, and 26, which is the same as the conditions we mentioned for equivalence between ZCS and Q-learning, this algorithm will become equivalent with Q-learning with a residual gradient algorithm.

For convergence, a residual gradient algorithm requires a condition regarding the decay of the learning rate. By denoting the learning rate at a time step t as α_t, the condition is defined as:

$$\sum_{t=0}^{\infty} \alpha_t = \infty \quad \text{and} \quad \sum_{t=0}^{\infty} \alpha_t^2 < \infty, \tag{47}$$

which is required for the action value function to converge with probability 1. In the case of ZCS, learning rate α corresponds to $\beta/|A_{-1}|$, which means that it is dependent on the size of a previous action set $[A]_{-1}$ regarding the previous state-action pair (x_{t-1}, a_{t-1}). As this dependency is not assumed in the original convergence proof, an additional condition is introduced that makes learning rate β independent of a state-action pair. Let α be a learning rate that decays independently of a state-action pair, satisfies Equation 47, and uses $\beta = \alpha \cdot |A_{-1}|$ as the learning rate in ZCS. This additional condition and the condition retaining static classifier population $[P]$ are sufficient conditions for convergence[10].

7   Discussions and Conclusions

In this chapter, as the first step toward developing a foundation of LCSs, we focus on the relation between LCSs and RL methods regarding both generalization

---

[10] Baird proposed residual gradient algorithms in [31, 32] as a technique to attain convergence proof for RL methods with linear FA in general. However, the proof in these works is not regarded as *rigorous* enough in the RL field. Although such proofs are still awaited, recent contributions concerning RL theories partially obtained mathematically rigorous proofs for residual gradient algorithms for linear FA [33], where the result is limited to methods adopting: (1) value estimation, where the policy is kept unchanged through learning; and (2) synchronous update, where the change in parameter values are applied periodically and synchronously.

1 Initialize [P]

2 Repeat (for each episode):

3    $x \leftarrow$ initial state of episode

4    $[M] \leftarrow \{cl_i \in [P] \mid x \in condition_i\}$

5   Invoke COVERING
       until [M] satisfies covering condition

6   For all $a \in \mathcal{A}(x)$:

7      $[M]|_a \leftarrow \{cl_i \in [M] \mid action_i = a\}$

8      $payoff_a \leftarrow \sum_{cl_i \in [M]|_a} s_i$

9   Repeat (for each step of episode):

10   Choose action $a \in \mathcal{A}(x)$ using policy
          derived from $payoff_a$ (e.g., ϵ-greedy)

11     $[A] \leftarrow \{cl_i \in [M] \mid action_i = a\}$

12   Take action a, observer reward r,
          and next state x'

13     $\delta \leftarrow r - payoff_a$

14     $[M] \leftarrow \{cl_i \in [P] \mid x' \in condition_i\}$

15   Invoke COVERING
          until [M] satisfies covering condition

16   For all $a \in \mathcal{A}(x')$:

17       $[M]|_a \leftarrow \{cl_i \in [M] \mid action_i = a\}$

18       $payoff_a \leftarrow \sum_{cl_i \in [M]|_a} s_i$

19     $a^* \leftarrow \arg\max_a payoff_a$

20     $\delta \leftarrow \delta + \gamma payoff_{a^*}$

21   For all cl_i in [A]

22       $s_i \leftarrow s_i + \beta\delta/|A|$

23     $[M|_{a^*}] \leftarrow \{cl_i \in [M] \mid action_i = a'\}$

24   For all cl_i in $[M|_{a^*}]$

25       $s_i \leftarrow s_i - \beta\delta\gamma/|A|$

26   Invoke GENETIC ALGORITHM
          in probability $\rho/2$

27   until x' is terminal

Fig. 3. An algorithmic description of ZCS with residual gradient algorithm.

ability and rule condition generalization in LCSs and FA method in RL. In Sections 4 and 5, two common LCSs, ZCS and XCS are compared with Q-learning with FA, which revealed: (1) an equivalence of ZCS and Q-learning with linear FA; and (2) an inconsistency of XCS with Q-learning with FA, both regarding their reinforcement processes. Based on these results, in Section 6, a residual gradient algorithm is applied to ZCS, which resulted in a LCS with generalization ability that guarantees convergence with its conditions clarified.

Although these results are limited under the condition of LCS's rule discovery process being suppressed, LCS's reinforcement process can be discussed on the strong basis of RL methods with FA. One promising topic utilizing this basis is the generalization of continuous state-action space. For RL methods, such generalization techniques are available such as tile coding and radial basis functions (RBF), and applying these techniques to LCSs is expected to increase sophistication of the handling of continuous state-action space in LCSs.

At the same time, the design of the rule discovery process regarding the basis of RL should be addressed. Despite the consistency of ZCS compared with Q-learning with FA, ZCS's original rule discovery process has a problem called *overgeneralization* [8, 34, 35]. On the contrary, XCS avoids this problem by adopting accuracy as its fitness criteria, but this modification including prediction updates breaks the consistency between Q-learning with FA. The unification of the advantages of ZCS's consistency between Q-learning with FA and XCS's advantage of the effective rule discovery process will be one of our future works.

Acknowledgments

This research was conducted as part of 'Research on Human Communication' with funding from the National Institute of Information and Communications Technology of Japan and the Okawa foundation for information and telecommunications.

References

1. Holland, J.H.: Adaptation in Natural and Artifical Systems. The University of Michigan Press, Michigan (1975)
2. Holland, J.H.: Adaptation. Progress in Theoretical Biology IV (1976) 263–93
3. Goldberg, D.E.: Genetic Algorithms in Search, Optimization, and Machine Learning. Addison-Wesley, MA. (1989)
4. Holland, J.H.: Adaptive algorithms for discovering and using general patterns in growing knowledge bases. International Journal for Policy Analysis and Information Systems 4 (1980) 245–268
5. Booker, L.B.: Do We Really Need to Estimate Rule Utilities in Classifier Systems? In: Learning Classifier Systems. Springer (1998) 125–142
6. Riolo, R.L.: Lookahead planning and latent learning in a classifier system. In: From Animals to Animats: Proceedings of the First International Conference on Simulation of Adaptive Behavior. (1991) 316–326

7. Smith, R.E., Dike, B.A., Ravichandran, B., El-Fallah, A., Mehra, R.K.: The fighter aircraft LCS: A case of different LCS goals and techniques. Lecture Notes in Computer Science **1813** (2000) 283–300
8. Wilson, S.W.: Classifier fitness based on accuracy. Evolutionary Computation **3** (1995) 149–175
9. Sutton, R.S.: Learning to predict by the methods of temporal differences. Machine Learning **3** (1988) 9–44
10. Holland, J.H.: Escaping brittleness: the possibilities of general-purpose. Machine Learning, an artificial intelligence approach **2** (1986)
11. Sutton, R.S.: Generalization in reinforcement learning: Successful examples using sparse coarse coding. In Touretzky, D.S., Mozer, M.C., Hasselmo, M.E., eds.: Advances in Neural Information Processing Systems. Volume 8., The MIT Press (1996) 1038–1044
12. Watkins, J.C.H.: Learning from Delayed Rewards. PhD thesis, Cambridge University (1989)
13. Singh, S.P., Jaakkola, T., Littman, M.L., Szepesvári, C.: Convergence results for single-step on-policy reinforcement-learning algorithms. Machine Learning **38** (2000) 287–308
14. Watkins, J.C.H., Dayan, P.: Technical note: Q-learning. Machine Learning **8** (1992) 279–292
15. Dorigo, M., Bersini, H.: A comparison of Q-learning and classifier systems. In: Proceedings of From Animals to Animats, Third International Conference on Simulation of Adaptive Behavior. (1994)
16. Lanzi, P.L.: Learning classifier systems from a reinforcement learning perspective. Soft Computing **6** (2002) 162–170
17. Wilson, S.W.: ZCS: A zeroth level classifier system. Evolutionary Computation **2** (1994) 1–18
18. Bull, L., Hurst, J.: ZCS Redux. Evolutionary Computation **10** (2002) 185–205
19. Kovacs, T.: Evolving optimal populations with XCS classifier systems. Master's thesis, School of Computer Science, University of Birmingham (1996)
20. Wilson, S.W.: Generalization in the XCS classifier system. In: Genetic Programming 1998: Proceedings of the Third Annual Confe rence, Morgan Kaufmann (1998) 665–674
21. Butz, M.V., Pelikan, M.: Analyzing the evolutionary pressures in XCS. In: Proceedings of the Genetic and Evolutionary Computation Conference (GECCO-2001). (2001) 935–942
22. Butz, M.V., Goldberg, D.E., Lanzi, P.L.: Bounding learning time in XCS. In: Proceedings of the Genetic and Evolutionary Computation Conference (GECCO-2004). (2004)
23. Butz, M.V., Wilson, S.W.: An Algorithmic Description of XCS. In: Advances in Learning Classifier Systems. Volume LNAI 1996. Berlin: Springer-Verlag (2001) 253–272
24. Wada, A., Takadama, K., Shimohara, K., Katai, O.: Analyzing generalization in learning classifier system: From the aspect of function approximation method for reinforcement learning. In: The Forth Meeting for Youth COMmunity 2003 (MYCOM2003). (2003) 74–75
25. Wada, A., Takadama, K., Shimohara, K., Katai, O.: Comparison between Q-learning and ZCS Learning Classifier System: From aspect of function approximation. In: The 8th Conference on Intelligent Autonomous Systems. (2004) 422–429
26. Wilson, S.W.: Get real! XCS with continuous-valued inputs. Lecture Notes in Computer Science **1813** (2000) 209–222

27. Butz, M.V., Goldberg, D.E., Lanzi, P.L.: Gradient descent methods in Learning Classifier Systems: Improving XCS performance in multistep problems. In: Proceedings of the Genetic and Evolutionary Computation Conference (GECCO-2004). (2004) 751–762
28. Gordon, G.J.: Stable function approximation in dynamic programming. In Prieditis, A., Russell, S., eds.: Proceedings of the Twelfth International Conference on Machine Learning, San Francisco, CA, Morgan Kaufmann (1995) 261–268
29. Singh, S.P., Jaakkola, T., Jordan, M.I.: Reinforcement learning with soft state aggregation. In Tesauro, G., Touretzky, D., Leen, T., eds.: Advances in Neural Information Processing Systems. Volume 7., The MIT Press (1995) 361–368
30. Tsitsiklis, J.N., Roy, B.V.: Feature-based methods for large scale dynamic programming. Machine Learning **22** (1996) 59–94
31. Baird, L.C.: Reinforcement Learning Through Gradient Descent. PhD thesis, Carnegie Mellon University, Pittsburgh, PA 15213 (1999)
32. Baird, L.C.: Residual algorithms: Reinforcement learning with function approximation. In: International Conference on Machine Learning. (1995) 30–37
33. Merke, A., Schoknecht, R.: Convergence of synchronous reinforcement learning with linear function approximation. In: Proceedings of the Twenty-first international conference on Machine learning (ICML2004). (2004)
34. Cliff, D., Ross, S.: Adding temporary memory to zcs. Adaptive Behavior **3** (1994) 101–150
35. Kovacs, T.: Strength or Accuracy? Fitness Calculation in Learning Classifier Systems. Volume 1813. Springer-Verlag (2000)

Section III

Problem Characterization

On the Classification of Maze Problems

Anthony J. Bagnall and Zhanna V. Zatuchna

University of East Anglia, Norwich, NR4 7TJ, England
ajb@cmp.uea.ac.uk
z.zatuchna@uea.ac.uk

1 Introduction

A maze is a grid-like two-dimensional area of any size, usually rectangular. A maze consists of *cells*. A cell is an elementary maze item, a formally bounded space, interpreted as a single site. The maze may contain different obstacles in any quantity. Some may be significant for learning purposes, like virtual *food*. The agent is randomly placed in the maze on an empty cell. The agent is allowed to move in all directions, but only through empty space. The task is to learn a policy to reach food as fast as possible from any square. Once the food is reached, the agent position is reset to a random one and the task repeated.

Maze environments have been widely used as testbed problems in machine learning research [4], especially in the learning classifier system literature [27, 13, 2, 1]. There have been made some different approaches for the examination of problem complexity for learning agents [23, 8], some of them for the maze problem domain [15, 25]. The aim of the paper is through a thorough survey and analysis of research using alternative maze structures to define metrics to quantify the complexity of maze problems. For the purpose we collected 50 different mazes, 44 of them have been used in at least 55 publications.

Frequently mazes have been given different names in different publications. Where possible, we assign a maze the name most commonly used in the literature. If there is no commonly accepted name, the maze is named after the author of the first publication referring to it, suffixed by the year of publication. The techniques applied to mazes can be categorised as follows: XCS [27, 11, 12, 7] (13 mazes in 17 papers); ZCS [26, 5, 7, 1] (7 mazes in 10 papers); ACS [22, 3, 18] (18 mazes in 11 papers); Other methods (Witness algorithm [4], Q-learning with added memory [14], ATNoSFERES [10], others) (25 mazes in 20 papers). A more complete description can be found in [28].

The characteristics of mazes that determine the complexity of the learning task fall into two classes: those that are independent of the learning algorithm, such as number of squares and density of obstacles (described in Section 2) and those that are an artifact of the agents ability to correctly detect its current state, such as the number of aliasing squares. In Section 3 we examine the effect of agent perception on the problem complexity, and describe how alternative types of aliasing may effect complexity of mazes. Generalization and noise problems in maze environments are considered in the Section 4. Section 5 outlines the areas of future work. In Section 6 we summarize our conclusions.

A.J. Bagnall and Z.V. Zatuchna: *On the Classification of Maze Problems*, StudFuzz **183**, 307–316 (2005)
www.springerlink.com

2   Agent independent maze attributes

Size. The number of cells a maze contains obviously effects the complexity. The mazes range from 18 cells (Cassandra4 [4], Fig.1(a)) to 1044 cells (Woods7 [26]). Mazes smaller than 50 cells are classified as *small* (19 mazes). *Medium* mazes, such as MiyazakiC [19], Fig.1(b), have between 50 and 100 cells, and *large* mazes have more than 100 cells. We denote the number of cells of a maze m as s_m.

(a)

(b)

Fig. 1. (a) Cassandra4, $s_m = 18, \phi_m = 1.33$; (b) MiyazakiC, $s_m = 64, \phi_m = 3.37$

Distance to food. The average distance to food (ϕ_m) in a maze is an important characteristic of complexity. The bigger the value, the more difficult the maze is. The range of values in the mazes considered varies from $\phi_m = 1.29$ for Koza92 [9] to $\phi_m = 14.58$ for Nevison-maze3 [20]. We classify a maze as having a *short* distance to food if $\phi_m \leq 5$, a *medium* distance if $5 < \phi_m < 10$ and a *long* distance if $\phi_m \geq 10$.

(a)   (b)

Fig. 2. (a) Russell&Norvig, $o_m = 10, \delta_m = 0.48$; (b) Gerard-Sigaud-2000, $o_m = 12, \delta_m = 0.71$

Obstacles. Mazes may contain walls, partitions or both. A wall is complete cell that the agent cannot occupy or see through whereas a partition is a barrier between cells. For example, Russell&Norvig maze [21] (Fig.2(a)) is a *wall maze*, Gerard-Sigaud-2000 [6](Fig.2(b)) is a *partition maze*, and MiyazakiC (Fig.1(b)) is a *wall-and-partition* maze. Mazes like E2 (Fig.3(b)), that contain only surrounding walls, are *empty* mazes. The number of obstacles in a maze, o_m, is defined as the total number of internal wall cells plus the total number of partitions plus half the total number of surrounding walls. Thus, for mazes with a surrounding wall, we adjust the number of obstacles to allow for the fact that they may only ever present an obstacle from a single direction.

Density. The density of a maze is the proportion of obstacles to squares, $\delta_m = \frac{o_m}{s_m}$. A maze is *spacious* when $\delta \leq 0.3$ and a maze is *restricted* when $\delta \geq 0.6$. Mazes with intermediate values of δ_m are mazes of *average density*. Spacious mazes, for example Woods7, may be extremely difficult for an agent because the maze does not provide enough reference points for the agent to distinguish the environment state. Toroidal mazes are mazes without a border of obstacles. Under this measure, a toroidal maze will be classified as more difficult than the same maze enclosed by a wall.

We consider size, density and distance to food are the most important agent independent characteristics that effect complexity. Other features of mazes that influence the complexity of the problem include:

Type of objects. In addition to a target (food) state, some mazes contain a penalizing state, such as *enemy*. For example Russell&Norvig maze has an enemy marked as E. Enemy and enemy+food mazes present a different learning problem to food mazes, and have been used only by Russell and Norvig [21] and Littman [16]. Mazes may also have different types of obstacles as well as different kinds of food (Woods2 [27]). The number of types of object effects the agent's ability to perceive its environment, and hence influences the number of aliasing states.

Maze dynamics. Some mazes involve cells which change state. For example, a multi-agent maze will have cells that are sometimes empty and sometimes occupied. A dynamic maze with moving enemy will have uncertain position of the negative rewarding object. Other mazes such as Sutton98 [24] may include moving walls. On the whole, we can talk about three sources of non-static mazes: dynamics of indifferent objects (walls), dynamics of principal objects (food/enemy) and multi-agent systems. Dynamic mazes are obviously more difficult than static ones and represent a completely different kind of maze problems.

The complexity of the learning problem is only partially dependent on the physical complexity of the maze. Perhaps greater importance is the ability of the agent to perceive the environment.

3 Agent dependent maze attributes

The agent may not be able to distinguish one square from another, despite the fact that they are in different locations, because the environment signals the agent receives in the squares are the same. Cells that appear identical under a particular detector are commonly called *aliasing*, and a maze containing at least two aliasing cells, is called an *aliasing maze*. Aliasing mazes deserve special emphasis in the context of maze classification because they represent the most difficult to solve class of problem.

In [25] Wilson proposes a scheme to classify reinforcement learning environments with respect to the sensory capabilities of the agent. An environment belongs to Class 1 if the sensory capabilities of the agent are sufficient to determine the entire state of the environment. In Class 2 environments the agent has only partial information about the true state of the environment. Class 2 environments are said to be partially observable with respect to the agent, or equivalently are non-Markov with respect to agent's actions. Accordingly, the agent is said to suffer from the hidden state problem.

Littman in [15] presents a more formal classification of reinforcement learning environments, based on the simplest agent that can achieve optimal performance. Two parameters h and β characterize the complexity of an agent. An (h, β) environment is best solved by an (h, β) agent that uses the input information provided by the environment and at most h bits of local storage to choose the action which maximize the next β reinforcements. Hence, Class 1 environments correspond to $(h = 0; \beta = 1)$ and $(h = 0; \beta > 1)$ environments, while Class 2 environments correspond to $(h > 0; \beta > 1)$ (non-Markov) environments. Of the 50 mazes considered, 21 are Class 1 mazes and 29 are Class 2.

Whilst this classification is useful, there is still a large degree of variation in complexity within Class 2 problem and the nature of the aliasing may alter the difficulty of the learning problem.

Alternative types of aliasing. In reinforcement learning terminology, the presence of aliasing states is reflected in the characteristics of the transition matrix of the decision problem of an agent in a maze. A transition matrix describes the probability of moving from one state to another for any given action. Mazes with no aliasing squares have the characteristic that for any state action pair, there will be one state with a probability of transition of 1 (i.e. any action in any square will always move the agent to the same square).

We define an *aliasing state* as one where for at least one action the probability of moving to any other state is neither 0 nor 1. An *aliasing square* is a cell in the maze which is included in an aliasing state. Thus, two or more aliasing squares may appear to be a single aliasing state to the agent.

However, the complexity of the maze cannot be determined from just the transition matrix. Some mazes (e.g. Woods2) may produce a transition matrix with uncertainty but still be easily solved by a memoryless agent and are, according to Littman [15], Class 1 environments. The complexity of the problem is determined by not only the uncertainty, but also the optimal strategy. Woods2 is classified as Class 1 because the optimal strategy in the squares that appear the same is identical.

The complexity of a maze for a LCS agent with a particular detector can be quantified by how long, on average, an agent using a Q-table trained by the Q-learning algorithm takes to find food compared to the optimal steps to food. If

Q-learning can disambiguate all squares then, assuming it has been trained for long enough, it will find the optimal route to food. If, however, it has a detector that introduces aliasing, it will take longer if the aliasing effects the optimal strategy. We use a standard version of the Q-learning algorithm, $\gamma = 0.2, \alpha = 0.71$, with roulette-wheel action selection in exploration mode and greedy action selection (max Q) in exploitation mode, number of trials $n = 20000$. Let ϕ_m^Q be the average steps to food of a trained Q-learning agent that can only detect the surrounding squares. The complexity measure ψ_m is then defined as $\psi_m = \frac{\phi_m^Q}{\phi_m}$.

This measure gives us a metric that can quantify the effects of aliasing. For example, mazes E2 [18] (Figure 3(b)) and Cassandra4x4 [4] (Figure3(a)) both have aliasing squares and similar average steps to goal and density values. However, Cassandra4x4 is much easier to solve than E2, because the aliasing squares of Cassandra4x4 do not effect the optimal strategy. This is reflected in their widely different complexity values of $\psi_m = 251$ for E2 and $\psi_m = 1$ for Cassandra4x4.

(a)   (b)

Fig. 3. (a) Cassandra4x4, $\delta = 0.38$, $\phi = 2.27$, $\psi = 1$; (b) E2, $\delta = 0.25$, $\phi = 2.33$, $\psi = 251$

Lanzi [11] noticed that disposition of aliasing cells play a significant role in maze complexity. For most LCS agents there are two major factors that have a significant influence on the learning process: minimal distance to food, d, and correct direction to food, or right action, a. Let d_1 and d_2 be minimal distance to food from an aliasing cell 1 and aliasing cell 2 respectively, and a_1 and a_2 be the optimal actions for the cells. There are four different situations for that case:

- when the distance is the same and direction is the same ($d_1 = d_2$ and $a_1 = a_2$), the squares are *pseudo-aliasing*;
- when the distance is different but direction is the same ($d_1 \neq d_2$, $a_1 = a_2$) these are *type I aliasing squares*;
- when the distance is different and direction is different ($d_1 \neq d_2$, $a_1 \neq a_2$) these are *type II aliasing squares*;
- when the distance is the same but direction is different ($d_1 = d_2$, $a_1 \neq a_2$) the squares are *aliasing type III*.

Thus, there are three types of genuine aliasing squares and one type of pseudo-aliasing conditions. Woods2 is an example of a maze with pseudo-aliasing cells. It can be seen from Figure 4(a) that for Littman57 [16] the aliasing cells marked with 1, 2 and 3 have the same direction to food (aliasing type I). Figure 4(b) shows MazeF4 [22] with aliasing squares type II marked with 1. Both squares have different distances to food as well as different directions. Woods101 [17] (Fig. 6(a)) is an example of a maze with type III aliasing squares.

In some maze cells there are more than one optimal direction to the nearest food. Thus, there are two more additional subcategories which we consider as variants of the aliasing type I:

- when the distance is the same and direction sets are intersecting ($d_1 = d_2$ and $a_1 \cap a_2$), and
- when the distance is the different and direction sets are intersecting ($d_1 \neq d_2$ and $a_1 \cap a_2$).

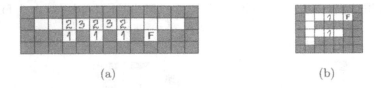

(a)   (b)

Fig. 4. (a) Littman57, aliasing maze type I; (b) MazeF4, aliasing maze type II

Influence aliasing types on maze complexity. Each aliasing type will produce distinctive kinds of noise in the agent's reward function and understanding the internal structure of the noise may help us to develop a mechanism for improving the learning of the agent. The obtained results show that the mazes with a large value of ψ_m ($\psi_m > 150$) all have type III aliasing squares (see Fig. 5). The majority of mazes that include aliasing type II squares as the highest aliasing have $10 \leq \psi_m \leq 150$. Mazes that include only aliasing type I produce a $\psi_m < 10$. Each maze can then be categorized by the type of aliasing cells it includes. For mazes that have *combined* aliasing (more than one aliasing type), we define the aliasing group a maze belongs to by the highest aliasing type it contains.

Thus, aliasing mazes type III may be considered as the most difficult group of aliasing mazes, mazes type II are of medium complexity and those type I are the easiest.

MASS system. The collected mazes have been assessed using created Maze Assessment Software System (MASS), capable of analyzing maze domains by: width; height; average steps to goal; max steps to goal; density; number of pseudo-aliasing and aliasing states and squares; average Q-learning steps, types and location of aliasing squares. MASS also produces the following outputs: transition matrix; step-to-food map; Q-learning coefficient map; Q-learning step map. A detailed description of the properties of all mazes considered can be found in [28]. The source code is available from either of the authors.

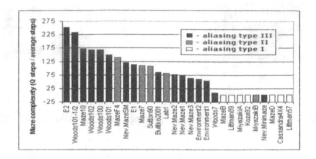

Fig. 5. Maze complexity chart.

Further aliasing metrics. Solving an aliasing non-Markov maze implies bringing it to the condition where it becomes a Markov one and hence predictable for the agent. Thus, it is the agent's structure and abilities that make an aliasing maze Markov or non-Markov, while dynamic mazes are completely agent-independent in their non-Markov properties.

Different learning systems may have different attributes that influence on complexity. For example, agents that belongs to the class of predictive modelling systems, like Anticipatory Classifier Systems (ACS) [22], predict not only reward, but also the next environmental state s'. Aliasing can thus be more complex and a wider classification is suitable:

- $d_1 = d_2$, $a_1 = a_2$, $s'_1 = s'_2$ — pseudo-aliasing, and
  $d_1 = d_2$, $a_1 = a_2$, $s'_1 \neq s'_2$ — pseudo-aliasing, predictive mismatch
- $d_1 \neq d_2$, $a_1 = a_2$, $s'_1 = s'_2$ — type I, and
  $d_1 \neq d_2$, $a_1 = a_2$, $s'_1 \neq s'_2$ — type I, predictive mismatch
- $d_1 \neq d_2$, $a_1 \neq a_2$, $s'_1 = s'_2$ — type II, and
  $d_1 \neq d_2$, $a_1 \neq a_2$, $s'_1 \neq s'_2$ — type II, predictive mismatch
- $d_1 = d_2$, $a_1 \neq a_2$, $s'_1 = s'_2$ — type III, and
  $d_1 = d_2$, $a_1 \neq a_2$, $s'_1 \neq s'_2$ — type III, predictive mismatch

In addition, some aliasing mazes may have *aliasing chains*, like Woods102 [5] (Fig.6(b)) with adjacent aliasing squares 1 and 2. Other mazes may have *communicating aliasing cells*, like Woods101 (Fig.6(a)) with two aliasing cells bordering on the same neighbour cell. The chains may be composed of different aliasing states, or, on the contrary, of the same aliasing states (e.g. E2). The environments may present a task of increased complexity for some kind of predictive modelling agents, compared to the aliasing mazes that do not have such conditions.

According to the maze complexity chart (Fig. 5), some aliasing mazes are much harder for Q-learning than could be expected. For example, among aliasing type III, such small short distanced mazes as Woods100 and Woods101-1/2 produce extremely high ψ_m coefficients, coming up with in much bigger mazes featuring numerous aliasing states, such as Maze10 and E2. Among aliasing type II the same position employs MazeF4, surpassing quite intricate maze Sutton90, featuring 7 different aliasing states in 23 aliasing squares.

(a)   (b)

Fig. 6. (a)Woods101, communicating aliasing cells ; (b)Woods102, aliasing chains

Upon examination it can be noticed that MazeF4, Woods100 and Woods101-1/2, as well as some other mazes of higher complexity, have the property that to reach food an agent has to pass through a wall-isolated aliasing square situated close to the food object from the majority of starting positions. The presence of the *alias gate* may make the maze significantly harder for some Q-based agents. Quantifying and specification of the alias gate effect may be necessary for further research.

4   Generalization and uncertainty issues

Generalization. In terms of LCS, generalization is reducing the number of significant bits used to represent an environment situation. The process groups similar types of states together in a less specialized state based upon common attributes and substitutes 'zero' and 'one' with the don't care symbol. The goal of generalization is to extend the range of the states that can be represented by a smaller population without being too crowded or too sparse. The main question is how the right generalization can be differentiated from overgeneralization.

Any generalization process applied to a maze introduces aliasing. As far as the generalized states have the same distance and the same directions to food (i.e. if they fall into the pseudo-aliasing category), the generalization is correct and beneficial. Generalization leading to the aliasing type I (the same directions, different distance) also can be beneficial, although the error-based classifier systems may be sensitive to the continuous changes in the reward, thus, some disturbance to the learning process should be expected. Any generalized state that contains aliasing type II or III is overgeneralized, because the squares concealed in the state always demand completely different actions.

Noise. Noise in LCS is a disturbance of a random nature in the agent's information system, bringing an uncertainty either to its actions or to the environment signals it receives.

The *detector noise* means the every perceived state s_{per} is a probabilistic function of the original environment state s. Thus, each environment state will correspond to a set of perceived environment states: $s \Rightarrow s_{per1}, s_{per2}, s_{pern}$. The size of the set depends on the noise function that is used, and is limited to the number of states the detector is able to perceive.

As a result, the number of states significantly increases as well as the learning time. The outcome of the learning will depend on the result sets of perceived states. If for each two states s_i and s_j in a non-aliasing maze the sets of perceived states do not intersect, the maze can be solved by the same agent and with the same amount of classifiers, provided that an appropriate generalization technique is used. Otherwise, if the sets of perceived states are intersecting, the noise function introduces aliasing and the outcome will depend on some characteristics of it and of how big the intersections are. In any case, the performance of the learning agent is considerably affected.

The *effector noise* means the conducted action a_{cond} of the agent is a probabilistic function of the original action a. Thus, for each action-state pair (s^{t-1}, a) in the environment, there will be a set of next environment states $\{s_1^t, s_2^t, \ldots, s_n^t\}$, but the size of the set cannot be greater than the number of actions available for the agent. The effector noise always introduces aliasing, although it seems to be simpler than the aliasing introduced by the detector noise because the overall number of states in the maze remains the same. The outcome of the learning will also depend on the noise function.

5 Future research

Future research may include investigations into maze complexity for predictive modelling systems, and testing different LCS agents on the mazes to define their sensitivity to alternative aliasing types. The future research may also examine the influence of further aliasing metrics (aliasing gates, chains, communicating aliasing cells) on the learning process. Investigation of different generalization techniques and specific noise functions also can be beneficial. Finally, study of maze topology and specific-purpose maze generation seems to be an essential direction of maze research in the nearest future.

6 Conclusion

Maze problems are useful and popular test problems for reinforcement learning algorithms, particularly LCS. The research covered 50 different mazes of a wide range of complexity that have been used or can be used for LCS research.

We examined agent independent and agent dependent maze attributes and proposed a set of metrics for measuring maze complexity. We considered the present definitions of aliasing, highlighted the effect of the nature of aliasing squares on the maze difficulty and introduced alternative aliasing types for a Q-based learning agent with a detector only able to perceive the surrounding squares. In addition, we proposed an approach to different aliasing types for predictive modelling systems, considered further aliasing metrics and had a short look at influence of generalization and noise on maze complexity.

The introduced metrics will provide a clearer mechanism for assessing the learning ability of new algorithms. The research also offered appropriate tools for analyzing the correlation between a learning agent and the kind of mazes it experiences difficulties with, that may provide a better understanding of its weaknesses and facilitate improvements into the agent's structure.

References

1. Bull, L.: Lookahead And Latent Learning In ZCS. GECCO-2202 (2002) 897-904
2. Bull, L., Hurst, J.: ZCS: Theory and Practice. Tech.Report UWELCSG01-001, University of West England (2001)
3. Butz M.V., Goldberg D.E., Stolzmann W.: Introducing a Genetic Generalization Pressure to the Anticipatory Classifier System. GECCO-2000 (2000)
4. Cassandra A.R., Kaelbling L.P., Littman M.L.: Acting Optimally in Partially Observable Stochastic Domains. Proc. of the 12th Nat. Conf. on Art. Intel. (1994)
5. Cliff, D., Ross, S. : Adding memory to ZCS. Adaptive Behavior 3(2) (1994) 101-150
6. Gerard, P., Siguad, O.: YACS: Combining Dynamic Programming with Generalization in Classifier Systems. Advances in Learning Classifier Systems. Springer (2001) 52-69
7. Hurst J., Bull L.: A Self-Adaptive Classifier System. Advances in Learning Classifier Systems. Springer (2000) 70-79
8. Kovacs, T., Kerber, M.: What makes a problem hard for XCS? Advances in Classifier Systems. (2001) 80-99 Springer
9. Koza, J.R. Evolution of Subsumption Using Genetic Programming. Proc. of the 1st European Conference on Artificial Life. (1992) 110-119
10. Landau, S., Picault, S., Sigaud, O., Gerard, P. : A Comparison Between ATNoS-FERES And XCSM. GECCO-2002 (2002) 926-933
11. Lanzi, P.L.: Solving Problems in Partially Observable Environments with Classifier Systems. Tech. Rep. 97.45 (1997) Politecnico di Milano
12. Lanzi, P. L. and Colombetti, M.: An extension to XCS for stochastic environments. GECCO-99 (1999) 353-360
13. Lanzi, P. L., Wilson S. W.: Toward optimal classifier system performance in non-Markon environments. Evol. Comp. 8 (4) (2000) 393-418
14. Lanzi, P.L.: Adaptive Agents with Reinforcement Learning and Internal Memory. 6th Inter.Conf. on the Simulation of Adaptive Behavior (SAB2000) (2000) 333-342
15. Littman, M. L.: An Optimization-Based Categorization of Reinforcement Learning Environments. 2nd Inter.Conf. on Simulation of Adaptive Behavior, MIT (1992)
16. Littman, M. L., Cassandra, A. R., Kaelbling, L. P.: Learning policies for partially observable environments. The 12th Intern. Conference on Machine Learning (1995)
17. McCallum, R. A.: Overcoming Incomplete Perception with Utile Distinction Memory. Proc. of the 10th Intern. Machine Learning Conference (1993)
18. Metivier, M., Lattaud, C.: Anticipatory Classifier System using Behavioral Sequences in Non-Markov Environments. For 5th Intern. Workshop, IWLCS-2002
19. Miyazaki, K. and Kobayashi, S.: Proposal for an Algorithm to Improve a Rational Policy in POMDPs. Proc. of Int. Conf. on Systems, Man and Cybernetics (1999)
20. Nevison, C.: Maze Lab 1: Event Loop Programming. Colgate University (1999)
21. Russell, S., Norvig, P.: Artificial Intelligence: A Modern Approach. Hall (1994)
22. Stolzman, W.: An introduction to Anticipatory classifier system. Learning Classifier Systems, From Foundations to Applications. Springer (2000) 175-194
23. Smith S. J., Wilson S. W.: Rosetta: Toward a Model of Learning Problems. Proc. of the Third ICGA (1989) 347-350
24. Sutton R. S., Barto, A. G.: Reinforcement Learning: An Introduction. MIT (1998)
25. Wilson, S. W.: The animat path to AI. Proc. of the 1st Intern. Conference on the Simulation of Adaptive Behaviour. MIT (1991)
26. Wilson, S. W.: ZCS: A Zeroth Level Classifier System. Ev. Com.2 (1) (1994) 1-18
27. Wilson. S.W.: Classifier Fitness Based on Accuracy. Evol. Comp. 3(2) (1995)
28. Zatuchna, Z.V.: To the studies on maze domains classification in the framework of LCS research. Technical Report CMP-C04-02, University of East Anglia (2004)

What Makes a Problem Hard for XCS?

Tim Kovacs[1] and Manfred Kerber[2]

[1] Department of Computer Science
The University of Bristol
Bristol BS8 1UB England
kovacs@cs.bris.ac.uk
http://www.cs.bris.ac.uk/

[2] School of Computer Science
The University of Birmingham
Birmingham B15 2TT England
M.Kerber@cs.bham.ac.uk
http://www.cs.bham.ac.uk/

1 Introduction

Two basic questions to ask about any learning system are: to what kinds of problems is it well suited? To what kinds of problems is it poorly suited? Despite two decades of work, Learning Classifier Systems (LCS) researchers have had relatively little to say on the subject. Although this may in part be due to the wide range of systems and problems the LCS paradigm encompasses, it is certainly a reflection of deficiency in LCS theory.

Delving into the subject more deeply, a host of other questions arise. What is it about a problem that makes it difficult or easy for a given system? In other words, what factors are involved in determining a problem's difficulty? How do these factors interact? What effects do the representation(s) used by the system have? What effects does the rule discovery system have? In short, we want to know about the dimensions of problem complexity for the system of interest. Clearly this is a difficult subject even for one particular learning system. At present we consider only Wilson's *XCS* [10] classifier system, although we hope much of the approach and some of the results will transfer to other LCS. To simplify matters, we restrict consideration to the standard ternary LCS language, to a binary action space and to single step reinforcement learning problems.

One reason to ask questions like those above is, of course, to find out when we can hope to successfully apply a given system to a given problem. Another reason is to better understand how to evaluate a system. Since we can't test our systems on all possible problems, we must consider only a subset. How can we choose this subset? There are at least two conflicting criteria. First, the subset should maximise the coverage of the dimensions of difficulty in order to more fully represent the space of all possible problems. That is, it should include as many as possible of the features which make a problem difficult for the system in question. Otherwise the test set may not detect deficiencies in an algorithm which would become apparent on other problems. Second, we would like to

T. Kovacs and M. Kerber: *What Makes a Problem Hard for XCS?*, StudFuzz **183**, 317–336 (2005)
www.springerlink.com

minimise the number of tests which must be made in order to make testing more manageable. Optimising these two criteria requires a good understanding of the dimensions of problem complexity for the system in question.

We address questions of problem complexity by considering the space of all possible test functions (i.e. learning problems) given our various restrictions. Based on our insights into this space we suggest a simple ternary single step test suite for LCS, and provide some results for XCS on it. To begin with, however, we consider how to approach the study of problem complexity in XCS.

2   Methodological Considerations

In this section we briefly motivate our study, consider some representational issues, and outline our approach.

2.1   Why Study Single Step Tests?

Single step functions are those in which the LCS's actions have no influence on which inputs it receives in the future. This contrasts with *sequential* (also called *multi step*) functions, in which actions *do* influence future inputs.[1]

In previous work we have mainly studied single step functions and we continue that practice here for two reasons. First, some applications, e.g. data mining, are single step, so it is of interest to understand and to optimise LCS for single step problems. Second, single step functions avoid many complications of sequential ones. Even if we're interested only in sequential problems, it seems reasonable to first evaluate, understand, and possibly improve our systems in the simpler single step case. When we have a good understanding of the basic LCS mechanisms we can go on to look at issues which are specific to the sequential case. This seems easier than starting with the more complex case and having to face a host of additional problems at the outset. We would argue that present understanding of LCS is limited enough to justify this approach. (Clearly, however, useful work *is* being done with sequential tests, not all of it addressing problems exclusive to them.)

Of course we have to be careful when evaluating potential improvements to ensure that we're not overfitting; optimising performance in single step problems at the expense of sequential ones. If we are interested in sequential problems we need sequential tests in our test suite.

2.2   Representing and Manipulating Functions

In Reinforcement Learning (RL), feedback to the learner about the value of its actions consists only of numeric values, called *rewards*, and the goal of any reinforcement learner is to maximise (some function of) the rewards it receives. Rewards are defined a priori by the experimenter in the form of a *reward function*, which (for our purposes) maps input/action pairs to integers. The reward

---

[1] We'll refer to states and inputs interchangeably.

function is an essential component of any RL problem specification; changing the reward function changes the problem.

As an aside, XCS – like most other reinforcement learners, but unlike traditional LCS – learns an approximation of the entire reward function. That is, every input/action pair is represented. Traditional LCS, in contrast, are only concerned with representing the more rewarding parts of this space [10, 6].

Specification of SL and RL Problems. Working with LCS, we often refer to test problems in terms of an input/output mapping. One example is the 6 multiplexer, which has often been used as an LCS test. Figure 1 shows the 3 multiplexer, a related but simpler function. This exhaustive listing of input/output cases is called a *truth table*.

The 3 multiplexer is defined on binary strings of length 3, and treats the string as being composed of an index segment (the first bit) and a data segment (the remaining two bits). The value of the function is the value of the indexed bit in the data segment, so, for example, the value of 000 is 0, the value of 001 is 0, the value of 101 is 1 and so on. Knowing the value of the string, we can do Supervised Learning (SL); that is, we know which action the LCS must respond with in order to be correct.

However, the input/output mapping of figure 1 alone is not a complete specification of an RL problem, since it does not specify rewards. In adapting a boolean function to an RL paradigm we need to extend it by defining a reward function. So the 3 multiplexer is not a complete RL problem – we need to extend it with a reward function, which we have done in figure 2. (Horizontal lines have been inserted between different inputs in this and other long figures simply as a visual aid.) Note that this figure refers to *actions* rather than the *output* of the function, because we are now dealing with a learning agent which acts (predicts the output of the function). Note also that we specify the reward for both the correct and incorrect action for each input, since we must specify a reward for each possible input/action pair. By "correct" we mean the action which receives the higher reward for that input.

In RL Rewards Determine Input/Output Mappings. We associated certain rewards with input/action pairs in figure 2 but clearly could have used other values. Other rewards will produce a 3 multiplexer problem as long as the correct action in each state is the output specified for that state in figure 1. If this is not the case, we no longer have a 3 multiplexer problem since it is the rewards which determine what input/output mapping will be learnt.

Even when rewards *are* consistent with the 3 multiplexer input/output mapping, 3 multiplexer RL problems differ when their reward functions differ. We'll see an example of this shortly.

Representing Generalisations. In the original and standard representation used with XCS [10], each rule has a single fixed-length l-bit condition which is

Input	Output
000	0
001	0
010	1
011	1
100	0
101	1
110	0
111	1

Fig. 1. The 3 multiplexer function.

Input	Action	Reward
000	0	1000
000	1	0
001	0	1000
001	1	0
010	0	0
010	1	1000
011	0	0
011	1	1000
100	0	1000
100	1	0
101	0	0
101	1	1000
110	0	1000
110	1	0
111	0	0
111	1	1000

Fig. 2. The 3 multiplexer, with one possible reward function.

Input	Action	Reward
00#	0	1000
00#	1	0
01#	0	0
01#	1	1000
1#0	0	1000
1#0	1	0
1#1	0	0
1#1	1	1000

Fig. 3. The 3 multiplexer, with rewards and generalisations expressed with the ternary syntax.

Input	Output
000	1
001	0
010	0
011	1
100	0
101	1
110	1
111	0

Fig. 4. The even 3 parity function.

a string from $\{0, 1, \#\}^l$, and single action which is a string from $\{0, 1\}^a$. In this work $a = 1$. A condition c matches a binary input string i if the characters in each position in c and i are the same, or if the character in c is a $\#$, so the $\#$ is the means by which conditions generalise over inputs. For example, the condition 00$\#$ matches two inputs: 000 and 001. Actions do not contain $\#$s and so, using this representation, XCS cannot generalise over actions.

The input/action/reward structure of the 3 multiplexer in figure 2 admits a number of accurate generalisations over inputs, and XCS seeks to find them. Figure 3 shows how XCS will learn to represent the problem defined in figure 2 using ternary classifier conditions to express generalisations over inputs. Using generalisation allows us (and XCS) to represent functions with fewer rules.

Notice that a table like figure 3 can be used in two ways, both as a specification of the function to be learned by XCS, and by XCS to representation its hypotheses about the function it is learning.

Input/Output Functions Constrain Generalisation. The amount of generalisation possible depends on the input/output mapping and the representation used. Consider the even 3 parity function, whose output is 1 when there are an even number of 1s in the input, and 0 otherwise (figure 4). While the 3 multiplexer admits considerable generalisation using the ternary LCS language, a parity function admits none whatsoever. That is, any condition which contains 1 or more $\#$s is overgeneral.

Consequently, to represent the even 3 parity function (regardless of what reward function is associated with it) XCS must use the full set of rules with fully specific conditions (i.e., conditions which have no $\#$s) of the appropriate length. Note that this set includes rules with identical conditions but different actions. The 3-bit version of the fully specific rule set (with different rewards) can be found in figures 2 and 5. Note that we can represent *any* 3-bit Boolean function with this same rule set by introducing an appropriate reward function.

Rewards Can Constrain Generalisation. Input/output mappings and representations constrain what generalisation is possible, and adding rewards to an input/output mapping can further constrain what generalisations XCS can make. For example, if we choose a reward function in which the rewards for each state are sufficiently[2] different, as in figure 5, XCS will be unable to generalise at all.[3] In this case XCS needs to use the set of all rules with fully specific conditions to represent the function. That is, XCS will need a rule for each row in the table in figure 5 (16 rules) rather than a rule for each row in figure 3 (8 rules).

Note that the altered reward function still returns a higher reward for the correct action in each state, and a lower reward for the incorrect action. For

---

[2] What constitutes *sufficiently* different rewards depends on XCS's tolerance for differences in rewards, which in turn depends on how XCS has been parameterised.

[3] This reward function was constructed by allocating a reward of 1000 for the correct action and 0 for the incorrect action in the first (topmost) state and incrementing the rewards for correct and incorrect actions by 100 for each subsequent state.

Input	Action	Reward
000	0	1000
000	1	0
001	0	1100
001	1	100
010	0	200
010	1	1200
011	0	300
011	1	1300
100	0	1400
100	1	400
101	0	500
101	1	1500
110	0	1600
110	1	600
111	0	700
111	1	1700

Fig. 5. The 3 multiplexer with rewards which may cause XCS to learn to represent it without generalisation.

example, for input 000, action 0 receives more reward than action 1, and so is to be preferred by a system whose goal is to maximise the rewards it receives. This reward function is consistent with the input/output mapping we call a 3 multiplexer function. If we changed the rewards so that action 0 received less reward than action 1 for input 000 then we would no longer be dealing with a 3 multiplexer function.

Although the reward function in figure 5 is consistent with the 3 multiplexer function, from XCS's point of view, however, the change in rewards means the problem is equivalent to the 3-bit parity problem, since with the new reward function XCS cannot generalise accurately over inputs at all. That is, even though the input/output mapping remains that of a multiplexer problem, XCS cannot generalise and must represent each input/action pair individually, as in a parity problem. Thus the representational complexity of this particular 3 multiplexer problem is equivalent to that of a parity problem (again, assuming XCS is parameterised such that it cannot generalise). This demonstrates that referring to input/output functions (e.g. multiplexer and parity functions) can be misleading when we are really referring to RL problems.

To summarise, the representational complexity of a single step RL problem depends not only on the input/output mapping but on the rewards associated with actions, the representation used, and XCS's parameterisation.

Optimal Rule Sets. Notice how we represented the 3 multiplexer in figure 3 using the language of classifier conditions to express generalisations over inputs. That is, each line in the table can be interpreted as a classifier, and the function can be represented by a set of classifiers. We could have used other sets of rules to represent the function, e.g., the set of fully specific rules used in figures 2 and 5. The set we used was chosen because it has certain properties. It is:

1. **Complete.** The rule set maps the entire input/action space.
2. **Accurate.** For our purposes this means each rule maps only to a single reward.
3. **Non-Overlapping.** No input/action pair is described by more than one rule.
4. **Minimal.** The rule set contains no more rules than are needed to satisfy the other three properties.

A set of rules with these 4 characteristics is called an *optimal population* or *optimal rule set*, denoted [O] and pronounced as the letter O [4]. We'll see in section 3.2 that it is interesting to know the size of an optimal population, denoted $||[O]||$.

We find it convenient to represent test functions by their optimal populations because we can easily manipulate them to produce other optimal populations with their own corresponding functions. Working directly with the target representation makes it obvious what effects a transformation has.

Minimality and Default Hierarchies. XCS does not support *Default Hierarchies* (see, e.g., [1]) so we do not consider solutions involving them when calculating minimality. Also, default hierarchies inherently violate the constraint that solutions be composed of non-overlapping rules, so no solution containing a default hierarchy can be an [O].

More Compact Representations. We can represent functions more compactly if we assume we have binary rewards, that is, a reward function in which all correct actions receive the same reward r_1, and all incorrect actions receive another reward r_2, where, again, correct actions are simply those which return more reward, i.e. $r_1 > r_2$. This allows us to omit the reward column and the incorrect actions when specifying a test function: if we know the correct action for a state, we know what reward it will receive (r_1), and also what reward the other action in that state will receive (r_2). (Alternatively, we can omit the correct actions instead of the incorrect ones.) These conventions allow us to represent complete RL problems in the input/output form of figures 1 and 4.

If we go further and also omit the action column we cannot specify unique functions, but we can, using only conditions, specify classes of functions. The advantage of this approach is slight, being only that we can omit actions in our specification, and to obtain fully specified functions from this representation requires some computation. To do so, we systematically assign correct actions to the conditions in such a way as to avoid making it possible to replace conditions with more general ones. Effectively, we require that the functions obtained from the condition set all have the same number of rules in their minimal representation (i.e., the same $||[O]||$). (We ignore the capacity of non-binary reward functions to influence the minimal number of rules required, since we continue to assume binary rewards.) Adding actions to conditions yields fully specified input/output functions, which in turn specify full input/action/reward mappings if we assume some binary reward function.

For example, we can assign 1s and 0s as correct actions to the conditions in figure 6 as we like, as long as we avoid assigning the same correct action to both 00 and 01, which would make it possible to replace them with 0# and yield an [O] with only 2 rules, which consequently lies outside the class denoted by figure 6 (because it contains 3 conditions). By assigning 1s and 0s to a set of conditions in different ways, and avoiding the possibility of replacing conditions with more general ones, we can obtain different functions, all of which have ||[O]|| = 4. The 4 [O]s which can be produced from the condition set in figure 6 are shown in figure 7.

This approach of specifying only conditions is used in figure 12. Note that we are using these more compact representations for the specification of test functions, but XCS is not using them to represent its hypotheses about the function it is learning.

Fig. 6. A set of conditions used to specify a class of functions.

Input	Action		Input	Action		Input	Action		Input	Action
00	0		00	1		00	0		00	1
01	1		01	0		01	1		01	0
1#	0		1#	0		1#	1		1#	1

Fig. 7. The four [O]s represented by the conditions in figure 6.

2.3   Measuring Problem Difficulty

So far we've discussed problem difficulty without defining its meaning. Different measures are possible. The primary measure we'll use is %[O], the proportion of the optimal population present in the classifier system on a given time step, which is useful as a measure of the progress of genetic search. An alternative metric, more commonly used with XCS, is that used by Wilson in, e.g., [10], simply called 'performance'. This is defined as the proportion of the last 50 inputs to which the system has responded correctly. It measures the extent to which XCS has found a population of classifiers constituting *a solution* to the problem. %[O], in contrast, measures the extent to which XCS has found *the optimal solution*. The latter is naturally more difficult to find, and requires more trials (inputs to the system) to learn. Even after XCS has reached a point where it responds perfectly correctly to all its inputs it still needs more time to find the optimal solution.

We prefer the use of %[O] because it is more sensitive to the progress of genetic search. As demonstrated in [5], %[O] can reveal differences which do not show up using the other performance measure. %[O] also seems a natural choice of metric given the discussion in section 3.2.

Another metric we'll use is the *mean %[O] regret* (or simply *regret*) which is defined as the mean difference between 100% [O] and the observed %[O] averaged over all trials and all runs. This corresponds to the mean distance between the top of a %[O] graph and the %[O] curve. Perfect performance would produce a %[O] regret of 0, while the worst possible performance would produce a regret of 1.

2.4 Population Sizing

We could consider the population size required to efficiently learn a function as another measure of its complexity. Different test functions – even those of the same string length – can have rather different population size requirements. Because of this, it is important to take population size into consideration in any comparison of different test functions. Otherwise, differences in performance may be due to the suitability of the population size used, rather than to some other feature of the problem. Different search mechanisms may have different population size requirements too, so population size should also be considered when comparing them.

Population sizing experiments on a range of 6-bit functions showed that in each case performance plateaued around a certain point. We chose to use a population size limit of 2000 rules for all experiments as all 6-bit functions should have plateaued by this point.

2.5 Experimental Procedure

The tests used in the following sections followed the standard experimental setup defined in [10] and subsequently used in many other studies of XCS. The essence of a trial is that XCS is presented with a randomly generated binary string of a fixed length as input, it responds with either a 0 or a 1, and receives a numerical reward as specified by the reward function.

Unless otherwise stated, the settings used were the standard ones for the 6 multiplexer from [10]. We used the specify operator from [7], GA subsumption deletion [11], a niche GA in the action sets [11], and the t3 deletion scheme from [5] which protects newly generated rules (with a delay of 25 for all tests). This configuration represents our current best guess at a good implementation of XCS for the problems considered here (apart from the ommission of action set subsumption, which would probably generally improve performance). No attempt was made to optimise parameters for individual problems. Finally, we used uniform crossover rather than the 1 point crossover used in previous work with XCS in order to avoid any bias due to the length or position of building blocks [1] for the solution.

3   Dimensions of Problem Difficulty

Now that we have outlined our approach, we can finally ask the question: what dimensions of single step problem difficulty are there for XCS?

Wilson briefly investigated complexity in [11] using a series of boolean multiplexer functions of string length 6, 11 and 20. The size of the optimal populations |[O]| for these functions is 16, 32, and 64 respectively. By evaluating the number of inputs XCS needed to learn each function, Wilson worked out an expression relating |[O]| and difficulty, and another relating string length and difficulty [11]. Not surprisingly, functions of greater string length and |[O]| were more difficult. Our experience with multiplexer-based functions of various |[O]| [2] and multiplexer and parity functions [4] indicates that larger |[O]| is more difficult even when string length is constant. But, as we will soon see, string length and |[O]| are not the only dimensions of problem complexity.

3.1   String Length

The lengths of the input and action strings determine the size of the space of rules XCS searches in, so it seems reasonable that difficulty would generally increase with string length. But if we consider the Kolmogorov complexity of binary strings we can see that it is quite possible for a long string to have low complexity while a much shorter string has high complexity. (See [9] for an introduction to Kolmogorov complexity.) For example, it seems intuitive that a string of 10,000 zeroes has low complexity, while a binary string of 100 randomly distributed ones and zeroes has higher complexity. Technically this assumes the use of an appropriate language to represent the strings, but intuitively the longer string has much greater regularity than the shorter one. If the strings were computer files the longer string could be compressed into a smaller file than the shorter string.

Just as some strings are much more regular than others, some functions are much more regular than others (assuming a given language in both cases). If our language can capture the regularities in a function we can represent it much more compactly. We'll see in section 3.2 that how compactly we can represent a function correlates well with the difficulty XCS has in learning it, which means that string length does not have as great an effect on difficulty as we might think. That is, some functions of a given input length will be much easier than others, and will even be easier than some others of lesser input length.

3.2 ||[O]||

It makes some intuitive sense that ||[O]|| would be a major factor in problem complexity for XCS, because ||[O]|| is determined by how much useful generalisation XCS can express, and XCS is largely concerned with finding generalisations. Unlike earlier LCS, XCS was designed to learn a complete mapping from state/action pairs to their values [10, 6]. Further, it was designed with accurate generalisation over this mapping in mind, and accurate generalisation has been a focus of XCS research from the beginning (see [10, 2, 4, 3, 7, 11, 8]).

Since an optimal population for a function is – assuming the ternary language – a minimal representation of the function, ||[O]|| is a measure of its complexity using this language. Thus we can think of ||[O]|| as a representation-specific measure of Kolmogorov complexity.

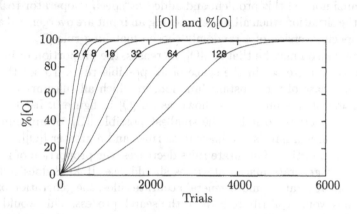

Fig. 8. %[O] for 6-bit functions with ||[O]|| equal to 2, 4, 8, 16, 32, 64 and 128. Curves are averages of 1000 runs.

Figure 8 shows the difficulty of a series of 6-bit functions of increasing ||[O]|| (the series is defined in section 5). Difficulty increases with ||[O]||, but the rate of increase of difficulty slows as ||[O]|| rises. That is, changes in ||[O]|| make less difference to difficulty at higher values.

Why exactly does difficulty increase with ||[O]||? One factor seems to be that as ||[O]|| increases, the rules in the population become, on the whole, more specific. More specific rules may be more difficult for XCS to find because XCS updates rules and affords them opportunities to reproduce only when they match an input. This has two consequences. First, more specific rules match fewer inputs and so are updated less frequently, and reproduce less frequently. The reproductive trial allocation scheme used in XCS (see [10]) balances out the difference between general and specific rules to some extent, but specific rules still reproduce less than more general ones. Now consider that a specific rule is likely to have been generated by the reproduction of another specific rule. This means that specific rules are generated less frequently because they reproduce less frequently.

The second reason more specific rules may be more difficult to find is that rules start with low fitness and only gain fitness as they are updated. It takes a number of updates for a rule to reach its full fitness, and more specific rules will take longer because they are updated less frequently. Because it takes such rules longer to reach their full fitness, they reproduce less than do more general rules.

A third reason is that genetic search in XCS seems to move mainly from specific to more general rules. Rules which are too general are unfit, whereas rules which are too specific are fit (although not as fit as rules which are as general as possible without being too general). So rules which are too general do not reproduce, while rules which are too specific do. If we think of a fitness landscape, more general rules have larger basins of attraction, while more specific rules have smaller ones. It should take longer for the GA to create a rule which falls in the basin of a specific rule – the GA seems inefficient at finding these rules. Lanzi noticed this problem and added a "specify" operator to XCS which detects the situation when all rules matching an input are overgeneral and creates a more specific version of a randomly chosen matching rule [7].

A final reason may be that as $||O||$ increases, the proportion of inaccurate to accurate rules increases in the space of all possible rules. To see this, consider the extreme case of the constant function, in which all rules are accurate. The 3-bit version of this function is shown as an [O] in figure 9. Its [O] has only 1 rule for each action, so it has the smallest possible $||O||$ of any function. Any other function will have some inaccurate rules and a greater $||O||$.

As the proportion of accurate rules decreases, the proportion of reproductive events which generate inaccurate rules should rise. If XCS does indeed search mainly from accurate to more general accurate rules, the generation of new inaccurate rules would contribute little to the search process. This would mean that reproductive events which generate inaccurate rules contribute little to search, so search would be slower on functions with more inaccurate rules, i.e. those of greater $||O||$.

Input	Action	Reward
###	0	1000
###	1	0

Fig. 9. The 3-bit constant function, with one possible reward function.

3.3   The Reward Function

As we saw in section 2.2, the reward function can constrain what generalisations are possible. Apart from this, however, we expected the form of the reward function to have little effect on problem difficulty, because in XCS fitness is based on the accuracy with which a rule predicts rewards, and not on the magnitude of the reward prediction. However, results show the range of rewards – that is, the difference between the highest and lowest reward – has a potentially strong effect on problem difficulty.

Input	Action	Reward
0000	0	100
0000	1	0
0001	0	0
0001	1	100
0010	0	0
0010	1	100
⋮	⋮	⋮
0110	0	100
0110	1	0
1111	0	x
1111	1	y

Fig. 10. The 4-bit parity function, with rewards.

To study the effect of reward range we used 4-bit parity functions in which the rewards for most of the function were binary and held constant, but the rewards for one state (1111) were varied (see figure 10). The population size limit was 500 and other settings were as defined in section 2.5 and used for the 6-bit tests shown in figure 8.

As the range in rewards increases, problem difficulty initially decreases, but then increases (see figure 11). To explain the means by which the reward range has its effects we must refer to the XCS update rules [10], in particular the prediction error update:

$$\varepsilon_j \leftarrow \varepsilon_j + \beta(\frac{|R - p_j|}{R_{max} - R_{min}} - \varepsilon_j)$$

and the rule accuracy update:

$$\kappa_j = \begin{cases} 1 & \text{if } \varepsilon_j \leq \varepsilon_o \\ 0.1 e^{(\ln \alpha)(\varepsilon_j - \varepsilon_o)/\varepsilon_o} & \text{otherwise} \end{cases}$$

where ε_j is the prediction error of rule j, $0 < \beta \leq 1$ is the learning rate, R is the reward, p_j is the prediction of rule j, R_{max} and R_{min} are the highest and lowest rewards possible in any state, κ_j is the accuracy of rule j, ε_o is a constant controlling the tolerance for prediction error and $0 < \alpha < 1$ is a constant controlling the rate of decline in accuracy when the threshold ε_o is exceeded (see [10]).[4]

---

[4] Wilson has since changed the accuracy update to:

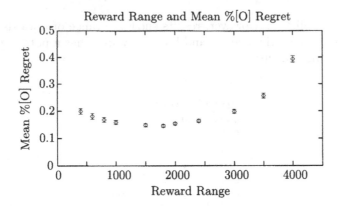

Fig. 11. Mean %[O] regret with 95% confidence intervals vs. reward range on the 4-bit parity problem (figure 10). Curves are averages of 400 runs.

We can see that in the prediction error update the error between prediction and reward $|R - p_j|$ is normalised to fall between 0 and 1 by dividing it by the range in rewards $R_{max} - R_{min}$. This means the error of a rule is inversely proportional to reward range; the larger the range the smaller the error.

The accuracy update says something like: the larger the error the lower the accuracy. Since fitness is based on accuracy, larger errors mean lower fitness. Putting all this together, larger reward ranges mean the fitness calculation is less sensitive to prediction error – it takes a larger error to produce the same effect. This means that, with larger reward ranges, XCS will *often* have more difficulty distinguishing accurate from inaccurate rules. We attribute the increase in difficulty as the reward range grows to this effect.

However, with larger reward ranges XCS will not *always* have more difficulty distinguishing accurate from inaccurate rules. In the extreme case a rule may be updated towards both R_{max} and R_{min}, and will have a large error. The errors of other rules will, in comparison, be small. This may allow XCS to more easily distinguish the overgenerality of some rules, and may account for the initial decrease of problem difficulty as reward range increases. We hypothesise that as reward range increases further this effect is then swamped by the more significant effect (that XCS becomes less sensitive to prediction error) and problem difficulty increases.

In order to avoid confounding effects of rewards, in this investigation we use only binary reward functions with the same reward range unless otherwise noted.

$$\kappa_j = \begin{cases} 1 & \text{if } \varepsilon_j < \varepsilon_o \\ \alpha(\varepsilon_j/\varepsilon_o)^{-v} & \text{otherwise} \end{cases}$$

where $0 < v$ is another constant controlling the rate of decline in accuracy when ε_o is exceeded.

3.4 Mean Hamming Distance

The hamming distance between two strings is defined as the number of characters which must be changed to transform one into the other. The mean hamming distance (MHD) of an [O] is the mean distance between each pair of condition/action strings in the population, including comparison of a given string with itself.

To study the effect of MHD on difficulty we used the 4 functions represented in figure 12. This figure shows 4 sets of conditions, each of which can be transformed into a fully specified [O] by assigning alternating 0s and 1s as the correct action for each condition, commencing with 0 as the correct action for the topmost condition and working downwards. Correct actions received a reward of 100 while incorrect actions received a reward of 0.

The 4 [O]s represented in figure 12 each have $\|[O]\| = 8$ and string length 6, but each has a different mean hamming distance. Figure 13 shows that difficulty increases with MHD on this set of functions.

	H1	H2	H3	H4
	000 ###	00# 0##	00# 0##	000 ###
	001 ###	00# 1##	00# 1##	001 ###
	010 ###	01# 0##	01# 0##	01# 0##
	011 ###	01# 1##	01# 1##	01# 1##
	100 ###	10# #0#	10# #0#	10# #0#
	101 ###	10# #1#	10# #1#	10# #1#
	110 ###	11# #0#	11# ##0	11# ##0
	111 ###	11# #1#	11# ##1	11# ##1
MHD	2	2.75	2.9375	3.125

Fig. 12. From left to right: [O]s with increasing mean hamming distance, but constant $\|[O]\|$ and string length.

We hypothesise that this effect is partly due to the greater ease of transforming more similar strings into each other by crossover and mutation. In [O]s with shorter mean hamming distances it is easier to move from one accurate general rule to another.

An additional factor may be involved. In section 3.2 we hypothesised that inaccurate rules contributed little to genetic search. If this is the case, mutation of an accurate rule into an inaccurate rule is a waste of effort, and slows the rate of genetic search. Rules which are more similar to other accurate rules are more likely to mutate into them, and less likely to mutate into inaccurate rules. Even if the accurate rule already exists, there may be more benefit in creating another copy of it than in creating an inaccurate rule. Optimal populations with smaller hamming distances between their rules are less likely to waste their efforts by producing inaccurate rules.

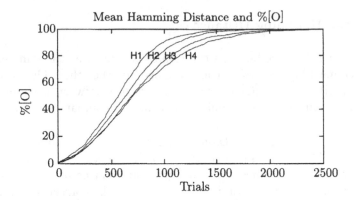

Fig. 13. %[O] for [O]s with different mean hamming distances. Difficulty increases with MHD. Curves are averages of 100 runs.

4   The Space of Single Step Functions

In the preceding sections we've identified several dimensions of single step problem complexity for XCS. In this section we consider some characteristics of the space of single step functions and how these dimensions structure it.

The space of single step functions grows rapidly with string length – there are 2^{2^l} possible binary functions for a binary string of length l. We know that some of these functions are more difficult for XCS than others. One dimension which affects difficulty is $||[O]||$ – results from section 3.2 show that difficulty increases with $||[O]||$. We can use $||[O]||$ to structure the space of l-bit functions: at one extreme, with $||[O]||$ maximised for a given l, we have a parity function. In this case [O] consists of $2^l \cdot 2$ fully specific rules. (There are 2^l inputs to match, and each maps to 2 actions.) At the other extreme, with $||[O]||$ minimised, we have a constant function. In this case [O] consists of 2 rules with fully generalised conditions (since the fully generalised condition maps to 2 actions).

Although the parity and constant function bound $||[O]||$ for a given string length, they are atypical: of the 2^{2^l} functions for a given l there are only 2 parity and 2 constant functions. The vast majority of functions have $||[O]||$ somewhere between the two.

If $||[O]||$ was the only dimension relevant to difficulty we would be justified in stating that the difficulty of a function $d([O_1])$ is greater than another $d([O_2])$ if its $||[O]||$ is greater. That is, if $||[O_1]|| > ||[O_2]||$ then $d([O_1]) > d([O_2])$. This would mean that, for l-bit functions, parity was the hardest and the constant function the easiest. This would give us bounding cases on complexity for l-bit functions and a unique ordering among them (by $||[O]||$). Further, it would give us only $2^l \cdot 2$ *complexity classes* (sets of functions of equivalent difficulty) in a much larger space of 2^{2^l} functions. That is, if we wanted to test XCS on all different levels of difficulty for l-bit functions, we would only have to test $2^l \cdot 2$ rather than 2^{2^l} functions.

However, we know that $\|[O]\|$ is not the only dimension of problem difficulty. Let's consider the others we've identified. Mean hamming distance, for a given string length l, covaries with $\|[O]\|$: the constant and parity functions – the bounding cases for $\|[O]\|$ of a given l – have fixed mean hamming distances. MHD is only variable away from these extremes, so we need only consider its effect away from them. For example, we need not worry about whether MHD can make one parity function more difficulty than another, since they must have the same MHD. This suggests that, unless MHD has a very strong effect – and our studies suggest it does not – then complexity for l-bit functions is indeed bounded by the constant and parity functions. This issue deserves further study.

Unlike MHD, the reward range is independent of $\|[O]\|$: we can use any rewards we like with any function. This suggests that in comparing the complexity of functions, we should hold reward range constant, unless reward range is itself the object of comparison, in which case we should hold all other dimensions constant. This was the approach taken in section 3.3.

The above suggests that, for a given string length and reward range, $\|[O]\|$ may be a reasonable metric for problem difficulty, and that the idea of dividing the space of functions into 2^l complexity classes defined by $\|[O]\|$ is also reasonable.

It is unfortunate that we have been unable to devise a more theoretically satisfying model of complexity than the "$\|[O]\|$ + noise" model proposed above. However, it is perhaps not surprising given the complexity of XCS: the classifier update rules, genetic algorithm, generalisation mechanism, deletion scheme, triggered operators and other mechanisms all affect the system's performance. While no simple precise model of all the above has been found, we are pleased that a single dimension, $\|[O]\|$, provides a simple and seemingly reasonable metric for problem difficulty. A somewhat more precise metric could perhaps be devised by combining $\|[O]\|$ and MHD, but we will not consider this here.

What other dimensions of single step problem difficulty exist for XCS, and what their significance is, remains to be seen. Because of this, it also remains to be seen whether $\|[O]\|$ is sufficient as a complexity metric.

5   A Ternary Single Step Test Suite

In section 3.2 we noted that generalisation is an important subject for XCS, and that $\|[O]\|$ is a measure of the degree to which generalisation is possible using the ternary language. We also saw that $\|[O]\|$ has a major effect on problem difficulty. In section 4 we saw how $\|[O]\|$ can be used to structure the space of functions of a given string length, and how using $\|[O]\|$ as a complexity metric divides the space of functions into a set of complexity classes. There are many fewer complexity classes than functions, which means we have to test XCS on only a small fraction of the function space to evaluate it at all levels of difficulty. However, there seems little need to go into such detail, particularly since higher values of $\|[O]\|$ make increasingly fine distinctions about difficulty.

Based on these observations, we propose a single step test suite which ranges over the dimension of $\|[O]\|$, and – to the extent that $\|[O]\|$ captures problem difficulty – ranges over problem difficulty. The suite is generated for l-bit strings as follows:

1 The first function in the series is the parity function for strings of length l. This function allows no useful generalisation.
2 Obtain the next function by making one of the l bits in the string irrelevant to the string's value. In effect we have a parity function for a string of $l - 1$ bits computed from a string of l bits. This function allows XCS to generalise over the irrelevant bit.
3 Repeat step 2 to cumulatively make more bits irrelevant and to obtain more functions until we reach the constant function, in which all bits are irrelevant.

This algorithm yields a set of $l + 1$ functions for l-bit strings. Recall that the number of functions grows hyperexponentially with the input string, and that the number of complexity classes defined by $||[O]||$ grows exponentially with it. Using this test suite, however, the number of tests we have to make grows only linearly with the the input string. In other words, it scales well to longer string lengths in terms of the effort required to perform the tests.

Note that this test suite is specific to the ternary LCS language, and not to XCS. That is, it may be used with any LCS, or indeed any system employing the ternary LCS language.

A disadvantage of the test suite is that it considers $||[O]||$ as the only dimension of problem difficulty. We would argue that the reward range can be considered separately to $||[O]||$ – we can use any reward range with the test suite. We would also argue that the bounds on $||[O]||$ provide bounds on MHD, since there is no variation in MHD at the bounds of $||[O]||$. There is the possibility that other, as yet unknown, dimensions of single step problem difficulty exist for XCS. Note, however, that the algorithm for generating the test suite does not specify how to select bits to ignore. By selecting bits in different orders we end up with different versions of the test suite. To cater for the possibility of unknown dimensions of problem complexity we could iterate the suite generation algorithm many times to produce many suites and average the results obtained from using them to test XCS.

The 6-bit tests shown in figure 8 were generated using this algorithm, with the leftmost relevant bit becoming irrelevant on each iteration of step 2.

6   Summary

We began with some methodological considerations, arguing that our approach of studying single step tasks is reasonable even if we're really interested in sequential ones. We then distinguished between the input/output functions we often speak of and the RL problems XCS is really applied to. Next we presented a way of representing RL problems which is particularly well suited to systems which use the ternary LCS language. Then we saw, for the first time in the literature, how population size affects performance in XCS and took measures to take its effect into account.

We've also taken some steps towards answering the questions posed at the start of the paper. We've examined a number of dimensions of problem complexity, some of them (reward range and MHD) previously unknown. We've illustrated how a significant dimension, $|[O]|$, structures the space of functions and defines complexity classes within it. Based on this we've presented a single step test suite template that's simple to describe and implement, and which scales to any length input string. We hope this test suite will prove useful, both by improving the way we evaluate LCS, that is, through its use, and by spurring the search for a better suite, and the knowledge needed to construct one.

The work begun here can be extended in many ways. To begin with, the search for additional dimensions of complexity for XCS seems important, as does evaluation of the many hypotheses introduced to account for the effects observed in section 3. To what extent our approach is appropriate for other LCS remains to be seen, as does their sensitivity to the complexity dimensions we've examined with XCS.

Finally, we've provided a great deal of additional empirical evidence to support the suggestion in [4] that XCS reliably evolves [O]s for boolean functions. We suspect that XCS can reliably learn *any* function from the class studied here, given enough resources.

Acknowledgements

We'd like to thank the two anonymous reviewers for their helpful comments and Stewart Wilson for his support and inspiration over several years.

References

1. Goldberg, D. E. *Genetic Algorithms in Search, Optimization, and Machine Learning.* Addison-Wesley, 1989.
2. Kovacs, T. Evolving Optimal Populations with XCS Classifier Systems. MSc Thesis, University of Birmingham. Also Technical Report CSR-96-17 and CSRP-96-17, School of Computer Science, University of Birmingham, Birmingham, U.K., 1996.
3. Kovacs, T. Steady State Deletion Techniques in a Classifier System. Unpublished PhD report, 1997.
4. Kovacs, T. XCS Classifier System Reliably Evolves Accurate, Complete, and Minimal Representations for Boolean Functions. In Roy, Chawdhry, and Pant, editors, *Soft Computing in Engineering Design and Manufacturing*, pages 59–68. Springer–Verlag, 1997.
5. Kovacs, T. Deletion schemes for classifier systems. In W. Banzhaf, J. Daida, A. E. Eiben, M. H. Garzon, V. Honavar, M. Jakiela, and R. E. Smith, editors, *GECCO-99: Proceedings of the Genetic and Evolutionary Computation Conference*, pages 329–336. Morgan Kaufmann, 1999.
6. Kovacs, T. Strength or Accuracy? Fitness Calculation in Learning Classifier Systems. In P. L. Lanzi, W. Stolzmann, and S. W. Wilson, editors, *Learning Classifier Systems: An Introduction to Contemporary Research*, pages 143–160. Springer–Verlag, 2000.

7. Lanzi, P. L. A Study of the Generalization Capabilities of XCS. In Thomas Bäck, editor, *Proceedings Seventh International Conference on Genetic Algorithms (ICGA-7)*, pages 418–425. Morgan Kaufmann, 1997.
8. Lanzi, P. L. Generalization in Wilson's XCS. In A. E. Eiben, T. Bäck, M. Shoenauer, and H.-P. Schwefel, editors, *Proceedings of the Fifth International Conference on Parallel Problem Solving From Nature*, number 1498 in LNCS. Springer–Verlag, 1998.
9. Li, M. and Vitányi, P. *An Introduction to Kolmogorov Complexity and Its Applications*. 2nd edition. Springer–Verlag, 1997.
10. Wilson, S. W. Classifier fitness based on accuracy. *Evolutionary Computation*, 3(2):149–175, 1995.
11. Wilson, S. W. Generalization in the XCS classifier system. In J. Koza et al., editors, *Genetic Programming 1998: Proceedings of the Third Annual Conference*, pages 665–674. Morgan Kaufmann, 1998.